T0340710

THE AUTOMOTIVE BODY MANUFACTURING SYSTEMS AND PROCESSES

THE AUTOMOTIVE BODY MANUFACTURING SYSTEMS AND PROCESSES

Mohammed A. Omar

Clemson University International Center for Automotive Research CU-ICAR, USA

A John Wiley & Sons, Ltd., Publication

This edition first published 2011

Registered office
John Wiley & Sons Ltd, The Atrium, Southern Gate, Chichester, West Sussex, PO19 8SQ, United Kingdom

For details of our global editorial offices, for customer services and for information about how to apply for permission to reuse the copyright material in this book please see our website at www.wiley.com.

Library of Congress Cataloging-in-Publication Data

Omar, Mohammed A.
The automotive body manufacturing systems and processes / Mohammed A Omar.
p. cm.
Includes bibliographical references and index.
ISBN 978-0-470-97633-3 (hardback)
1. Automobiles–Bodies–Design and construction. I. Title.
TL255.O43 2011
629.2'34–dc22
2010045644

A catalogue record for this book is available from the British Library.

Print ISBN: 9780470976333 [HB]
ePDF ISBN: 9780470978474
oBook ISBN: 9781119990888
ePub ISBN: 9781119990871

Set in 11 on 13 pt Times by Toppan Best-set Premedia Limited
Printed in Malaysia by Ho Printing (M) Sdn Bhd

To Rania and Yanal, my sources of inspiration.

Contents

Preface

This book addresses the automotive body manufacturing processes from three perspectives: (1) the transformational aspect, where all the actual material conversion processes and steps are discussed in detail; (2) the static aspect, which covers the plant layout design and strategies in addition to the locational strategies; and, finally, (3) the operational aspect. The transformational aspect is discussed in Chapters 2, 3, 4, 5, and 6; while the static aspect is given in Chapter 7 and the operational aspect with its two different levels—operational and strategic—is presented in Chapter 8.

The transformational perspective starts by covering the metal forming practices and its basic theoretical background in Chapter 2. It also addresses the potential technologies that might be used for shaping and forming the different body panels using lightweight materials with a lower formability window, such as aluminum and magnesium. The text discusses the automotive joining processes in Chapter 3, covering the fusion-based welding technologies, mainly the metal inert gas (MIG), the tungsten inert gas (TIG), and the resistance welding practices. These welding technologies are discussed to explain their applicability and limitations in joining the different body panels and components. The welding schedules for each of these technologies are explained and the spot-welding lobes and dynamic resistance behavior are also explained. Additionally, Chapter 3 describes the adhesive bonding practices and the different preparations and selection process needed to apply and decide on the correct adhesive bonds. The different strategies applied by automotive OEMs to enable their welding lines to accommodate different body styles using intelligent fixtures and control schemes are also discussed. Finally, the robotic welders and their advantages over manual applications, in addition to discussing potential joining practices such as friction stir welding, are addressed in this chapter.

Chapter 4 discusses the automotive painting processes and its different steps; starting from the conditioning and cleaning, then the conversion and E-coating, followed by the spray-based painting processes. Also, this chapter describes the automotive paint booths' design and operation, while addressing the difference between the solvent-borne, and power-coat-based booth designs. Other miscellaneous steps that include the sealant, PVC and under-body wax application and curing steps are presented. In Chapter 5, the final assembly area and the different processing applied to install the different interior and exterior trim parts into the painted car shell, are presented, in addition to the marriage area where the power-train joins the painted shell. The mechanical joining and fastening practices are given in detail, explaining the different strategies that automotive OEMs use to ensure the right tension loads are achieved in their mechanical joints.

In Chapter 6, the automotive manufacturing ecological aspects, from the materials used and their utilization in addition to the energy expended in the manufacturing process, are discussed. The ecological chapter includes comprehensive analyses of the energy and resources footprint for each of the transformational aspects. Additionally the painting process is discussed in detail to explain its air conditioning requirements, water usage and treatment, and finally its air emissions. Also, the effect of reducing the current automobiles' weight on their overall environmental footprint, especially in the usage phase, is presented.

Chapter 7 starts the discussion of the static aspect of the automotive manufacturing processes; it explains the different strategies used to plan the factory layout from the process-based, the product-based, and the cell-based layouts. Additionally, the different details in regard to the factors that affect the manufacturing plant location are presented; also the factor rating method, the center of gravity method, and the transportation table are described. Chapter 8 provides the operational and strategic management aspects of automotive manufacturing. This chapter explains the aggregate planning process and the master production scheduling, then it further discusses the material requisition planning (MRP) steps and its basic operation.

This book can be perceived to be composed of four basic modules: module one starts with Chapter 1 that provides a basic introduction to the automotive manufacturing processes from the assembly and the power-train manufacturing steps, in addition to explaining the basic vehicles' functionalities and performance metrics and the industry basic drivers and changers. The second module is focused on the transformational aspect, which is found in Chapters 2, 3, 4, 5 and 6. The third module is concerned with the static aspect of automotive manufacturing, which is found in Chapter 7. The fourth and final module is in Chapter 8, where the operational and strategic tools are discussed. Dividing the automotive manufacturing processes into these four modules enables the reader to gain a comprehensive understanding of the automotive manufacturing processes, control schemes, and basic drivers, in addition to their environmental impact.

Foreword

The automobile is the most complex consumer product on the market today. It affects every aspect of our lives. It also requires significant intellectual, capital and human investment to produce. The market related to automotive production is second to none, and vehicle production drives multiple sectors of national and international interest including areas related to energy, emissions and safety. Manufacturing a vehicle requires a multi-billion dollar investment, and is one of the highest tech operations in the manufacturing sector. Certainly, automotive plants are one of the largest wealth generators in the industrial world. Furthermore, there is also no doubt that the architecture of the automobile will change rapidly over the next several product generations. Such a rapid enhancements will induce a significant strain on vehicle production in the future. Much of the technology and concepts employed in vehicle manufacture will, by necessity, change to meet the growing demand for rapidly changing technology, higher quality, improved safety, reduced emissions and improved energy efficiency in new vehicles.

Mohammed Omar has significant experience in a variety of automotive manufacturing environments. He has taken these experiences and developed a number of thorough and innovative courses at the Clemson University—International Center for Automotive Research. This text is the culmination of over a decade of these industry, research and teaching efforts.

This text is presented to the reader in four main modules that clearly and concisely present automotive technology and vehicle manufacture. The first module provides an introduction to automotive engineering and the key manufacturing processes necessary to successfully product the modern vehicle. Basic vehicle functions and performance metrics are presented in this module, as well as typical drivers and changers in the automotive industry. Within this model the base processes such as welding/joining, paint/coat and assembly are presented. Such processes are critical not only in final product quality and capability, but also define the resource needs of the overall production process. This leads directly into the second module, which targets the transformational aspects of automotive production as they relate to the environment and the economy. In the second section, issues from material utilization to energy and resource consumption are analyzed and discussed. The text highlights these factors and their overall impact on the resource footprint of both the product and its manufacturing process. The third module shifts from the production processes to the static aspect of vehicle manufacture. Issues such as the overall plant design, manufacturing cell integration, operation and optimization strategies are presented along with several examples of successful implementations from various corporate strategies. Finally, the fourth module addresses operational and strategic tools used in automotive manufacturing. Issues such as aggregate planning process, master production scheduling and Material Requisition Planning (MRP) are discussed.

The integration of these four modules provides a fresh and innovative perspective on automotive manufacture that enables the reader to have a comprehensive understanding of the automotive production processes, control schemes, basic drivers, in addition to environmental impact. The text is a must have for the modern manufacturing engineer, and will provide the reader with a state-of-the art foundation for modern manufacturing. I highly recommend Dr. Omar's timely book. I believe it will benefit many readers and is an excellent reference.

Thomas Kurfess
Professor and BMW Chair of Manufacturing
Director, Automotive Engineering Manufacturing and Controls
Clemson University

Acknowledgments

This book reflects the work of thousands of mechanical and automotive engineers and researchers, whose dedication to their engineering profession has led to great advancements in science and mobility products that served and continue to serve us all.

Sincere and special dedication is due to Professor Kozo Saito (University of Kentucky) for his continuous academic and personal guidance. I would like also to thank the mechanical engineering professors at the University of Kentucky for their encouragement and mentoring during my PhD studies. Additionally, I would like to thank all my colleagues at the Clemson University automotive and mechanical engineering departments for their continuous support and enriching discussions. Special thanks are due to Professor Imtiaz Haque for his invaluable guidance and support. Also, I would like to recognize all my students for their dedication and hard work; especially Yi Zhou (my first PhD student) and Rohit Parvataneni (artwork) for their selfless work.

I would like to recognize; Jürgen Schwab, Brandon Hance, and Ali Al-Kilani for their technical contribution and discussions. Finally, I would like to thank my high school mathematics teacher, Mr. Mohammed Edrees.

Abbreviations

ACEEE	American Council for an Energy Efficient Economy
AHSS	advance high strength steel
AISI	American Iron and Steel Institute
AIV	Aluminum Intensive Vehicles
APQP	Advanced Product Quality Planning
ARB	accumulative roll bonding
BH	Bake harden-able
BiW	body in white
BoM	bill of material
BUT	Bending-Under-Tension
CAD/CAM	Computer Aided Design and Manufacturing
CAFE	Corporate Average Fuel Economy
CBS	Cartridge Bell System
CCD	charged coupled devices
CGA	circle grid analysis
CMM	Coordinate Measuring Machine
CNC	computer numerically controlled
CO	Change-Over
COPES	Conductive Paint Electrostatic Spray system
CT	cycle time
DBS	Draw Bead Simulator
DC	Deformation Capacity
DoC	Degree of Cure
DP	Dual-Phase
DQ	Draw Quality
DQSK	Drawing Quality Special Killed
DSC	Differential Scanning Calorimetry
EDDQ	Extra Deep Draw Quality
EGA	electro-galvanized Ze-Fe alloy
ELU	Environmental Load Unit
EPA	Environmental Protection Agency

EPI	Energy Performance Indicators
EPS	Environmental Priority Strategy
ERP	enterprise requisition planning
FE	Finite Element
FLC	forming limit curve
FLD	forming limit diagram
FMEA	Failure Mode and Effect Analysis
FSW	friction stir welding
FTIR	Fourier Transform Infrared Spectroscopy
GD&T	Geometric Dimensioning and Tolerancing
GIS	Geographical Information System
GMAW	Gas Metal Arc Welding
GQA	general quality agreement
HAP	Hazardous Air Pollutants
HAZ	Heat Affected Zone
HDGA	hot-dip galvanneal
HSLA	High Strength Low Alloy
HSS	high strength steel
HVLP	high volume of air supplied at low pressure
IF	Inter-terrestrial Free
IOI	Industrial Origami Incorporated
JIT	Just In Time
LDH	Limiting Dome Height
LDR	Limiting Draw Ratio
LED	Light Emitting Diode
LIEF	Long-Term Energy Forecasting
LM	Lean Manufacturing
MIG	metal inert gas
MMCs	Metal Matrix Composites
MPS	master production schedule
MRP	material requirement or requisition planning
MTBF	Mean Time Between Failures
NC	Numerically Controlled
NVH	noise, vibration and harshness
OCMM	Optical Coordinate Measuring Machines
OEMs	original equipment manufacturers
PLCs	programmable logic controllers
PPAP	production parts approval process
PUEL	post-uniform elongation
RDC	remaining deformation capacity
RH	relative humidity
RHT	Ring Hoop Tension

SHA	Systematic Handling Analysis
SLP	Systematic Layout Planning
SMCs	sheet molding compounds
SMED	Single Minute Exchange of Die
SoP	Start of Production
SPM	Strokes Per Minute
STS	Shape Tilt Strength
% TE	% total elongation
TFE	Tube Free-Expansion
TGA	Thermo-Gravimetric Analysis
TME	Temper Mill Extension
TPS	Toyota Production System
TRIP	TRansformation-Induced Plasticity
TSA	thickness strain analysis
TWB	Tailor Welded Blank
TWB/C/T	tailor-welded blanks, coils, and tubes
TWC	Tailor Welded Coil
UEL	uniform elongation
UTS	Ultimate Tensile Strength
VOC	Volatile Organic Compounds
VSM	Value Stream Mapping
WHP	Work Hardening Potential
WHR	Work Hardening Rate
WIP	Work In Process
YS	Yield Strength

1

Introduction

1.1 Anatomy of a Vehicle, Vehicle Functionality and Components

Customers today perceive the value of an automobile based on its structure, its mobility function, its appearance, and other miscellaneous options such as infotainment. This fact motivates the automotive engineers to develop engineering metrics to judge each of these perspectives in a quantitative manner to help them improve their design, benchmark their vehicles against their competitors and, more importantly, meet the legal regulations. For example, the performance of a vehicle structure is dependent on the following criteria: Crash-worthiness (or passive safety), service life (or durability), its noise, vibration and harshness (NVH) characteristics, in addition to new metrics that have recently been viewed as value-adding, such as structure recyclability and weight efficiency.

Crash-worthiness defines the vehicle structure ability or capacity to absorb dynamic energy without harming its occupants in an accident, while the durability is the probability that the structure will function without failure over a specified period of time or frequency of usage. The NVH describes the structure performance in absorbing the different vibration levels and providing a desired (designed) level of comfort. The noise is defined as vibration levels with low frequency (<25 Hz), while harshness is the term for vibrations at (~25–100 Hz). All the above structural requirements are controlled by intrinsic (density, Young's modulus) and extrinsic (thickness, geometry, and shapes) material properties and joining strategies. So the material, its shape selections, and the manufacturing process control the overall performance of vehicular structures.

The vehicle mobility function is controlled by its ride and handling dynamics in addition to the drive-line and power-train systems' reliability. Again, the choice of material (weight and stiffness) and the design geometries (center of gravity location) affect the vehicle's performance. The vehicle appearance can be described in its styling, which is controlled by the panels' shape, its geometrical fit (gaps, flush

---

The Automotive Body Manufacturing Systems and Processes, First Edition. Mohammed A. Omar.

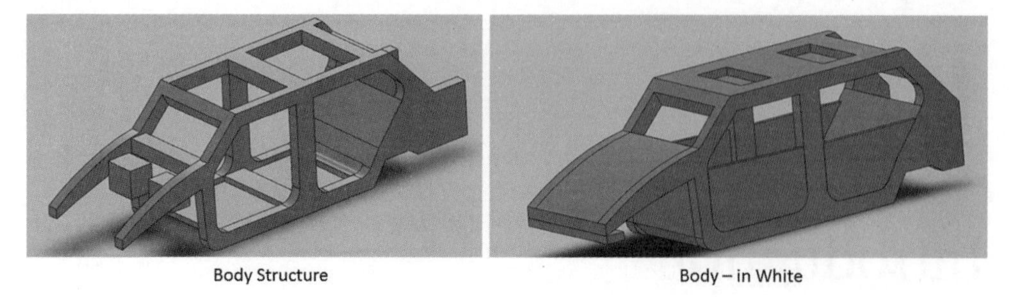

Figure 1.1 Left: the vehicle body structure without closures, right: the complete vehicle BiW

setting, etc.), and its final paint finish. Human visual perception evaluates the vehicle's finish in terms of specific visual qualities: Color properties, encompassing three color attributes, in addition to color matching between the different vehicle parts such as the steel body and the plastic trim. Also, surfaces' spatial properties as well as the vehicle's geometric attributes such as gloss, texture, and haze control the customer's perception. For example, if the paint on a vehicle suffers from orange peel, i.e. the paint looks like the peel of an orange, the customer might mistakenly observe this as a defect (variation) in the sheet metal roughness.

The vehicle's main components and sub-systems can be categorically listed as: Power-train, chassis, exterior and interior trims, and the body in white (BiW) or vehicle body-shell. The power-train is composed of the prime-mover (the internal combustion engine, or electric motor), the gear system, and the propulsion and drive shafts, while the chassis includes the suspension and steering components, in addition to the wheel, tires, and axles. The interior and exterior trims compose the front and rear ends, the door system, and the cockpit trim. Finally, the body in white is made up of the closures (doors, hood, tail-gate) and the frame, see Figure 1.1). The frame can be of a uni-body design (Figure 1.2 (a) uni-body), a body-on-frame (Figure 1.2 (b)), or a space-frame (Figure 1.2 (c)). The uni-body design features stamped panels, while the space-frame is made up of extrusions and cast parts. The BiW closures are selected based on the vehicle's constituent material dent-resistance properties (i.e. yield strength) while the frame is designed to provide specific torsional and bending stiffness.

1.2 Vehicle Manufacturing: An Overview

After reading Section 1.1, we can conclude that vehicle performance is judged based on design strength, stiffness, energy absorption, dent resistance, and surface roughness. However, before designers select a material or design a specific shape, they should consider manufacturability. The manufacturability from an automotive body structure's point of view is described in terms of the design formability, the joining

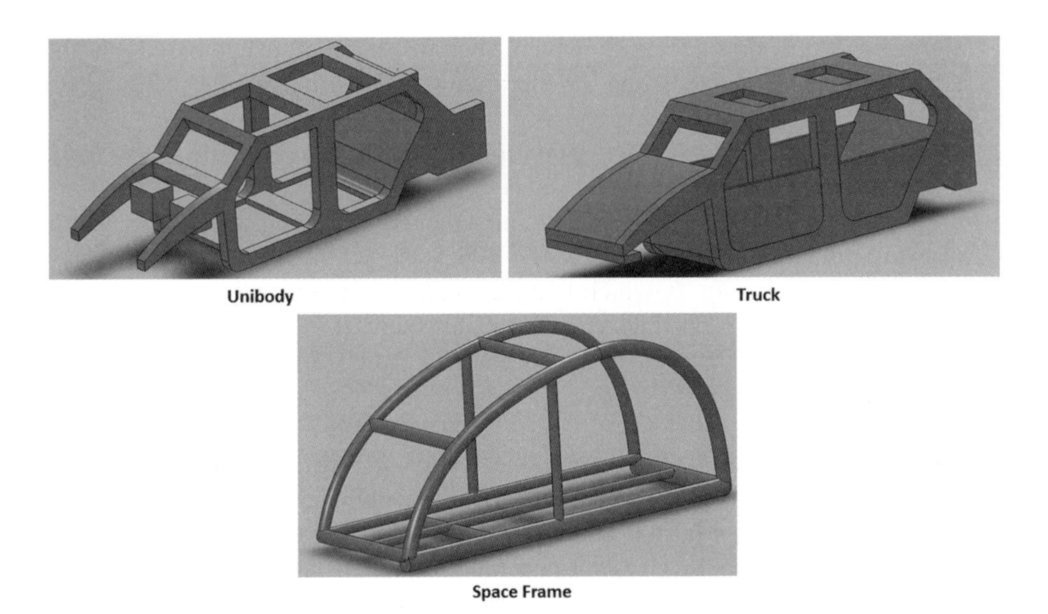

Figure 1.2 Top left: (a) a uni-body design, top, right: (b) truck platform; and bottom right: (c) space-frame design

ability (weldability and hemming ability), the achieved surface finish and surface energy, and its overall cost. This fact motivates a deeper understanding of the automotive manufacturing processes and systems, because it will ultimately decide the design's overall cost, final shape, and functionality, that is, the design validity.

The automotive manufacturing activities can be analyzed on two levels: the manufacturing *system* and *process* levels. The manufacturing system view is typically investigated from three different perspectives: the production line (the *structural aspect*) which covers the machinery, the material handling equipment, the labor resources, and its allocations to the different activities. The *transformational aspect* includes the functional part of the manufacturing system that is the conversion of the raw materials into finished or semi-finished products. The transformational activities include all the stamping, casting, welding, machining and painting efforts within the plant. The third aspect is the *procedural aspect* which describes the operating procedures and strategies, which is further viewed from two different levels; the strategic level which identifies the product type, and volume (product planning), given the operating environment conditions (customer demands and regulatory issues). Additionally, the strategic plan includes the resources' allocation in the manufacturing enterprise. The second level is the operational level which is focused on production control, i.e. meeting the strategic plan objectives through planning, implementing, and control and monitoring activities. These operational activities are further categorized by [1]:

1. *aggregate production planning* which suggests product plans based on the required product volume, using a generic unit such as the vehicle platform not type, to increase the level of confidence from the forecast information;
2. *production process planning* which controls the production techniques to be used, in addition to process routes and sequence;
3. *production scheduling* to determine an implementation plan for the time schedule for every job in the process route;
4. *production implementation* which is the execution of the actual production plan according to the time schedule and allocated resources;
5. *production control* to measure and reduce any deviations from the actual plan and time schedules.

Another important view on the automotive manufacturing systems relates to the information, materials, and value-added (cost) *flows* within the plant. The raw materials and supplier parts flow from upstream to downstream through the material supply system, the material handling system and finally through the material distribution system. However, the information flows in the opposite direction, that is from downstream to upstream, to synchronize the rhythm of production and control its quality; this information flow is typically called the pull production system to indicate that the customer side controls the quantity and quality (product type) of the production. On the other hand, the old push system meant that the manufacturing plant outputs vehicles according to a mass production scheme without any feedback from the customer side.

The automobile manufacturing processes are divided into two plants: the assembly plants and the power-train plants. Both of these plants specialize in different transformational processes and convert different raw materials into final parts. However, both are synchronized in time to integrate their final outputs into complete vehicles.

1.2.1 Basics of the Assembly Processes

An automotive assembly plant is responsible for the fabrication of the complete vehicle BiW, starting from a steel and/or aluminum coil and ending with a complete painted car shell. Additionally, the power-train, chassis components, interior and exterior trims are all integrated into the BiW at the end of the assembly process in the final assembly area.

The assembly sequence starts with the receiving area for the coil (which is typically made out of steel and aluminum), which also includes a testing laboratory to check material thickness and surface characteristics. After passing the testing, the coil is either stored or staged for blanking. The blanks are then transferred to the stamping press lines to form the different vehicular panels. A typical BiW consists of about 300–400 stamped pieces, however, only a few main panels affect the overall

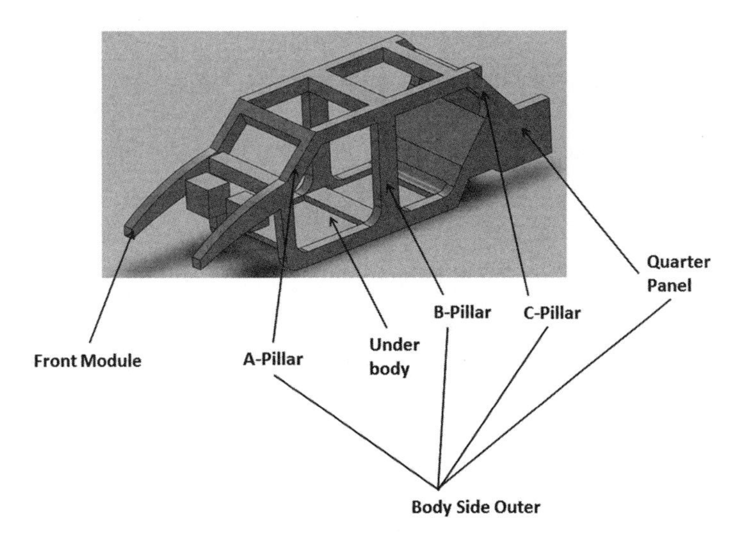

Figure 1.3 The different panels of the vehicle structure

geometry, fit and finish. These panels are the roof, the trunk (inner, outer, and pan), the hood (inner and outer), the under-body, the wheel-house, the body-side, A and B pillars, the floor pan, the front module (engine cradle, crush zones, shock towers), the quarter panels, and doors (inner, outer). Some of these panels are displayed in Figure 1.3.

After the stamping process, some of the panels are joined to create sub-systems in specialized cells, as in the case for the doors where their inners and outers are adhesively bonded, hemmed and spot-welded. Additional cells exist in the stamping area for other components which are then fed to the body-weld or body-shop area. The stamping process utilizes mechanical and hydraulic presses with different tonnage, accessories and dies, so it can handle different panels ranging in shape and size, from 0.1–6.5 mm in thickness and with dimensions as small as 1 x panel thickness to as large as 500 x panel thickness.

In the body-shop area, the different panels are joined to form the car shell, starting with the under-body and then the body-side (left- and right-hand) outers. The joining of such panels is first done using tack welding to hold the pieces in place, followed by permanent spot welds. A typical vehicle shell has around 5000 spot welds, achieved through robotic welders working in designated cells and programmed offline. The completed body shells will also go through a dimensional check process using laser illumination with a charged coupled devices (CCD) camera system to monitor the shell gaps, flush setting and fit. The body-weld also features metal inert gas (MIG) welding for the under-body.

The robotic welding cells are controlled and monitored through separate programmable logic controllers (PLCs) which are then connected through a main controller to enable the complete line control through a master PLC.

The completed BiW is then transferred to the paint-line. The paint booth area cleans the car shells in immersion tanks and applies a conversion coating layer (iron phosphate or zinc phosphate) followed by an electro-coat or e-coat layer. The subsequent paint layers require drying or curing, through a combination of convection and radiation-based ovens. Spray paint booths follow the immersion stages, to apply the primer, top coat and clear coat layers. Also the paint booth area features other important steps such as applying the under-body wax and sealants followed by their curing process. Inspection for paint quality in terms of thickness, color match and contaminants is also important in the paint-line. In the paint-line, the vehicles might be taken out of the overall production sequence to create color batches, thus reducing the paint color change time. However, at the end of the line, all vehicles are arranged back in sequence.

After the paint-line, vehicles are transferred to the final assembly area, where the interior (cockpit, seats, etc.) and exterior trims are installed. The final assembly area consists mainly of manual labor using power-tools and fixtures for the ergonomics, in addition to autonomous carriers that transfer the power-train components (engine, transmission, etc.) for assembly work (installing the cables, fuel hoses, and controllers) and then to the marriage area. The marriage area is where the power-train is installed in the vehicle body. The final assembly area features a variety of mechanical fastening and riveting operations to install the different trim components in the vehicle shell. Additionally, a variety of sensory systems is used to check the dimensional fit of the different components, in addition to ensuring the proper torque for each joint.

The final step in the assembly process tests the vehicle operation and build, using a chassis dynamometer and a water-test chamber.

The assembly plants require a sophisticated control system that not only monitors the different areas' performance (stamping, body-weld, paint and final assembly) but also synchronizes these activities with the reception of parts from the suppliers' network and with the power-train facility.

The flow of parts and semi-finished vehicles within an assembly plant go through different layouts within each assembly area. In the stamping area, the parts are distributed between the different stamping presses depending on the press tonnage and the dies assigned to that press. Also the staging and storing of stamped pieces are done on racks and then transferred to the body-shop or to specialized cells separately, see Figure 1.4. In other words, the layout in the stamping area is similar to a product-based layout not a process-based one. A product-based layout is similar to the ones found in small workshops or a carpenter shop, where the flow of pieces (panels) and equipment allocation (dies) changes according to the product type (vehicle type).

In the body-weld, there is a main assembly line where the sub-assemblies are fed to be joined to the main body frame. So the body-shop layout is a process-based layout, because the focus is on repeating the same process for all product (vehicle) types. The body-weld overall layout is similar to a spine, where the specialized cells that create the door sub-assemblies (joining inner and outer), the hood, the under-

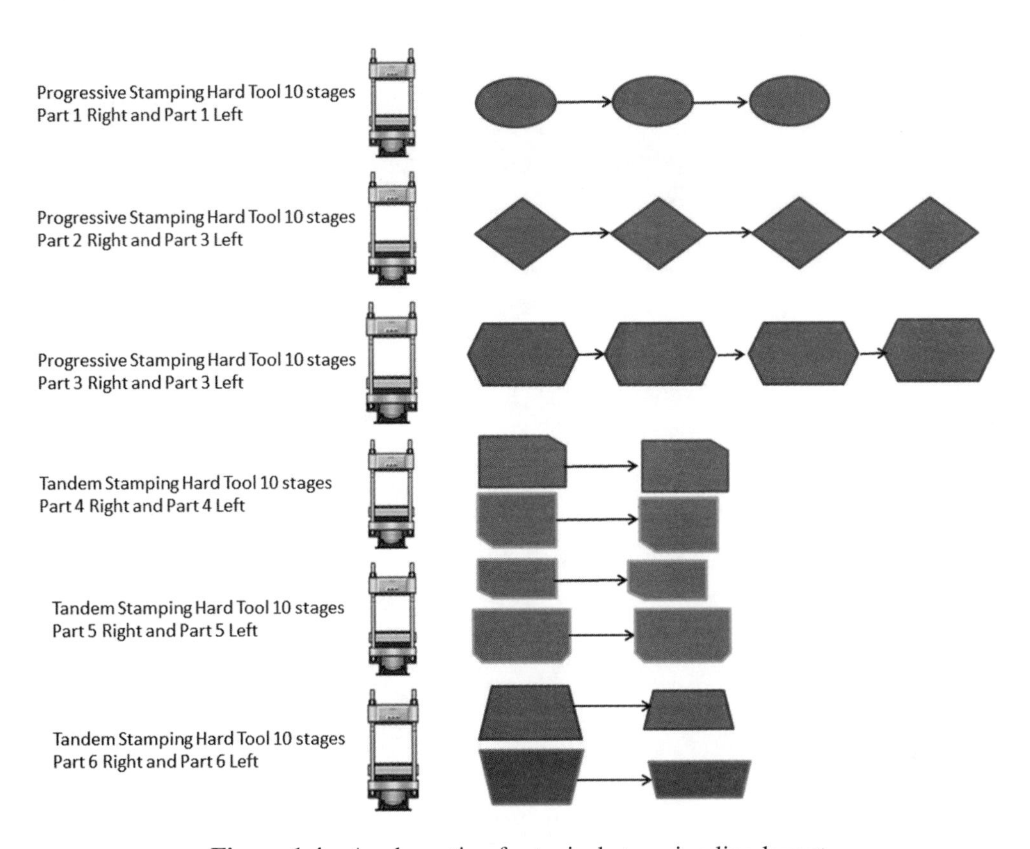

Figure 1.4 A schematic of a typical stamping line layout

body, feed the main line that joins them to the body main shell, see Figure 1.5. The paint-line layout starts with a single straight line for the cleaning and the conditioning steps, the conversion coating (phosphate), and the e-coating immersion tanks. Then the vehicles are sent to a selectivity bank area (with a flexible conveyor system) so batches of vehicles of the same color are created for the spray booths. Some original equipment manufacturers (OEMs) like Toyota do not use the color-batching strategy but instead developed their paint-line booths to use a cartridge color system, where the robots can switch between different cartridges to change colors, thus eliminating the need to clean the paint supply line every time the color is changed. The overall flow within the paint-line is displayed in Figure 1.6, illustrating the different painting steps and layout. The final assembly area follows a process-based layout using a straight or a horseshoe-shaped assembly line.

1.2.2 Basics of the Power-train Processes

The power-train facilities are mainly responsible for building the vehicle power-train and drive-line components such as the engine and transmission. The power-train

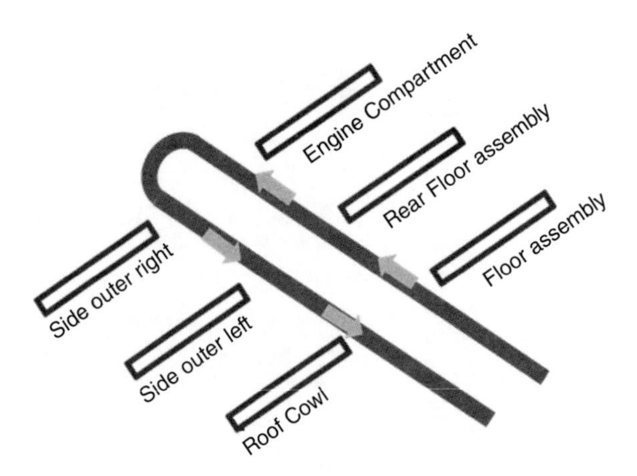

Figure 1.5 The layout of a body-weld line

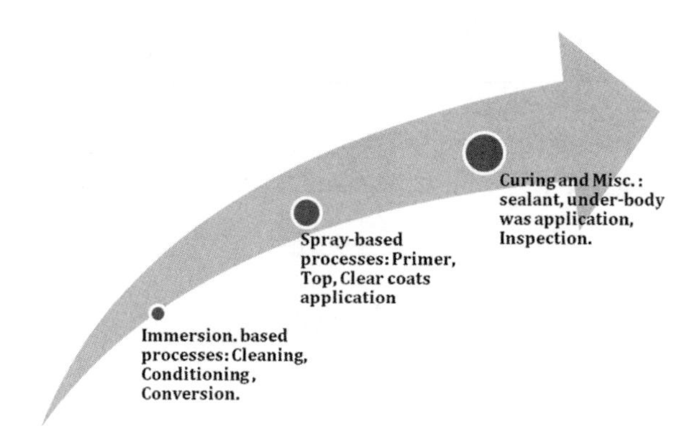

Figure 1.6 The basic processes in an automotive paint-line

plants feature different transformational manufacturing processes from those found in the assembly plants. The power-train plants use a variety of forging, casting and machining operations to fabricate the engine components and the transmission. For example, the engine cylinder blocks are made of cast iron or are cast out of aluminum or in some cases from aluminum with a magnesium core to reduce the total weight of the engine. After the entire engine and the transmission components have been manufactured, they are assembled manually. For example, after casting and machining the engine cylinder block and the exhaust manifold, forging the pistons and the crankshaft, and finishing the valves, the crankshaft is installed manually in the cylinder block and secured by the bearing caps, which are torqued automatically. Then the pistons are lubricated and installed in the cylinder block carefully to prevent scratching the cylinder lining. Then the cylinder head is mounted and torqued to hold

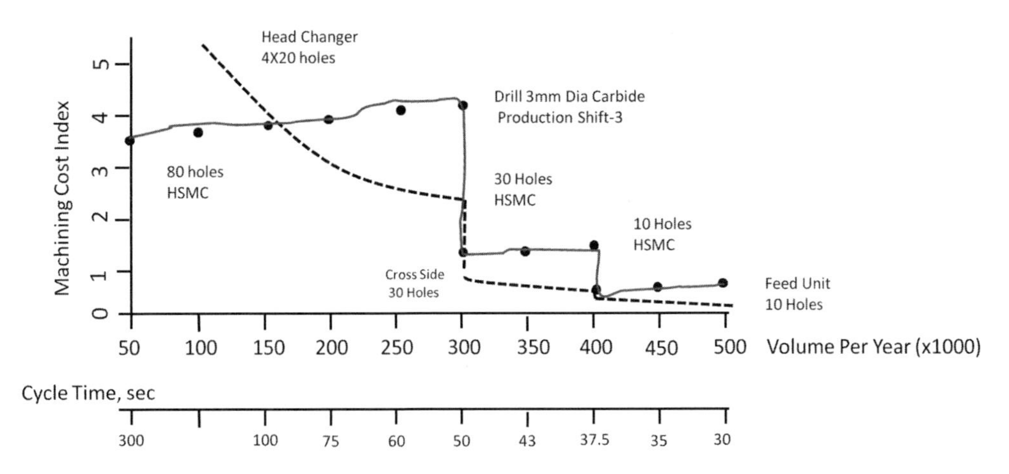

Figure 1.7 An estimated cost comparison between the multi-spindle drilling and high speed machining centers

the valves assembly. The inlet and exhaust manifolds are installed next and fastened mechanically. The testing of the engine operation is done next, using an engine dynamometer.

The power-train plant relies mainly on in-house parts and components, in contrast to the assembly plants. However, new trends in the power-train manufacturing have reduced the number of parts manufactured in-house, so most OEMs now manufacture using basic engine components: the cylinder block and head, camshaft, crankshaft, and the connecting rods. The cylinder block goes through several machining processes that consist of rough and final millings, in addition to a variety of drilling, reaming and tapping processes. In general, a typical cylinder block will go through around 70 processes. The typical cycle time in an assembly process is around 60–80 seconds per station, however, the typical cycle time for a power-train operation is around 3 minutes, which highlights the need for advanced or improved material transfer technologies in addition to high speed machining centers and multi-spindle drilling. This fact has motivated the use of specialized tooling and fixture systems along with multi-spindle head-changer and multi-slide accessories. Additionally, the three-axis computer numerically controlled (CNC) machining centers have increased the power-train flexibility and agility. Figure 1.7 shows an estimated cost-based comparison between the high speed machining centers and the multi-spindle drilling at different vehicle production levels. Additional advances in power-train machining include the use of super-abrasive tooling for the boring, milling and honing operations. For example, the crank-boring operation used to rely on tungsten carbide tooling to achieve close tolerances in the range of 0.02 mm, however, the use of a poly-crystalline diamond tool has led to better quality at higher boring speeds.

To manufacture the cylinder head, around 50 drilling operations are used to apply around 70 holes, so flexible transfer lines and cell-to-cell automation help to reduce

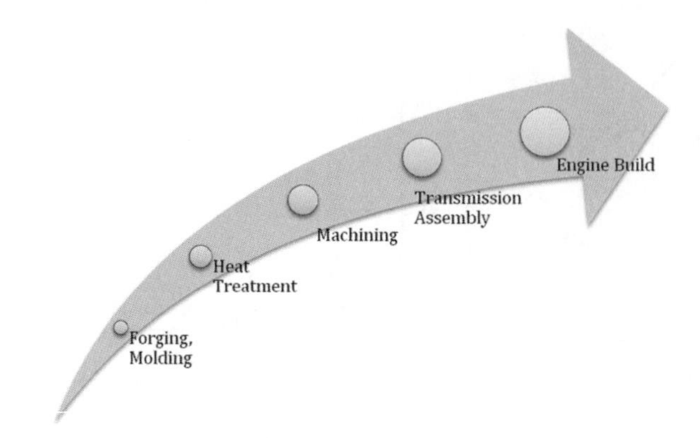

Figure 1.8 The basic processes within a power-train facility

the cycle time. Other specialized operations like the cam boring include the use of a long-line boring bar with custom fixture, to lower or raise the cam. On the other hand, to manufacture the crankshaft, the OEMs have to apply a series of different operations with tight tolerances, that include balancing the mass of the forged steel material, turning of both edges for clamping, and turning for the main and pin bearings, drilling the oil holes, finish grinding for the main and pin bearings, then superfinish the main and pin bearings. Finally, the crankshaft is washed, balanced and inspected. The balancing is done using an intelligent fixture that rotates the shaft and compensates for any imbalances by drilling holes.

The camshaft follows a similar processing sequence to that of the crankshaft, with changes in the tooling used and with the addition of a hardening process, where the shaft is heated using induction coils, then cooled rapidly.

The power-train manufacturing processes are also responsible for making the transmission components, mainly the gear system. The typical material for the different automotive gears is based on alloyed steel that provides the hard finish for the gear teeth while the core is soft and tough, so that it resists continuous use in terms of fatigue and wear resistances. These requirements motivate the use of different heat treatment steps to achieve the hard teeth and ductile core, which include a carburizing step to increase the carbon content within a controlled depth within the gear surface, a quenching process to increase the hardness, and a tempering step to improve the core toughness.

The basic operation used to form the gear is based on hot forging, followed by variety of hobbing and shaping cutting steps to generate the gear teeth. In the shaping process, a cutting gear with the designed profile is used to generate a similar tooth profile in the blank gear, however, in the hobbing process, a worm-like cutter cuts teeth on a cylindrical blank to generate the teeth, hence the hobbing process cannot be used to generate internal gears. Other subsequent operations include gear shaving,

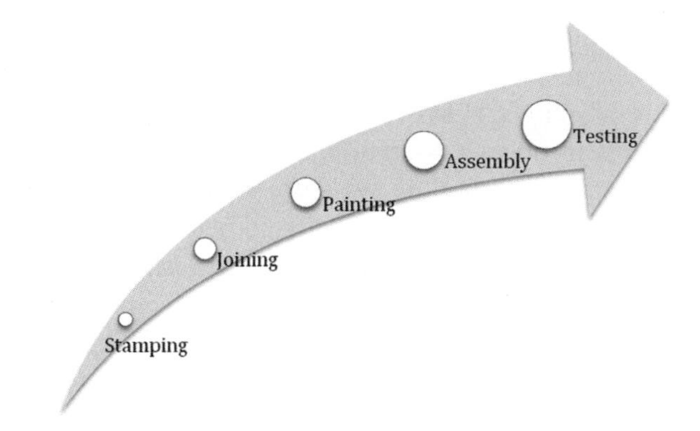

Figure 1.9 The basic processes in an automotive assembly plan

where a helical gear-like cutter, with closely spaced grooves, meshes with the gear so that a controlled material is removed from the gear teeth surfaces.

The standard processes within a power-train facility are displayed in Figure 1.8. So, the overall functional look of the vehicle manufacturing processes can be shown in Figure 1.9.

1.3 Conclusion

The automotive manufacturing processes (assembly and power-train) play a major role in deciding on the vehicles' design characteristics and the overall cost. Thus it is very important for designers and engineers to understand the current manufacturing infrastructure available in their company's production lines. This will pinpoint the manufacturing capabilities and limitations. At the end of the day, the designer will specify the design tolerances but the machine will control the achieved tolerances. Additionally, designers should consider the materials' compatibility from the joining process welding point of view, to avoid galvanic corrosion issues.

Additionally, designers should be aware that the vehicle design complexity in terms of number of parts and intricate shapes results in additional manufacturing steps (added cost and processing time). Also, the number of robotic welders and stamping presses should be taken into account due to their direct impact on the production rate. However, one should recognize that different OEMs make their decisions with regard to the vehicle type, volume and design based on their business models, which might be based on one of the following;

1. competing based on differentiation;
2. competing based on cost; or
3. competing based on time to market.

A company competing based on different and distinguished vehicle types might add complications to their manufacturing systems to achieve new added features or provide a wide range of options. However, an OEM competing on cost aims to reduce the manufacturing cost through less complicated designs and options. The OEMs who compete based on response time typically utilize common components and designs shared between different vehicle types and between old and new models of the same vehicle type. For example, the engine cradle design can be shared between old and new models without affecting the customer's perception of the vehicle as a new model, at the same time it helps the OEM to cut the development time and cost, in addition to reducing the set-up changes in the manufacturing plants.

Additionally, recent changes in the automotive market have forced the automotive OEMs to increase their product portfolio to accommodate new demands from emerging markets, mainly in Brazil, Russia, India, and China or the BRIC countries. This increase in vehicle models has shifted the OEMs' manufacturing models from the *economy of scale* to an *economy of scope*, which motivates further understanding of the manufacturing environment because such a shift adds complications in the following areas: the sequencing of the different models, the production capacity forecasting, and parts (suppliers) and sub-systems' diversity. So new manufacturing and design strategies should be implemented and explored to alleviate some of these challenges such as the use of modular systems and subsystems between the different models, which reduce the parts' diversity and the variations within the processes. At the same time, the modularity might have negative impact on the OEMs' overall flexibility [2].

The impact of the above challenges on the automotive industry have led the automotive OEMs to revise their production and business strategies through mergers with other OEMs and by implementing efficient manufacturing procedures such as lean manufacturing and its different derivatives and versions created to suit each company style and product type. The number of automotive OEMs has dropped from 36 in 1970, to 21 in 1990, and to 14 in 2000. However, the number of automobiles produced is around 55 million vehicles [3] with the majority of production taking place in Asia (around 18 million vehicles), followed by Western Europe (17 million vehicles) and then the USA (around 11.5 million vehicles).

Exercises

Problem 1

In your own words, describe the current metrics used to judge the automobiles, by a typical customer and by an automotive engineer.

Problem 2

What is the difference between the noise and vibration within the context of the automotive NVH requirement?

Problem 3

Explain the three manufacturing system perspectives.

Problem 4

What is the difference between the operational and strategic operations, within the procedural aspect?

Problem 5

List the basic manufacturing processes involved in the making of the automobile BiW, and comment on each process layout and drivers.

Problem 6

List the main differences between the automotive power-train and assembly processes, from the following perspectives: the dependence on suppliers' parts, and the nature of the processes utilized.

Problem 7

What is the difference between the gear hobbing and shaping processes?

Problem 8

What is the impact of economy of scale and the economy of scope on the automotive manufacturing process?

Problem 9

Automotive OEMs might compete based on different criteria, what are they?

Problem 10

List three challenges that have impacted the automotive industry in the past decade.

2

Stamping and Metal Forming Processes

The formability of sheet metals is one of the most important steps in the automobile manufacturing activities, because it is the decisive factor in the vehicle shell shape (styling), geometry (fit and finish), and performance (wind noise, water leakage). Additionally, the development and approval of the stamping dies are the most expensive and time-consuming efforts during a new vehicle design and launch. The die approval process can consume about 50 weeks before the Start of Production (SoP) to achieve the final dimensional validation.

The stamping of sheet metals can be defined as the process of changing the shape of the sheet metal blank into a useful shape in the plastic deformation state, using a die and a mechanical press; stamping is considered a net shaping process. However, the stamping engineering efforts are not limited to production engineering (i.e. the stamping process) but also include the development of the required tooling (i.e. stamping engineering). Such tooling includes the die making in addition to the fixtures and the automation tools such as the transfer mechanisms typically equipped with suction or electromagnetic cups. For the die making process, stamping engineering starts with the desired panel shape provided by the designer in a CAD file, in addition to the sought panel mechanical properties such as dent resistance (i.e. yield strength). Then, the engineers start with the material selection, i.e. selecting the steel grade, thickness and heat-treatment from what is typically provided by the steel mill. Feasibility analyses follow for each selected material, which lead to a process plan (process settings). After that, the die surface design starts with Finite Element (FE) simulation and numerical trials, followed by the actual (experimental) testing. Successful die designs will then be constructed and validated through a series of try-outs in the die-maker facility and then at the stamping line, using different number of parts (prototypes) and dimensional validation strategies (functional build, event-

The Automotive Body Manufacturing Systems and Processes, First Edition. Mohammed A. Omar.

functional build). Finally, the automation and auxiliary tooling are constructed for each approved die for series production.

On the other hand, the stamping process starts with the steel and aluminum coils provided by the mills with specific thickness, surface topography, widths, and heat treatments. Additional inputs to the stamping press are: the die (toggle, progressive), the lubricants (water or oil), the tonnage conditions, and other process settings such as clearances. Generally, the stamping process constitute following main operations; blanking (or blank preparation), stamping (forming), and assembling activities.

A typical material flow in the stamping area is shown in Figure 2.1. The sequence in Figure 2.1 describes the following main operations:

1. *Blank preparation:* involves a cutting action about a closed shape that is the piece retained for future forming (i.e. the blank). The blank shape is composed of any number of straight and curved line segments. A more detailed look at blanking shows that it is further composed of slitting and shearing. Slitting is the process of cutting of lengths (usually coils) of sheet metal into narrower lengths by means of one or more pairs of circular knives. This operation often precedes shearing or blanking and is used to produce exact blank or nesting widths. Shearing is done by a blade acting along a straight line. The sheet metal is placed between a stationary lower blade and a movable upper blade and is sheared by bringing the blades in contact.

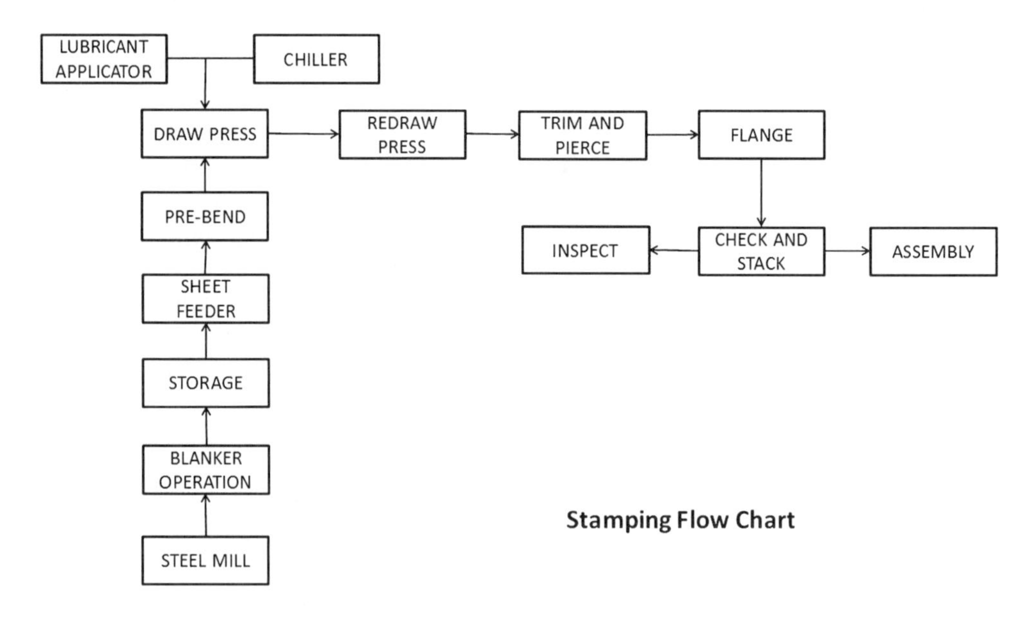

Figure 2.1 The sequence and basic steps in the stamping line

Other cutting operations exist in the automotive stamping for developed blanks, such processes include: (1) *piercing* which is the forming of a hole in sheet metal with a pointed punch without metal fallout; (2) *lancing* that creates an opening without completely separating the cut piece from the body of the sheet metal, such as the case for louvers; (3) *trimming* is the process of removing unwanted metal from the finished piece that was required for some previous stamping operation, such as binder areas, or was generated by a previous stamping operation, such as the earing zone on the top of a deep drawn cup; and (4) *parting* operations are used to separate two identical or mirror image stampings that were formed together (typically for the expediency of making two parts at one time or to balance the draw operation of a nonsymmetrical part). Parting also is an operation that involves two cut-off operations to produce contoured blanks from strip.

2. *First forming operations:* which aim at forming the blank into a semi-developed blank that has the initial shape. First forming operations include bending and flanging through either a shrink flanging, where the length of the flange shrinks as it is formed, or stretch flanging where the material is stretched as it is flanged.
3. *Drawing operations:* in the automotive press shop, most dies are called draw dies because the metal is drawn into the die cavity. However, most of the deformation modes are based on biaxial stretch over the punch or a bend-and-straighten from the flange. Drawing, sometimes known as cup drawing, radial drawing, or deep drawing, has a very specific set of conditions which differentiate it from other operations. The unique attribute of deep drawing is the deformation state of the flange. As the blank is pulled toward the die line, its circumference must be reduced. This reduction in the circumference generates a compressive stress in the circumferential direction, resulting in a radial elongation as the metal is extruded in the opposite direction.
4. *Subsequent operations:* most of the automotive panels require a sequence of forming steps because the degree of forming (flange angle, etc.) cannot be accomplished in a single step. Such operations include the re-strike step which comes after the metal has been stretched over a large radius punch (to avoid splitting), to spread the metal into the desired shape without any additional tension in the stamping line. Another typical subsequent forming operation is the redrawing. Limits are imposed on the blank diameter which can be drawn into a cup of a given diameter (this will be discussed in further detail in Section 2.1.3). Should a deeper cup be required, an intermediate diameter cup is drawn first; then the cup is redrawn in one or more subsequent stages to achieve the final diameter and height.
5. *Assembling activities:* these include variety of specialized cells for combining panels to form BiW components such as joining the door inners and outers. Additionally, other assembling might be done in die-joining strategies.

2.1 Formability Science of Automotive Sheet Panels: An Overview

The formability can be defined as the extent to which a sheet metal can be formed or worked into a specific shape without failure (cracking) and/or forming other undesirable features (e.g. Lueder bands). Formability is neither a material property nor a process property but it is a system property, dependent on the intrinsic and extrinsic sheet metal properties in addition to the process conditions.

In general, the variables that control formability of sheet metals within automotive production are:

1. *materials variables:* such as its thickness, width, *n*-value, *r*-value, *m*-value, surface topography, coating type, tensile and yield strengths, etc.;
2. *blank variables:* size, location, contour, flatness, edge conditions, pre-bend, etc.;
3. *die variables:* surface finish, rigidity, clearance, draw-beads, wear-plate tolerance, punch and die radii, etc.;
4. *press variables:* ram and bed flatness, shut height, inner ram load, press type and action, punch guidance, punch speed profile, etc.;
5. *other variables:* material temperature, die temperature, atmospheric conditions, etc.

The stamped pieces quality are typically judged based on the panel appearance, the resulted strains (patterns, directions), and its final dimensions. Such criteria can be further quantified through stamped panels' final geometric characteristics, which include one or a combination of four main geometric shapes: plane, tunnel feature, dome element, and irregular features. Additionally, the resulted strains values and gradient describe the metal flow pattern. The most severe of stamped defects is the formation of a split or crack in shaped panels. Researchers have tried since the 1950s to develop formability metrics and theories with a focus on split or crack avoidance in sheet metal. According to Wang (in [4]), research teams focused on correlating the split occurrence with the material *n*-value and *r*-value through utilizing fracture mechanics. Further work by Keeler and Goodwin [5, 6] has established the splitting criterion in the plane stress states, i.e. in the bi-tension deformation state and the tension-compression deformation state. This led to the forming limit curve (FLC) or forming limit diagram (FLD), which describe the split tendency and the material deformation capacity in relation to two plane strains, called the major and minor strains. The application of the FLC will be discussed in more detail in Section 2.1.4.

Later work by Yoshida [7] focused on the stamping surface defects such as the splits, and the formed panel shape change through spring-back and/or distortion. Yoshida developed the first stamping indices, anti-fracturability and the shape-fixability. However, recent advances in the stamping process and engineering required further investigations due to the addition of new steel grades, mainly the high strength steel (HSS) and the advance high strength steel (AHSS), in addition to the use of more stamped aluminum in vehicle bodies. This chapter will focus on the Universal Formability Theory as proposed and developed by Xu [4] to describe the different formability indices; however, the text will not discuss the mathematical background or derivation in detail but will focus on the application of such theory to the evaluation of automotive stamped panels.

The main stamping defects analyzed by the Universal Formability Theory are as follows:

1. *Splitting in the stamped panel:* local necking rupturing in the stamped panel away from the edge.
2. *Splitting at the edge of the panel:* rupturing near the edge of the stamping due to the lower deformation capacity at the edge due to the shear zones (edge burrs and cracks).
3. *Wrinkling:* surface waviness resulting from compressive plastic instability.
4. *Shape change:* this is the elastic recovery within the panel caused by distortion and spring-back. The spring-back can be a first spring-back or a second spring-back, depending on its occurrence after the first or the second unloading of the panel from sequential stamping processes.
5. *Low stretch:* causing a lower work hardening performance of the formed panel, thus affecting its dent resistance.
6. *Surface soft or low oil canning load ability:* typically caused by the residual stresses from the different loadings in sequential stamping.

From the above, one can summarize the mechanisms that form defects as: defects due to extreme stresses and strains (as in the case of splits); the stress and strain gradients; the deformation history of the panel; and the residual stresses after unloading from the die cavity. The following discussion focuses on the formability indices developed to address each of the above stamping defects and their formation mechanisms.

The Universal Formability Theory suggests six formability indices to comprehensively address these stamping defects:

1. *The anti-fracturability index:* to address the splitting of the stamped components due to tensile stresses (away from the edge). The sheet metal passes through four stages of deformation before it splits. These stages are: (a) the elastic deformation when stressed within the yield-stress of the material; (b) the uniform elongation when the applied stress reaches the yield-stress; (c) then the diffuse necking stage; and finally, (d) the local necking which leads to fracture. Figure 2.2 illustrates these four stages on the stress-strain diagram. The deformation when the local necking occurs is the material maximum deformation capacity (DC) that can be safely utilized in automotive stamping. Knowing the material DC enables the

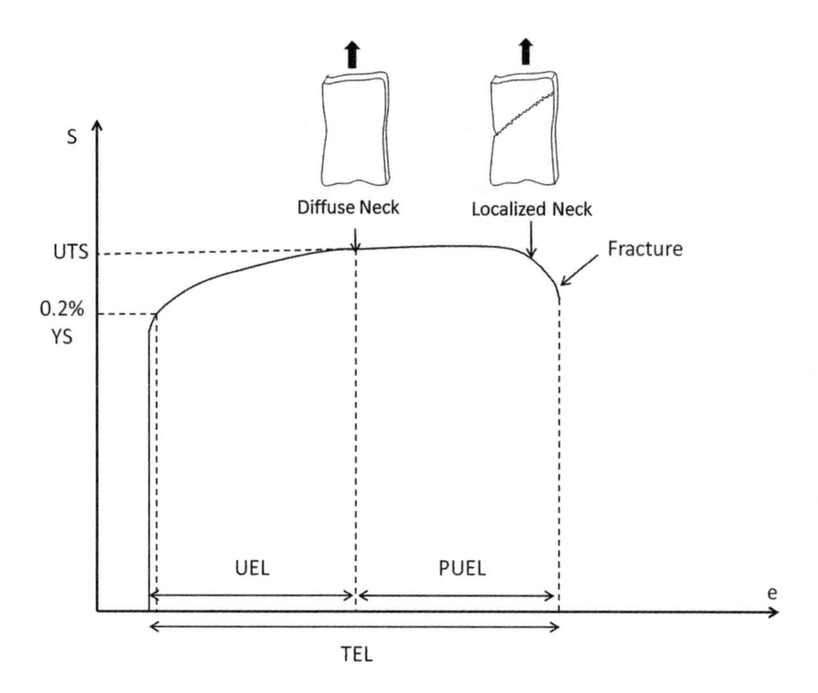

Figure 2.2 The stress-strain diagram

engineers to predict the remaining deformation capacity (RDC) of the shaped sheet metal after the first forming operation. This can be done by measuring the strain difference between the local necking point (i.e. DC) and the strain after the first forming.

Knowing the panel's RDC helps the stamping engineers to qualify their stamping practices as safe, marginal, or critical. Furthermore, the forming limit curve (FLC) helps the engineer visualize the stamping strain state relative to the material available DC. Figure 2.3 shows a typical forming limit diagram for a steel panel. The diagram includes the plane strains on the x and y axes, while the solid line represents the DC of the material. Thus one can measure the RDC at any point in the stamping by measuring the linear straining path or distance (measured from the origin) from that point to the solid line. A marginal region is typically established, around 10% for steel and 8% for aluminum samples to include a safety factor. This safety factor is typically measured as an increment in the major strain as shown in Figure 2.4 which shows the RDC and the ΔFLD or safety factor. However, one should note that the safety factor represents an actual value for the plane strain deformation; in other word, the case when the minor strain is zero, thus the RDC is a more accurate description of the strain state (safe, marginal, or critical). At the same time good stamping practice requires that the forming consumes at least 50% of the material DC. More details about the FLC and how to

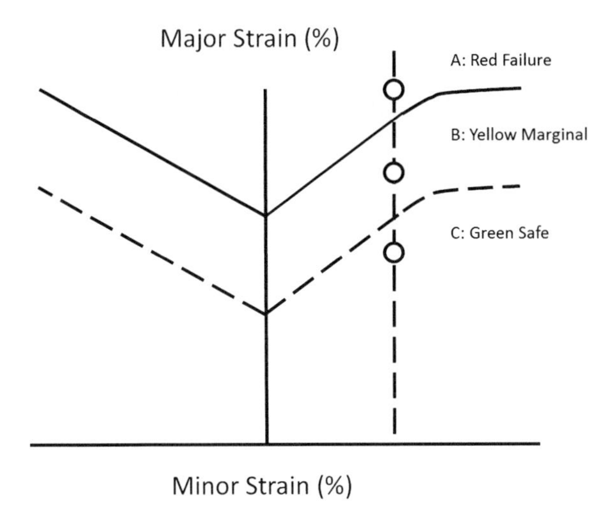

Figure 2.3 The forming limit diagram showing the different forming regions

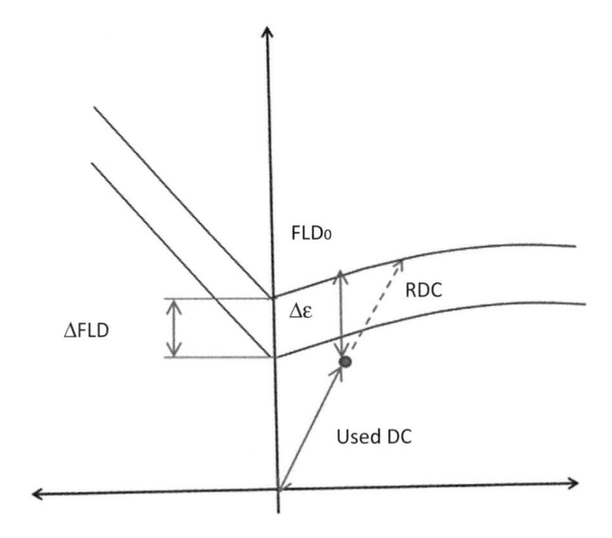

Figure 2.4 The FLD showing the RDC, DC and the safety factor involved

measure the major and minor strains using the Circle Grid Analysis is given in Section 2.1.4.

2. *The anti-edge fracturability index:* this index describes the weakened deformation capacity of the material near an edge. The shearing action along the material edge creates micro-cracks and burrs that tend to create a work-hardened region, that ranges in width from 50–70% of the material thickness. Additionally, the burrs and cracks work as fracture factors/initiators near the edge. The material RDC near the edge can then be evaluated in the same manner as in the case of the

anti-fracturability of the stamping away from the edge. However, the difference in straining path should be taken into consideration for the case of simple tension and one should also take the burr height into consideration. Simple calculations can assume that the burr height is equal to the panel thickness; thus, the RDC is the difference between the DC (in simple tension) and the major strain.

3. *The anti-wrinkle-ability index:* wrinkles resulting from unbalanced compressive stresses are typically evaluated based on their geometric characteristics as three types: Types I, II and III. Type I is the most severe with height greater than or equal to the gap between the upper/lower die and the sheet metal; and Type III with height that can only be measured using specialized optical illumination.
4. *Shape fixability and shape change index:* upon lifting the die surface, the formed panel still possesses some elasticity in the form of spring-back and distortion that can lead to changing the formed shape. The spring-back is typically considered a shape change that causes the product dimensions to be out of tolerance toward one side of the reference surface, while the distortion is when the product dimensions are out of tolerance with both sides of the reference surface.

   Spring-back is more severe for aluminum than steel, because spring-back decreases as the materials' Young's modulus increases. Steel has three times the Young's modulus that of aluminum, however, the spring-back increases as the yield strength increases. This means that the spring-back in the case of aluminum is less than three times that of steel. This also explains why the dies made for aluminum are typically larger than those made for steel, for panels with the same shape and dimensions. Also the spring-back in the case of high strength steel (HSS) is higher than that of low mild steel. The spring-back effect can be visualized in Figures 2.5 (a) and 2.5 (b).

   Typically, spring-back is more pronounced in channels and under-body structures and classified as an angular change, side-wall curl, and twist. The angular change happens when the bending edge line deviates from that of the forming tool and is typically caused by stress difference in the sheet thickness direction, when a sheet metal bends and unbends over a die radius. This stress difference in the sheet thickness direction creates a bending moment at the bending radius after dies are released, which results in the angular change. Sidewall curl is the curvature created in the side wall of a channel. This curvature occurs when a sheet of metal is drawn over a die/punch radius or through a draw bead. The main reason for this unevenness in the thickness direction is due to the stress generated during the bending and unbending process. The inside surface initially generates compressive stresses while the outer surface generates tensile stresses. During the bending and unbending sequence, the deformation histories for both sides of the sheet are unlikely to be identical. This usually manifests itself by flaring of the flanges, which is an important area for joining to other parts. The resulting sidewall curl can cause assembly difficulties for rail or channel sections that require close tolerances between the mating interfaces during joining. The twist

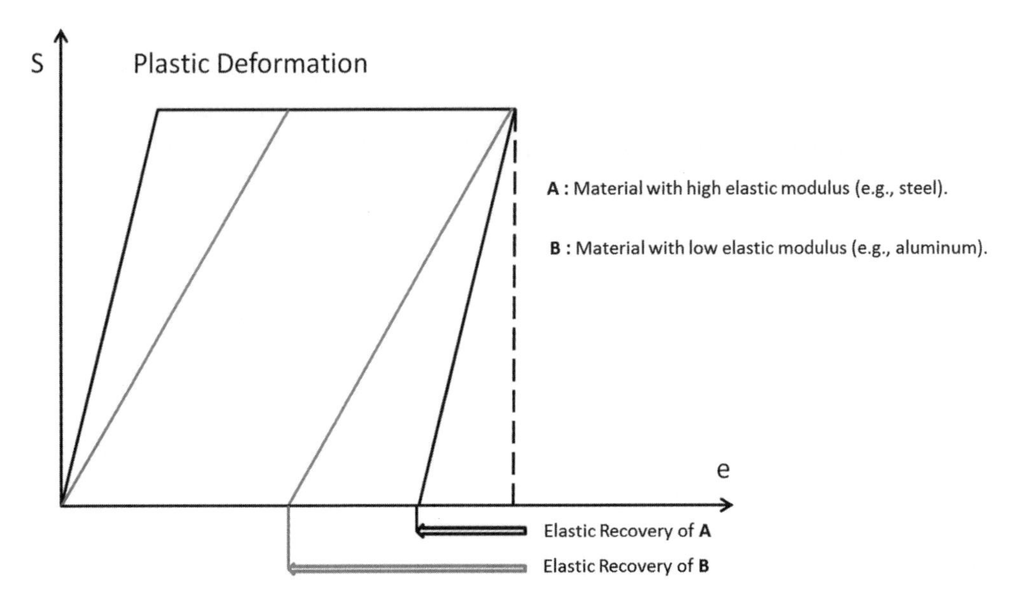

Figure 2.5 (a) The spring-back behavior for steel and aluminum.

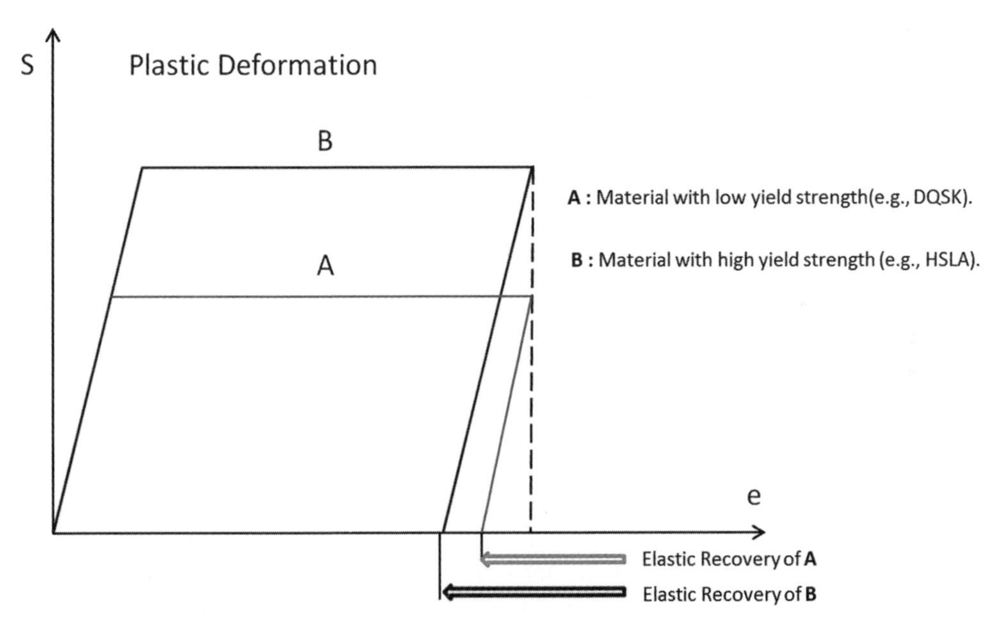

Figure 2.5 (b) The spring-back behavior comparison between two different grades of steel

spring-back is developed due to the torsional displacement developed in the panel cross-section as a consequence of the residual stresses acting in the part to create a force couple which tends to rotate one end of the part relative to another.

5. *Stretchability index:* this index is especially important in the case of flat, large exterior panels, because of the dent resistance and oil canning requirements. The oil canning tests for the maximum static, normal load cause surface elastic instability, while the denting for doors, as an example, can occur from stone impacts (dynamic denting). Denting can occur because the door surface is smooth and may not have sufficient curvature to resist "door slamming" (quasi-static denting) or along prominent feature lines where "creasing" can occur. The typically used standards, those defined by American Iron and Steel Institute (AISI), define a minimum dent resistance of 9.7 Joules. Based on testing using the practical techniques outlined, empirical formulae predicting the force and energy required to initiate a dent, can be shown in Equation 2.1:

$$W = k \times Y_s^2 \times t^4 / S \tag{2.1}$$

where W is the denting energy, K is a constant, YS is the material yield strength, t is the panel thickness and S is the panel stiffness. Panel stiffness depends upon the elastic modulus, the panel thickness, shape and geometry.

For such panels, the stamping process should achieve a critical strain value so a certain amount of work hardening is achieved. Work hardening is required so that the final flow stress is able to satisfy the dent requirements because the initial blank has lower yield strength; the larger the metal deformation, the higher the final yield strength of the shaped panel. The stretchability defect is sometimes called "low stretch."

6. *The anti-buckling index:* this is also related to the large, flat panels and its oil canning ability. The anti-buckling index focuses on the residual stresses that result from the unloading of the panels from the die cavity in sequential stamping. After the first unloading, compressive residual stresses exist in the flat panels and lead to a decrease in their oil canning. The main effect here is due to the stress gradients, not values.

So the Universal Formability Theory focuses on analyzing the stampings based on the final shapes (the four geometric characteristics) in addition to the metal flow patterns (strain gradients and values) and correlating these with the stamping process sequence of blanking, forming (metal flow patterns), unloading (residual stresses, spring-back), and then re-forming (work hardening). Also, the quality of the stamped panels is judged based on its allocated zone in the vehicle body and this zone relation to the customer perception of quality. The stampings or body-panels are typically classified as exterior and interior panels, with the exterior panels being further clas-

Automotive Body Panels	Exterior Panels	Exposed Panels	Permanently Exposed Panels (door outer skin, hood outer, roof), Class A
			Temporarily Exposed Panels (B-Pillar) , Class B
		Un-exposed Panels	Tactile (Touch) Regions
			Non-Tactile (Non-Touch) Regions
			Joining Interface (spot weld contact areas)
	Interior Panels		Non-Tactile (Non-Touch) Regions
			Tactile (Touch) Regions
			Joining Interface (spot weld contact areas)

Figure 2.6 Zoning of the automotive panels showing the different panels designation from surface finish perspective

sified into exposed and unexposed regions. The permanently exposed, exterior panels are known as the *Class A* surfaces. A *Class A* implies no stampings' defects are allowed even Type III wrinkles. The complete panels' allocation is shown in Figure 2.6 [4].

2.1.1 Stamping Modes and Metal Flow

Generally, the automotive stamping focuses on two modes of stamping: deep drawing and stretch forming. The deep drawing mode is when the sheet metal is formed (drawn) from the binder by the punch. In other words, if the clamping pressure of the holding down ring is high enough to avoid buckling and wrinkling in the flange (the part of the test-piece clamped between the holding down ring and the die), then the punch stroke will force the material down into the die, thereby pulling the material through the flange and into the die cavity. In this case, there will be very little deformation of the cup wall or cup bottom, if any, and most of the deformation will be in the flange. The deep drawing mode has the following characteristics:

1. sensitive to the r_m-value;
2. involves compressive surface strains and little or no thinning (actually thickening);
3. used in cases of severe deformation, such as cans, gas tanks, floor panels and oil pans;
4. extent or amount of drawing is determined by measuring the length of the impact lines from the binder to their final location on a drawn panel (e.g., cup height).

The stretch forming mode is when the clamping pressure of the holding down ring is very high, such that the friction forces in the flange are high enough to allow the material to flow in. The punch stroke will then force the material in the cup wall and on the bottom of the cup to deform instead. In this case, there will be very little deformation in the flange, but considerable deformation in the cup wall or cup bottom. The stretch forming mode has also the following characteristics:

1. sensitive to the n- and m-values;
2. as the material is locked in the binder, thinning is required to attain the desired part depth;
3. used for shallow, exposed parts such as doors, hoods and trunk lids;
4. degree of stretch measured by dimensional changes in inscribed circles (e.g., 100-mm).

The formability of sheet metal can be further analyzed from the metal flow patterns within the die cavity. In stamping, there are three main patterns of metal flow: elastic flow, plastic flow, and rigid movement. The rigid movement is when the position of sheet metal changes without actual deformation, while the elastic flow is the Hooke's-law based deformation that is temporary and disappears once the load is removed. The plastic flow is the permanent deformation of the sheet that should be constrained. To control the metal flow patterns, stamping engineers manipulate the process variables such as the binder pressure and panel alignment within the die cavity. Additionally, optimized metal flow can be achieved through the design of certain features in the stamping, such as the embossments. The embossing process is performed on a localized area so that the remainder of the blank is large compared to the deformation zone. This means that the blank is considered to be of infinite size and no metal flows from the blank into the deforming zone. Embossments are generally divided into three types (see Figure 2.7): beads and ribs, offsets, and decorative. The beads and ribs are characterized by a long, narrow depression which may be straight, circular, or a combination.

Most, if not all, automotive panels are the result of a combination of the different forming modes so that the different geometric characteristics and shapes can be formed, so to analyze a stamped panel, its different forming modes should first be identified and then analyzed according to the required metal flow pattern. Figures

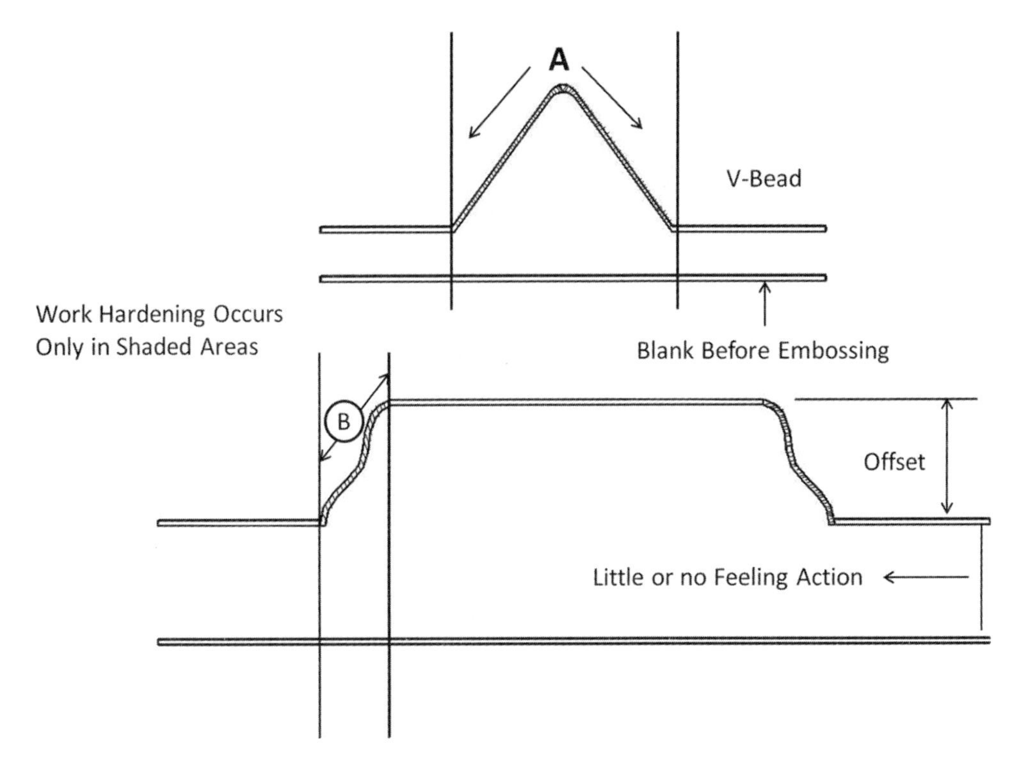

Figure 2.7 Examples of automotive embossments

2.8 (a) and (b) provide a pictorial example of such interaction between the different forming modes.

2.1.2 Material Properties and their Formability

Materials formability can be evaluated using their tensile test results by inspecting the following:

1. *Percentage of total elongation:* which is typically quantified through the amount of strain in a 2 inch (50 mm) gauge length of a tensile strength specimen; the higher number represents higher formability. Elongation is a direct measure of the material ductility.
2. *Percentage of uniform elongation:* The strain to maximum load in the tension test is called the uniform elongation. It is the limit of plastic strain that is uniformly distributed over the strained area.
3. *Work hardening exponent (n-value):* this indicates the relative stretch formability of sheet metals and the increase in strength due to plastic deformation, measured

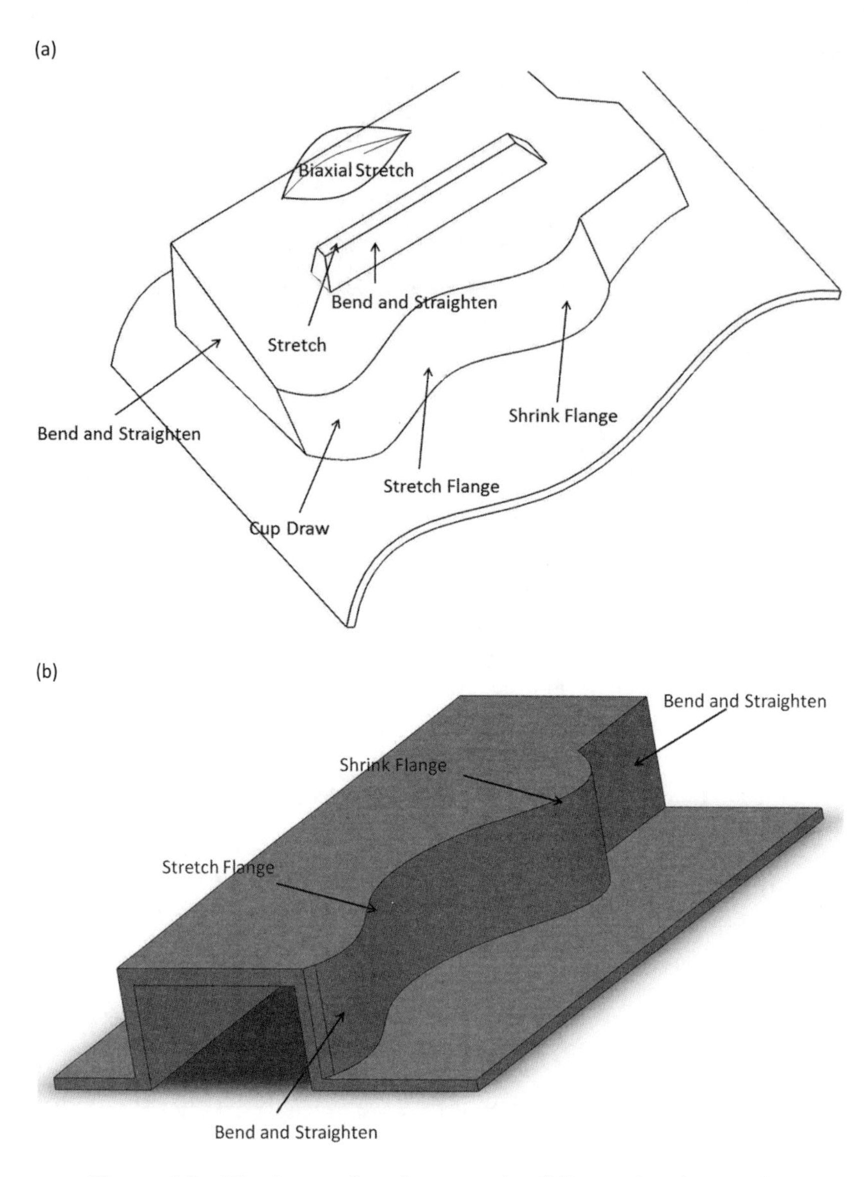

Figure 2.8 The interactions between the different forming modes

as the slope of the stress-strain curve between 10–20% strains. The n-value is not the work hardening rate, also a higher n-value does not imply a higher work hardening rate. For materials that exhibit power law strain hardening behavior, the n-value can be determined by plotting the flow stress curve behavior on a log-log scale and taking the slope to be the n-value or strain hardening exponent. For many automotive grades, $n10/20$ is normally specified. This particular n-value is the strain hardening exponent determined in the strain interval from 10–20%

elongation. Determination of the n-value from load elongation curves is described in [8].

The n-value can be defined numerically to be equivalent to the true strain at maximum load in a tensile test ($n = \varepsilon_u$) computed in Equation 2.2:

$$n = \varepsilon_u = \ln\left[1 + \frac{\%EL}{100}\right] \tag{2.2}$$

4. *Plastic strain ratio (r-value):* this describes the ability of a sheet metal to resist thinning or thickening when subjected to force, it is a measure of drawability, and equals the true strain in the width direction divided by the true strain in the through thickness direction. The procedure for measuring r-value can be found in ASTM E517.
5. *Strain rate sensitivity exponent (m-value):* this value describes the changes in strength and other properties of a metal as a function of the strain rate or deformation rate in tensile tests. At a given strength level, a material with a high m-value will generally have better formability because:
   i. during forming, the material that deforms most quickly will have the highest flow stress;
   ii. it is easier to deform the material around this region, which results in a more uniform strain distribution in the final part.

   Although a high m-value is generally regarded as having a positive effect on formability and strain distribution during forming, the primary focus of using materials with high m-values has been on "crash-worthiness." If a material has high positive strain rate sensitivity, it will be stronger during a crash.

   Determination of the m-value or strain rate sensitivity exponent can be done by: *the jump test*. In the jump test, a tensile test is run at one strain rate up to a prescribed strain level, then the strain rate is changed (usually increased), and the difference in flow stress is measured at the prescribed strain level at both strain rates. This test can be pictorially described in Figure 2.9, with the m-value computed using Equation 2.3.

$$m = \log\left(\frac{\sigma_1}{\sigma_2}\right) / \log\left(\frac{\dot{\varepsilon}_1}{\dot{\varepsilon}_2}\right) \tag{2.3}$$

One should also note that both the n-value and m-value affect formability: a high n-value can be attributed to a high uniform elongation (UEL), and the m-value is proportional to the amount of post-uniform elongation (PUEL). The n-value is related to strain hardening capacity, and the m-value is a measure of strain-rate sensitivity. Post-uniform elongation is proportional to the m-value because the appearance of necking during a tensile test will cause a sudden local increase in the strain rate.

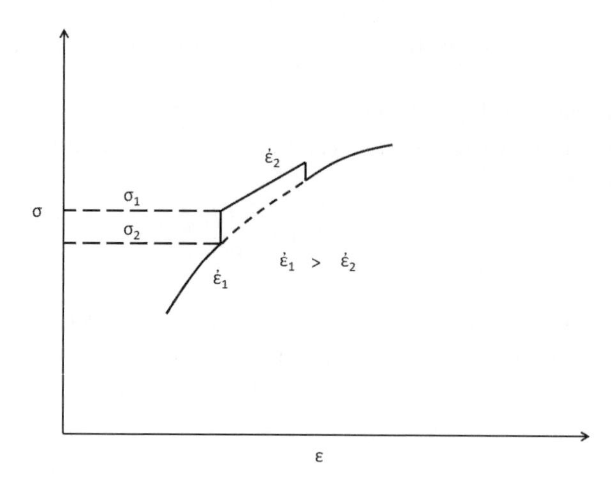

Figure 2.9 The *m*-value experimental procedure

6. *The strength (yield strength (YS) and ultimate tensile strength (UTS)):* this is generally not considered a measure of formability; however, strength can be a major influence in determining spring-back characteristics that can lead to shape distortion. Additionally, the YS affects the stamping process settings in terms of tonnage, in other words, engineers should ensure that they pass the YS to achieve permanent plastic shapes. Also the YS and UTS decide on the stamping performance in terms of dent resistance and oil canning, and a knowledge of YS and the *n*-value is important in predicting the final stamping flow stress after the work hardening and the bake-hardening effects. The UTS marks the end of uniform elongation and the beginning of post-uniform elongation. The UTS is reached when the strengthening effect of work hardening is matched by the weakening effect caused by material thinning.

   Figure 2.10 displays some of the typically used materials (steel grades and aluminum) in automotive panels, from a tensile test perspective and Figure 2.11 shows them from the stamping process engineer's point of view, which focuses on the yield strength to know the appropriate press tonnage to achieve plastic deformation and the final elongation achieved. In other words, it shows the process input and output. The inter-terrestrial free (IF) steel grade has the highest elongation at the lowest yield strength, which is the best choice from a stamping perspective. However, other performance characteristics (final strength) and cost should be taken into account.

The following section describes the different formability metrics and measures that are typically used to determine the material formability behavior for the different forming modes.

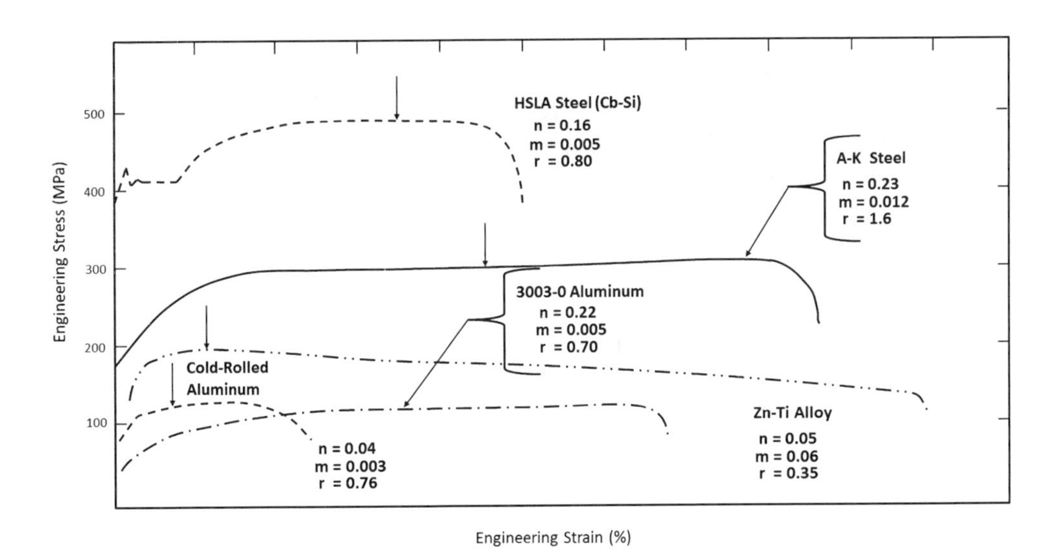

Figure 2.10 Tensile behavior of different materials, steel grades and aluminum

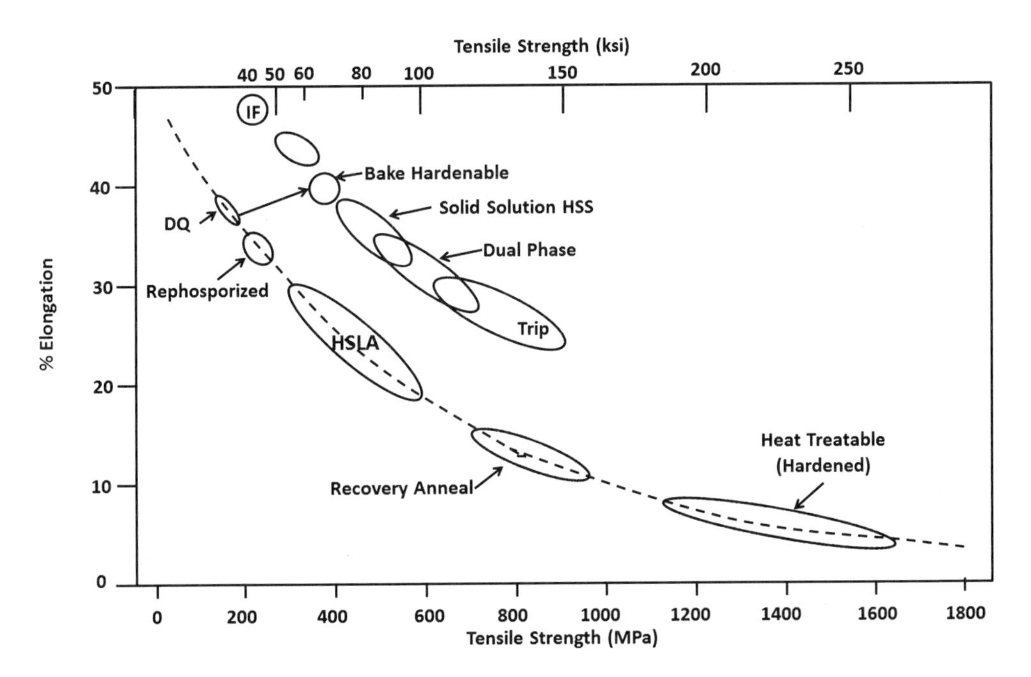

Figure 2.11 Different steel grades' elongation behavior versus their tensile strengths

2.1.3 Formability Measures

Formability measures describe the different testing schemes applied in the stamping lines to identify the material formability behavior under specific process conditions (e.g. lubricants) and forming modes. Some of the most commonly applied tests are: the anisotropy parameters (normal and planar), the limiting draw ratio (LDR), the limiting dome height (LDH), and the hole-expansion ratio (λ-value):

1. *Anisotropy parameters:* the normal anisotropy is used to measure mean r-value or the r_m-value, while the planar anisotropy evaluates the Δr that describes the variation of mechanical properties as a function of inclination to the rolling (applied at the steel mill) direction of the sheet metal. The normal anisotropy is measured during a uni-axial tensile test, where the longitudinal strain (ε_L) and width strain (ε_W) are measured, according to Equations 2.4 and 2.5:

$$\text{Length strain: } \varepsilon_L = \ln(1 + e_L) \tag{2.4}$$

$$\text{Width strain: } \varepsilon_W = \ln(1 + e_W) \tag{2.5}$$

   And recalling that the $\varepsilon_W + \varepsilon_T + \varepsilon_L = 0$, then the r_m-value can be computed from Equation 2.6 at 10% strain:

$$r = \varepsilon_W / [-(\varepsilon_W + \varepsilon_L)]. \tag{2.6}$$

   Planar anisotropy is described by a term called Δr, which is a measure of the variation in the r-value as a function of orientation, rather than a weighted average. A Δr-value close to zero suggests *planar* isotropy. By reducing $|\Delta r|$ the tendency for ear formation can be reduced, thereby minimizing the amount of trimming needed after deep drawing. Numerically Δr can be calculated from Equation 2.7:

$$\Delta r = (1/2)(r_{0^\circ} - 2r_{45^\circ} + r_{90^\circ}) \tag{2.7}$$

   Where, r_{θ° is the r-value at θ degrees from the rolling direction.

   Good formability can be achieved through a high value of r_m-value and a low $|\Delta r|$ value. Additionally, the Δr value affects the formation of ears in automotive stampings, if Δr value is <0, then the ears will form in a diagonal direction (i.e. 45 ° from the rolling direction), while if the Δr value is >0, then the ears will form in parallel and orthogonal directions (0 ° and 90 ° from the rolling direction). Such behavior is shown in Figure 2.12.

2. *The limiting draw ratio (LDR)* is the largest blank-to-punch diameter ratio that can be achieved in a simple cup (or can) drawing operation (without flange). The LDR values increase as the r_m-value increases and this is considered an important test for the material formability behavior with friction. Thus, the LDR test is considered a necessary test because it describes the friction conditions' impact on

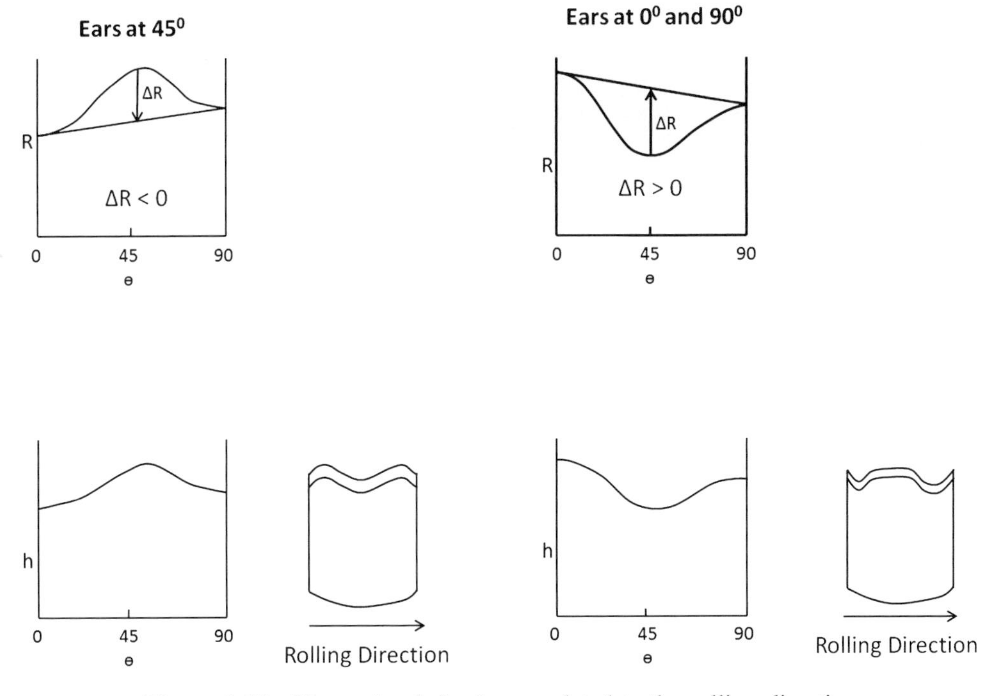

Figure 2.12 The earing behavior as related to the rolling direction

formability for wide range of stretch-forming operations. The LDR of steel is approximately equal to 2 or even higher.

3. *The limiting dome height (LDH) test* deforms fully constrained material specimens (of different widths) using a 100 mm diameter spherical punch, then the punch height or "limiting dome height" is recorded at fracture. The LDH is typically plotted against the samples' widths to obtain an LDH chart. Figure 2.13 shows the LDH and its chart. This test describes the die-contact and friction effect on materials' formability. Hence, the LDH chart represents a more realistic picture than a tensile test. To illustrate with an example, Figure 2.14 shows the LDH results for two low-carbon fully stabilized FS steels with similar tensile properties (*YS* ~ 160 MPa, *UTS* ~ 300 MPa, Tot. Elong. ~ 45 pct). The difference is that one is batch annealed and coated with an electro-galvanized Ze-Fe alloy (EGA), while the other has been hot-dip galvanneal (HDGA) coated on a continuous galvanizing line. The HDGA-coated material has a slightly higher *n*-value (~0.24) than its EGA-coated counterpart (~0.23). Based on tensile properties alone, one can predict that HDGA-coated material will result in better formability; however, the LDH curve of the EGA-coated material is higher. The LDH shows the interplay between mechanical properties and friction that governs the overall formability of a particular system. In this case, the EGA coating has a lower frictional response

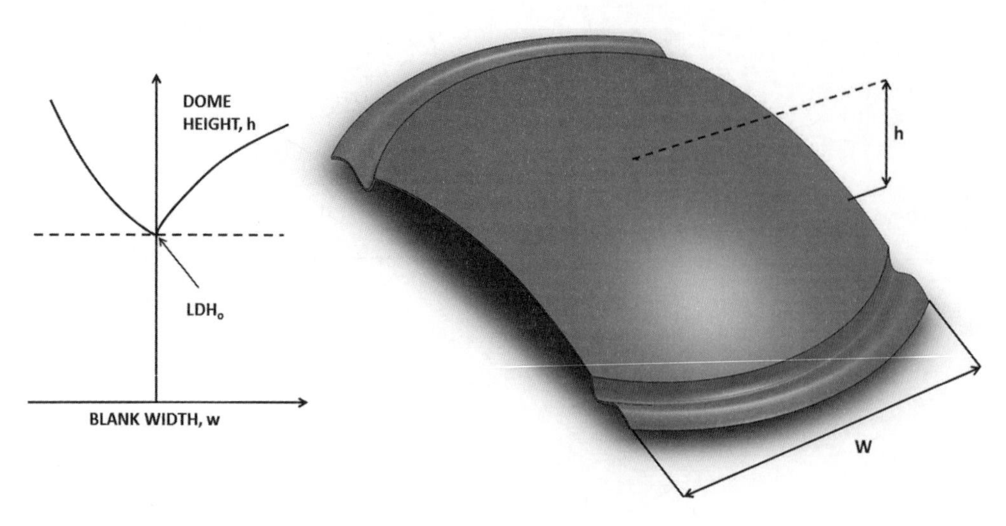

Figure 2.13 The limiting dome height test sample

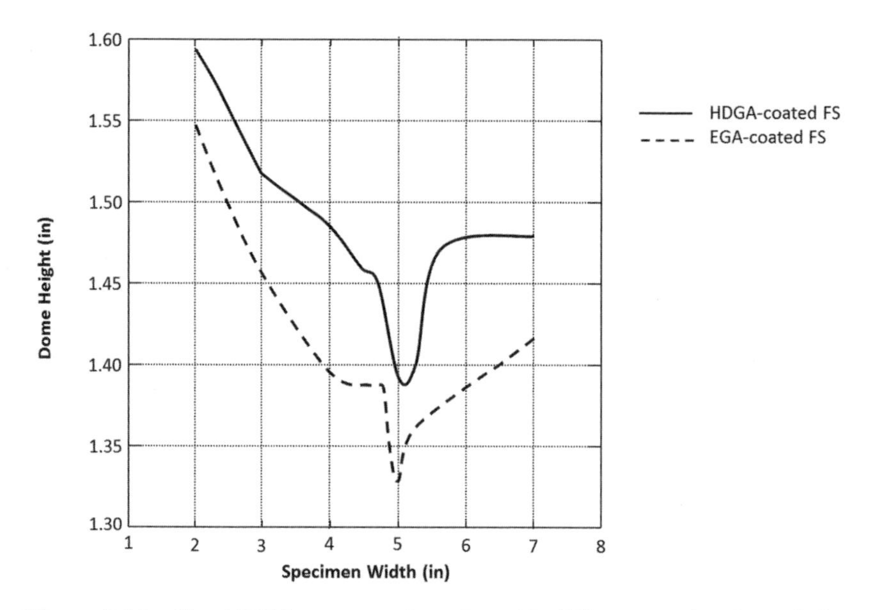

Figure 2.14 The LDH for two steel grades with different coating formulation

than the HDGA coating, and strain is distributed more uniformly, thereby resulting in a higher dome height in the LDH test.

The minimum in the curve is denoted LDH_0 where the subscript "0" refers to the fact that this minimum dome height typically corresponds to a plane-strain failure with a minor strain of zero. For some applications or comparisons, the LDH_0 value is all that is needed, as LDH_0 describes the general level of formability and suggests the position of the LDH curve with respect to the ordinate scale.

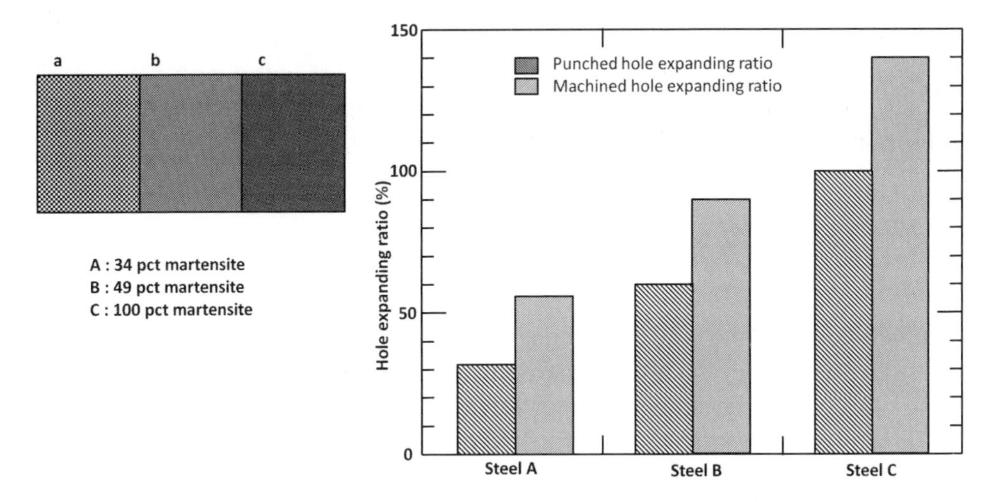

Figure 2.15 The expansion ratio for different steel grades with different martensite percentages in their microstructure

4. *The hole-expansion ratio* is considered a measure of "edge-stretchability" and it can simulate the case where material is blanked before forming and its sheared edges are subjected to deformation. It is considered very sensitive to edge quality, and the material microstructure, especially its microstructure uniformity.

   The λ-value is considered very important to describe the formability of advanced high strength steels (AHSS) because of their multi-phase micro-structure (Polygonal ferrite: α_P/ Bainite: α_B/ Martensite: α'/ and Retained Austenite: γ_R) and the heterogeneous distribution of such phases.

   For the hole-expansion test, a hole of prescribed diameter (usually 10-mm) is punched into a sheet steel sample with a prescribed clearance (usually between 10 and 15% of the thickness of the steel). Then a forming punch is used to expand the pre-punched hole until a visible crack is observed around the rim of the punched hole. Various tooling geometries are used for this test. The dimensions of the hole are measured before and after expanding, and the λ-value is simply expressed as:

$$\lambda = 100 \times (D_f - D_o) / D_o \tag{2.8}$$

   where λ is the hole-expansion ratio, and D_f and D_o are the final and original dimensions of the hole after punching and expanding, respectively.

   To illustrate the effect of the hole-expansion test results, Figure 2.15 shows the combined influence of microstructure and edge condition on hole-expansion capability. Shown are three types of steel with UTS around 1000-MPa, but with very different microstructures—all of which are various mixtures of ferrite (soft) and martensite (hard). From left to right, the martensite content increases from 34%

to 100%. Correspondingly, the hole-expansion ratio λ-value increases drastically. This is not to say that increasing the martensite content will improve formability in all situations. But the high degree of microstructural uniformity of material allows it to be expanded much more in this particular forming condition.

2.1.4 Circle Grid Analysis (CGA) and the Forming Limit Diagram (FLD)

The Circle Grid Analysis (CGA) is one of the most important tools that stamping process engineers apply to test the performance of the forming process, because this test describes not only the plane strain values but also the straining path, the metal flow direction, the deformation modes, and the strain distribution within the shaped panel. The CGA is conducted through etching a circle grid or array of certain size (depending on the panel) electrochemically at different locations within the stamping before it is formed, after the sheet is strained, the applied circles are deformed into ellipses with a major and a minor axes. The dimensional changes between the formed ellipses and the original circles can then be used to obtain the plane strains.

CGA is used to construct the forming limit diagrams (FLDs), which are very useful conceptual tools used in monitoring forming processes in actual production and can also be used in research environment. Figure 2.16 shows a typical circle grid array.

The change in dimension between the major axis of a deformed CGA and the original circle relative to the original circle diameter (i.e. strain) should always be positive and is an indication of the metal flow. The strain in the major axis direction

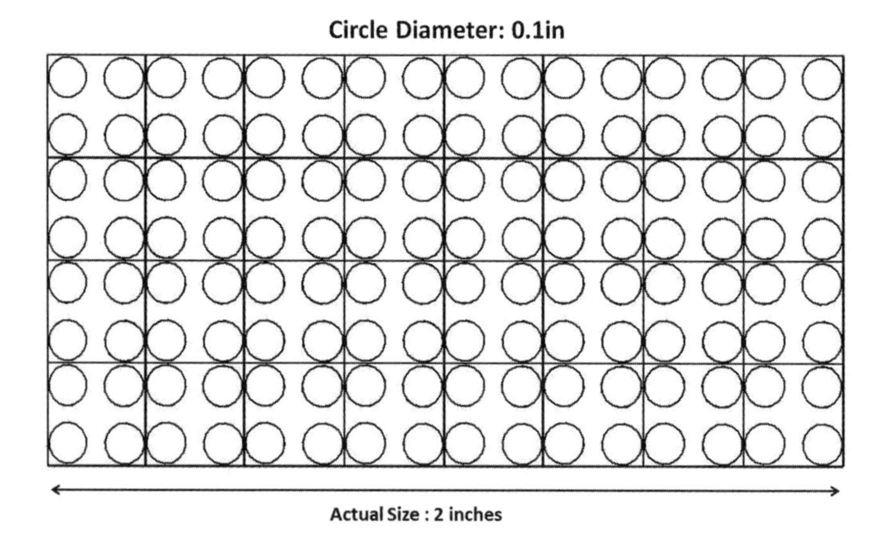

Figure 2.16 The circle grid analysis circle size and profile

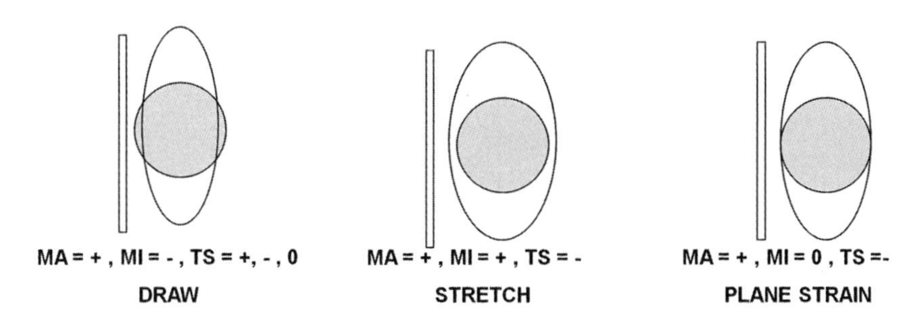

Figure 2.17 The three different forming modes and their relation to the deformed CGA

is called the major strain. On the other hand, the strain for the minor axis can be a positive, a negative value, or a zero, and is an indication of the forming mode at that location; the strain in the minor axis direction is called the minor strain. Based on the combination of the major and minor strain signs, one can decide on the deformation mode in the stamping according to:

- If the major strain is positive, while the minor strain is negative, then the deformation mode is drawing.
- If the major strain is positive, while the minor strain is positive, then the deformation mode is stretching.
- If the major strain is positive, while the minor strain is zero, then the deformation mode is plane strain.

These three cases are shown in Figure 2.17 along with the original circle plotted. Additionally, one can compute the thickness strain through the constant volume principle i.e. $\varepsilon_{major} + \varepsilon_{minor} + \varepsilon_{thick} = 0$ as in Equation 2.9:

$$\varepsilon_{thick} = -(\varepsilon_{major} + \varepsilon_{minor}) \tag{2.9}$$

Also, the thickness strain can be analyzed using an ultrasound transducer to measure the thinning percentage accurately. This can be done for regions with unacceptable strain values or metal flow patterns around features such as the draw-bead.

Once the major and minor stresses are computed, then one can use the FLD diagram to evaluate the stamping from the anti-fracturability point of view. This is done by first computing the FLD_0 using Equations 2.10 and 2.11 (for the case of steel), then plotting the major strain values on the y-axis and the minor strain values on the x-axis (see Figure 2.18).

$$FLD_0 = [(14.1 \times t) + 23.3] \times n / 0.21 \text{ (if the thickness } t \text{ is as received in mm)} \tag{2.10}$$

$$FLD_0 = [(359 \times t) + 23.3] \times n / 0.21 \text{ (if the thickness } t \text{ is as received in inches)} \tag{2.11}$$

With n being the n-value of the steel grade.

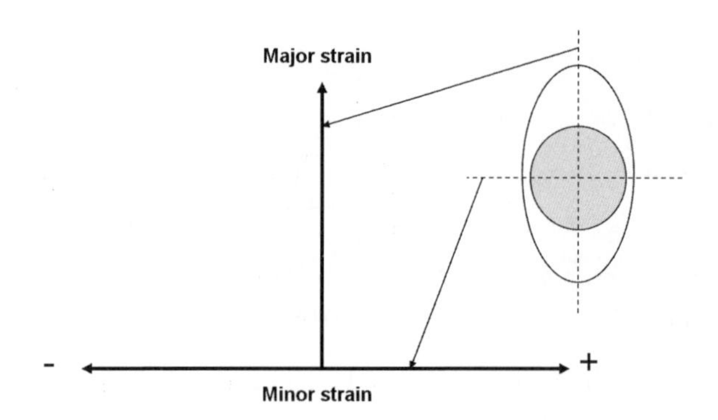

Figure 2.18 The plotting of the FLD strain points from the CGA; major and minor strains

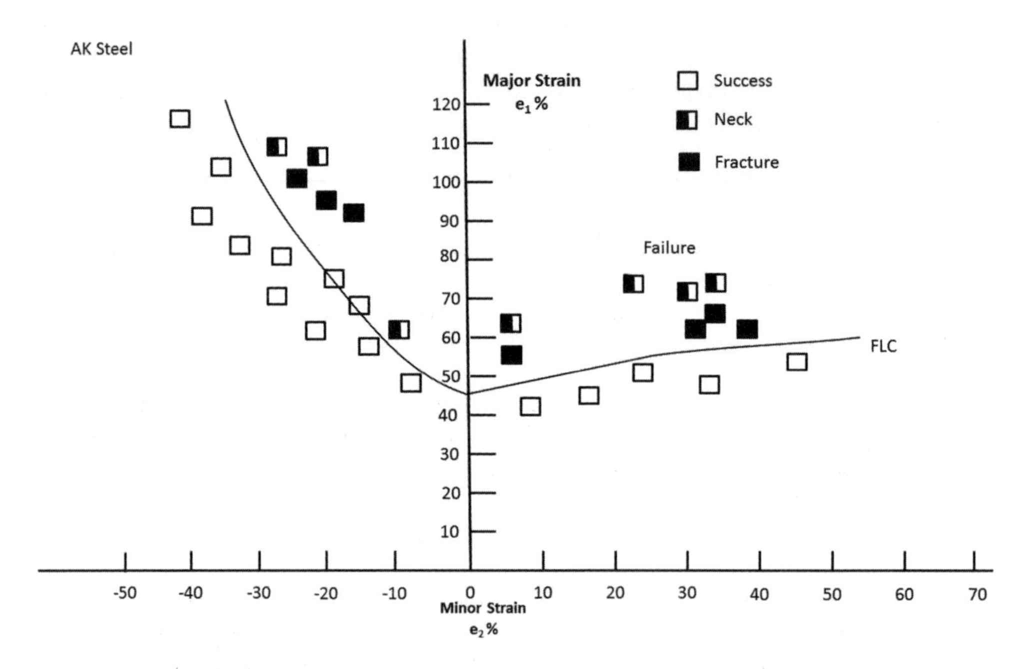

Figure 2.19 A typical FLD diagram for an automotive stamped panel

The shape of the FLD curve is the same for all types of low carbon steel as shown in Figure 2.19. The curve will shift up and down depending on the value of the FLD_0. The higher the value of the FLD_0, the larger the deformation window, i.e. the better the formability. So one can increase the deformation window by increasing the thickness of the panel or by selecting a steel grade with higher *n*-value. This trend can be seen in Figure 2.20.

In the case of aluminum coils, the FLD_0 relationship with thickness is not yet well defined, so it is not as readily computed as in the case of steel. But the aluminum

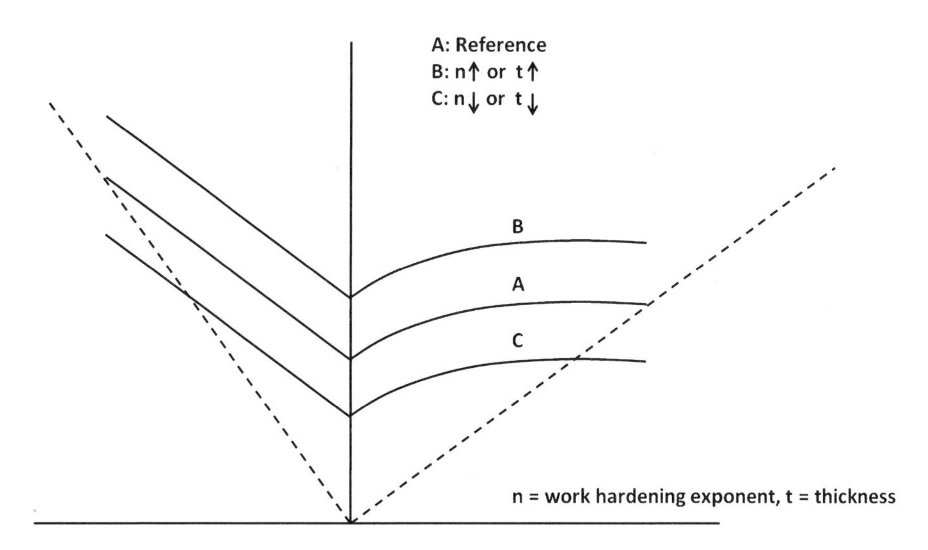

Figure 2.20 The FLD behavior with the steel grade *n*-value and panel thickness

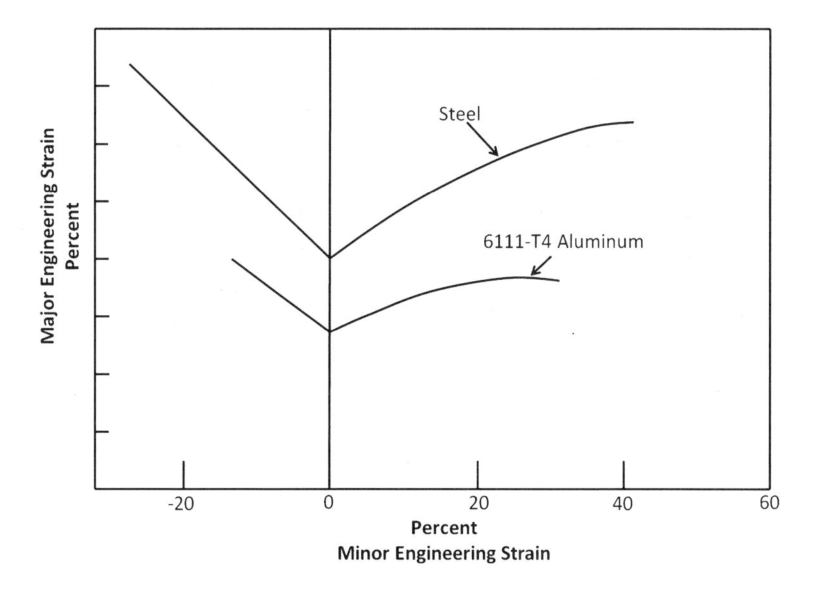

Figure 2.21 The FLD diagrams for steel and aluminum

FLD will have a smaller deformation window against that of steel, as shown in Figure 2.21.

The FLD can also reveal the straining path for any location within the stamping. The straining path is the sequence of strain combinations followed by a material as it is deformed further and further. In simple forming processes, the strain path is

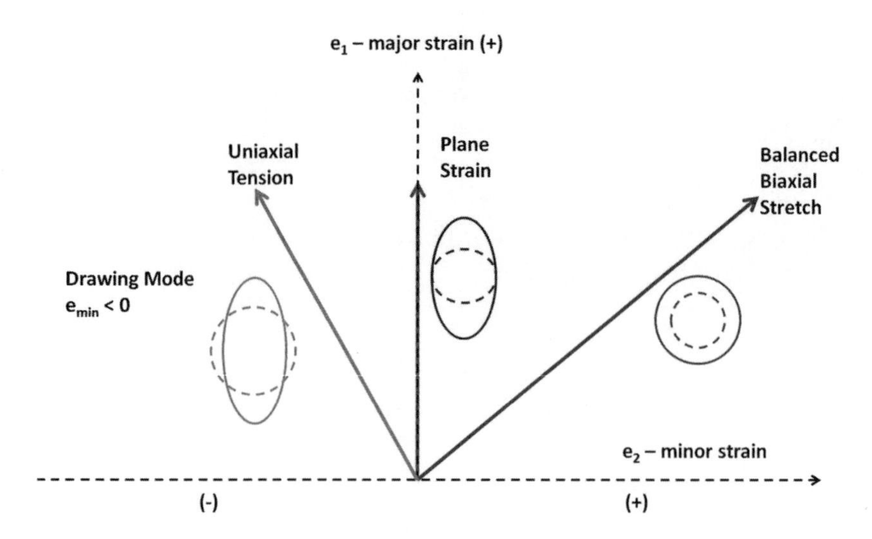

Figure 2.22 The different straining paths as plotted on the FLD

often assumed to be linear—starting from the origin at (ε_{major}, ε_{minor}) = (0, 0). A few important ideal strain paths are shown in Figure 2.22: (1) uni-axial tension (ε_{major} / ε_{minor} = −2); (2) plane strain (ε_{major} / ε_{minor} = 1/0 = ∞); and (3) balanced biaxial stretching (ε_{major} / ε_{minor} = 1).

For the areas with unacceptable strain values or metal flow patterns, engineers typically compute the critical thinning of the sample that causes a split. To compute the critical thinning percentage, the original panel thickness should be known, and the final thickness of the deformed panel at the suspect location should be measured using an ultrasonic transducer. Additionally, the critical fraction of the original thickness that might cause a split can then be computed through Equation 2.12:

$$\text{Critical fraction of original thickness} = 1/[1 + (\text{FLD}_0/100)] \quad (2.12)$$

Then the critical thickness is found by multiplying the critical fraction percentage by the original material thickness.

Furthermore, the critical percentage thinning can finally be determined through Equation 2.13:

$$\text{Critical percentage thinning} = 100\,[(t_{\text{original}} - t_c)/t_{\text{original}}] \quad (2.13)$$

This sequence of calculations is typically called the thickness strain analysis (TSA), and is typically used for specific locations within the stampings that exhibit a certain history of failure (splitting).

2.2 Automotive Materials

Automotive body panels are made from steel sheets and extrusions, aluminum sheets (as in the hoods in some vehicle models) and extrusions, in addition to injection-molded plastic parts (as in the fenders in some models). This text will focus on the different steel and aluminum grades and alloys that constitute the vehicle body's members and panels. The motivation behind selecting such materials for automotive applications is mainly due to their advantageous performance in terms of their dent resistance, energy absorption, strength (and their retained strength at different operating temperatures), and their stiffness. Additionally, steel and aluminum have favorable manufacturability characteristics i.e. their formability, weldability (and hemming ability), and finally their paintability. Other materials such as titanium might exhibit better performance in certain areas such as corrosion resistance and strength, however, the material cost (around 20–30 times that of steel), in addition to its processing difficulty (due to temperature sensitivity), limits its wide applicability in vehicle bodies. Other materials that are used in automotive body panels include sandwich structures and composites, mainly the sheet molding compounds (SMCs).

The following sections will classify the steel grades into three main categories: (1) the hot-rolled, cold-rolled low carbon steel; (2) the conventional higher strength steel (HSS); and (3) the advanced (or ultra) high strength steel (AHSS). The HSS and AHSS strength is achieved through special chemical composition, or special alloying elements, in addition to special mill processing. These steels are distinguished based on their strength values, with strength value >600 MPa being the AHSS threshold.

The low carbon formable steel grades are typically described from a formability point of view as in draw quality (DQ) and extra deep draw quality (EDDQ) grades because formability is the prime consideration in making a panel; whereas in high strength steels the yield and tensile strength level are the prime considerations. Hence such grades are classified accordingly. Higher strength steels are desirable for dent resistance, increased load-carrying capability, improved crash energy management, or for mass reduction through a reduction in sheet metal thickness, or gauge. Additionally later sections will discuss the tailor welded blanks/coils/tubes technology that allow the OEM to have customized coils by combining different thickness of the same or different grades in one coil or blank. The aluminum alloys will be discussed based on their formability and surface characteristics, with the main focus being on the aluminum 2XXX, 5XXX, 6XXX, and the 7XXX alloy series.

2.2.1 Automotive Steel Grades; Traditional Steel Grades

This section focuses on the following formable low carbon steel (traditional) grades; interstitial free (IF), deep drawing quality (DDQ), drawing quality special killed (DQSK), bake harden-able (BH), high strength low alloy (HSLA), and dent resistance.

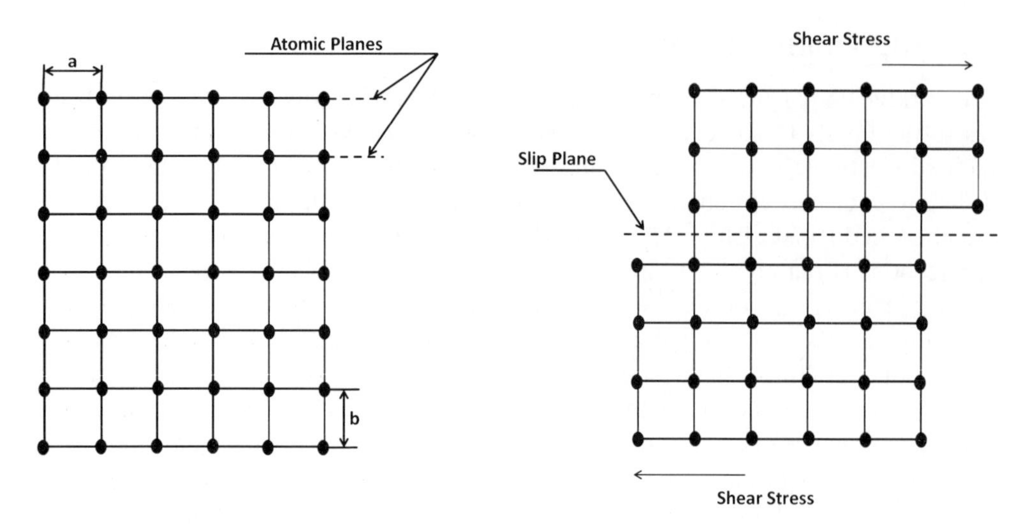

Figure 2.23 The slip and slip planes effect on material formability

The IF steel includes small amounts of elements, such as titanium or niobium, to combine with the interstitial elements carbon and nitrogen, to remove their strengthening effect; thus resulting in defect-free steel (at the interstitial level) with higher formability. This can be explained by knowing that any interstitial defects (carbon, nitrogen) hinder the propagation of the dislocation planes within the steel. To explain this further, by knowing that the plastic deformation under static load occurs when two groups of crystals under a shearing force move in a parallel fashion along what is called the slip or dislocation plane, causing the two groups to slip. However, the dislocation is first initialized by an edge dislocation that moves across the lattice under a shear stress to slip-planes. So if any impurities exist in the lattice, then it will hinder or block the dislocation movement by occupying the empty spaces. The slip plane movement while in deformation is shown in Figure 2.23. Removing the interstitial elements requires additional processing steps in the steel mill, which increases the cost of IF steel.

The DQSK is a steel grade that is deoxidized in the mill using some elements (typically aluminum thus sometimes called DQAK) to increase its formability.

The BH steels are specially processed to retain small amounts of carbon in a solid solution. During the automobile paint bake cycle, some of this carbon comes out of solution to increase the strength and dent resistance of the formed part. Therefore these steels can be produced in a relatively low strength condition and are easily formed into parts. However, after forming and paint baking, a significantly stronger part is produced.

The dent resistant steel contains increased levels of phosphorus (up to 0.10%), and possibly manganese and silicon for low carbon steels and IF steels. The high strength

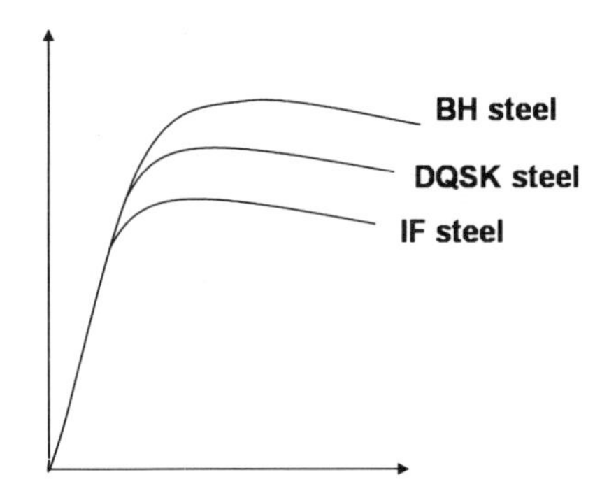

Property	IF	DQSK	BH	HSLA
Initial YS (ksi)	22	27	32	45
TS (ksi)	44	46	50	61
Tot. Elong (%)	47	42	39	30
n-value	0.24	0.21	0.19	0.16
r-value	2.0	1.7	1.6	1.1
YS (after 2%)	NA	NA	42-44	NA

Figure 2.24 The stress strain behavior for different steel grades with their mechanical properties described

solution strengthened steels contain increased levels of carbon and manganese with the addition of phosphorous and/or silicon.

Figure 2.24 shows the stress-strain behavior of the above steel grades and displays their material properties that relate to their formability.

The SAE J2329 classification for these automotive steel grades is given in Table 2.1.

Additionally, the cold-rolled and the hot-rolled steels are further classified in Tables 2.2 and 2.3.

Stamping engineers should take the available coil widths and thicknesses for each grade as provided by the steel mills, because selecting customized widths or thicknesses results in higher costs of raw materials. Steel mills specify the pricing of steel grades based on the number of production steps, special alloying elements, and the coating (zinc) required. Table 2.4 shows some of the available coils' widths and thicknesses for some of the steel grades.

Cold reduced uncoated and metallic coated sheet steels are produced in three surface conditions:

1. *Exposed* (E) is intended for the most critical exposed applications where painted surface appearance is of primary importance. This surface condition will meet requirements for controlled surface texture, surface quality, and flatness, i.e. Class *A* finish.

Table 2.1 The SAE J2329 classification for conventional automotive steels

AISI Designation	SAE Designation and grades	Characteristics
Commercial Quality (CQ)	J2329 Grade 1	Cold Rolled: YS 140-240 MPa, n-value 0.16 Hot Rolled: YS 180-290 MPa, n-value 0.16
Draw Quality (DQ)	J2329 Grades, 2-3	Cold Rolled: YS 140-260 MPa, n-value 0.18 Hot Rolled: YS 180-290 MPa, n-value 0.16
Deep Draw Quality (DDQ)	J2339 Grades, 3-4	Cold Rolled: YS 140-185 MPa, n-value 0.20 Hot Rolled: YS 180-240 MPa, n-value 0.18
Extra Deep Draw Quality (EDDQ)	J2329 Grade 5	Cold Rolled: YS 110-170 MPa, n-value 0.22

Note: n-value is based on the minimum value for the 5–15% strain.

Table 2.2 Further designations of the hot and cold rolled conventional automotive steels

Hot rolled grades	Yield strength (MPa)	Ultimate tensile strength (MPa)	Total elongation (%)	$n_{5\text{-}15}$-value
SAE J2329 Grade 2	140–250	270	34	0.16
SAE J2329 Grade 3	140–200	270	38	0.18
SAE J2329 Grade 4	140–190	270	40	0.21

Table 2.3 SAE designations for stronger hot and cold rolled grades

Hot rolled grades	Yield strength (MPa)	Ultimate tensile strength (MPa)	Total elongation (%)	$n_{5\text{-}15}$-value
SAE J2329 Grade 2	180–280	270	35	0.16
SAE J2329 Grade 3	180–240	270	39	0.18

Table 2.4 Availability of coils widths and thicknesses for each of the hot and cold rolled grades

Steel grade designations			Strength		Width range		Thickness		Characteristics			
AISI	SAE	Rolling	Yield (MPa)	Ultimate (MPa)	mm	inch	mm	inch	*n*-value	*r*-value	Elong. (%)	Hardness
CQ	J2329-1	Hot	269	386	610–1829	24–62	1.00–9.53	0.07–0.50	0.19	1.00	35	60
CQ	J2329-1	Cold	296	331	610–1829	24–62	0.38–3.30	0.02–0.12	0.20	1.10	35	50
DQ	J2329-2	Hot	248	338	610–1829	24–72	1.00–9.53	0.055–0.375	0.19	1.10	37	54
DQ	J2329-2	Cold	186	317	610–1829	24–72	0.38–3.30	0.015–0.130	0.22	1.50	38	42
DQ	J2329-3	Cold	186	317	610–1829	24–72	0.38–3.30	0.015–0.130	0.22	1.50	42	42
BH	J2340-180B	Cold	200	320	610–1829	24–72	0.64–2.79	0.025–0.11	0.20	1.70	39	52
BH	J2340-210B	Cold	221	352	610–1829	24–72	0.64–2.79	0.025–0.11	0.19	1.60	41	54

Reproduced by permission of American Iron and Steel Institute and Auto/Steel Partnership © 2010.

2. *Unexposed* (U) is intended for unexposed applications and may also have special use where improved ductility over a temper-rolled product is desired. Unexposed can be produced without temper rolling.
3. *Semi-exposed* (Z) is intended for non-critical exposed applications. This is typically a hot dip galvanized temper rolled product.

Additional steps in the steel mill are also taken to affect the surface texture and characteristics especially for the panels with Class *A* requirements (permanently exposed exterior panels). One of these processes which also impacts the material formability is temper rolling. Temper rolling (or "temper reduction") serves to improve the shape (flatness) of sheet metal. Another term that relates to temper rolling is "skin pass." This term refers to the fact that temper rolling is a very fine or light cold reduction process where only the skin or outer surfaces of the material is deformed, the amount of temper reduction is usually measured by the temper mill extension (TME), which measures the extension or the change in length over a certain portion along the length of a coil.

The temper rolling has significant effects on the mechanical properties. Temper rolling degrades both the n-value and the % total elongation (% TE) of the material. However, it is considered a necessary requirement for exposed automotive panels to avoid the yield point elongation phenomenon during the panel deformation. Temper rolling helps to minimize or eliminate yield point elongation or discontinuous yielding, because discontinuous yielding can lead to strain lines on the sheet surface, which negatively affects the surface appearance. The effect of the temper rolling on the n-value and % TE is displayed in Figure 2.25.

2.2.2 Automotive Steel Grades: High Strength and Advanced (Ultra)

The HSS steel grades are typically classified into two main types: conventional with an ultimate tensile strength <600 MPa, and advanced HSS, with ultimate tensile strength >600 MPa. The conventional HSS grades include the following main types:

- Bake-Hardenable (BH) steels which achieve a hardening effect inside the paint curing (at around 200 °C) ovens in addition to the typical work hardening from the deformation process. This is achieved through the carbon that is retained in solution through the annealing treatment; the carbon migrates to dislocations upon deforming the BH steel which locks in such dislocations, hence strengthening the steel (increasing its dent resistance). The BH work hardening is shown in Figure 2.26.
- High strength, low alloy (HSLA) steels which provide adequate formability for many simple forming processes (simple shapes), and have relatively low work

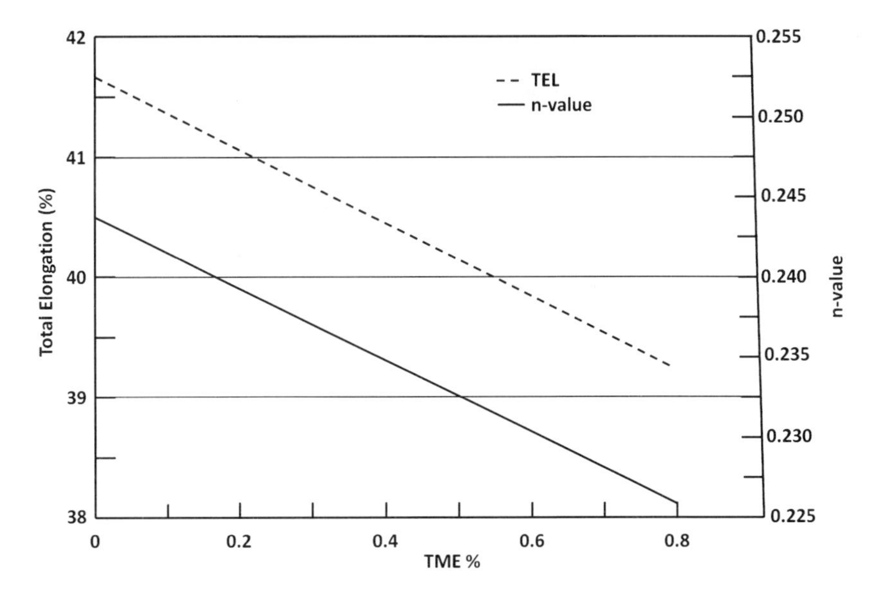

Figure 2.25 The effect of the temper rolling on the material's n-value

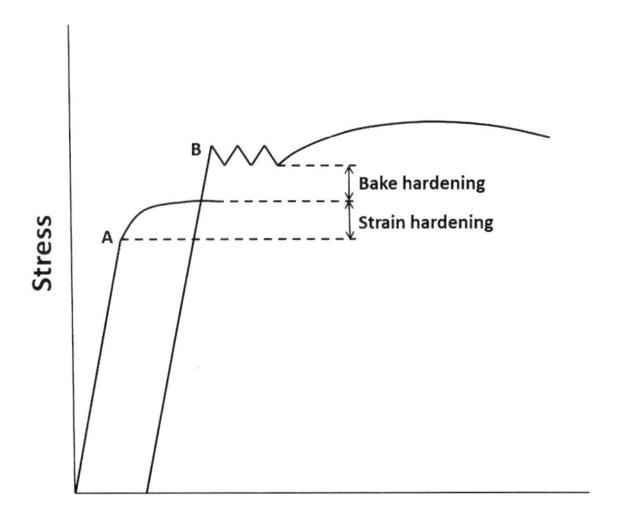

Figure 2.26 The bake and strain hardening effects on the final panel strength

hardening rate. The HSLA steels generally contain micro-alloying elements such as titanium, vanadium or niobium, which increase the steel strength by precipitation hardening. Additionally, the HSLA is strengthened by its fine grain size. The fine grain size and the fine distribution of the alloying elements hinder the dislocation edge movement, which means higher flow stress, i.e. stronger steel.

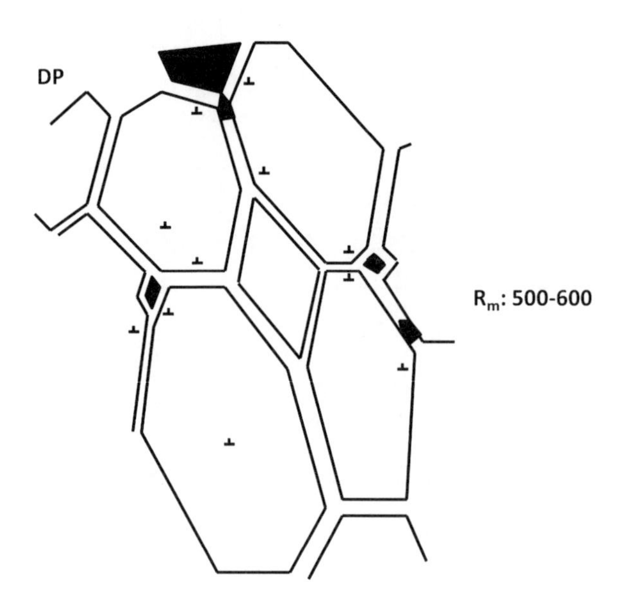

Figure 2.27 Dual phase steel microstructure

The Advanced HSS includes the following steel grades:

- *Dual-phase (DP) steels* have more enhanced formability over HSLA steels but with higher cost. Also, DP steels possess a very high work hardening rates at initial strains and can provide a uniform strain distribution during the forming. DP steels have high buckling resistance during plastic deformation. The making of DP steel requires processing at the micro-structural level through an inter-critical heating to achieve the ($\alpha + \gamma$) phases, then rapid cooling is applied to achieve the two phases: islands of relatively very hard martensite (α'), and a matrix of ductile polygonal ferrite (α). Figure 2.27 shows the DP micro-structure [9].

  The DP steel's high work-hardening rate can be explained through the DP micro-structure; the ductile ferrite allows the dislocation edge to propagate freely while the martensite islands hinder its progress, so the dislocations will pile up at the martensitic edges. So the ductile ferrite enables good formability and elongation, while the martensite increases the work-hardening effect, hence the DP steel has good energy absorption characteristics at high strength. Some bake-hardening is also achieved for the DP grade. The DP alloying process may include manganese, chromium, molybdenum, vanadium and nickel, which might be added individually or in combination. The alloying of DP grade is carefully done to avoid challenges in the spot-welding process, especially for stronger DP grades such as the DP (700/ 100) MPa.
- *Transformation-induced plasticity (TRIP) steels*: Typically used when formability of DP steels is not adequate to achieve certain complex shapes, but TRIP steels

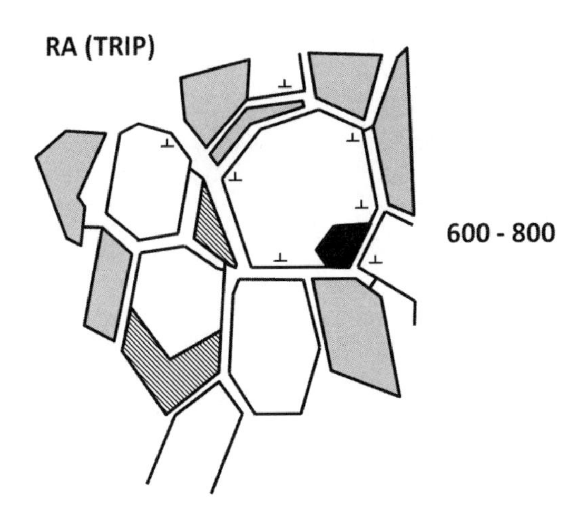

Figure 2.28 TRIP steel microstructure

require more processing at the steel mill in addition to higher alloying cost, all this leads to higher cost (more than DP). The TRIP steels can also sustain a high work hardening rate even at high strains. TRIP has similar performance to the DP steel in regard to crash-worthiness.

The TRIP steel microstructure includes a continuous ferrite matrix containing a dispersion of hard second phases of martensite and/or bainite. These steels also contain retained austenite in volume fractions greater than 5%. The retained metastable austenite transforms into martensite as the deformation process progresses, thus increasing the work-hardening effect even more than that of the DP, and at higher strain values. Furthermore, the new transformed martensite acts to hinder the dislocation, increasing the flow stress even further. Figure 2.28 shows the TRIP micro-structure.

Figure 2.29 displays the work-hardening values for TRIP, DP and HSLA grades and Table 2.5 describes some typical AHSS properties with the following classifications: XX aaa/bbb where XX = Type of Steel: aaa = minimum YS in MPa, and bbb = minimum UTS in MPa. The types of steels are: DP = Dual Phase, CP = Complex Phase, TRIP = Transformation-Induced Plasticity, Mart = Martensitic. Other AHSS types include: Complex-Phase (CP) Steels, Martensitic Steels, and Hot Stamping Steels.

Due to the AHSS complex micro-structure and work-hardening behavior, their stress-strain behavior cannot be mathematically explained using the standard power-law (Equation 2.14) as is the case for the conventional steel grades:

$$\sigma = K\varepsilon^{n} \tag{2.14}$$

where n is the n-value and K is the strength coefficient, and σ being the flow stress.

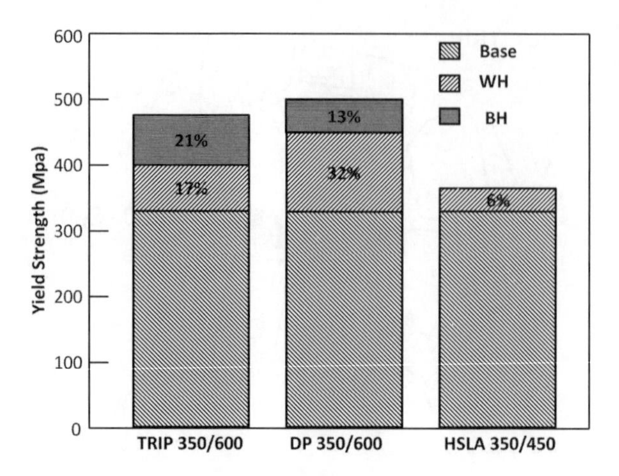

Figure 2.29 The hardening effects for the TRIP, DP, and HSLA grades

Table 2.5 The main characteristics of the advanced high strength steels

Steel grade	Yield strength (MPa)	Ultimate tensile strength (MPa)	$n_{5\text{-}15}$-value	r-value*	k-value (Strength coeff.)	Total elongation (%)
DP 300/500	300	500	0.21	1.00	1100	35
DP 400/700	300	700	0.14	1.00	1030	25
TRIP 450/800	450	800	0.24	0.90	1700	30
Martensitic 950/1200	950	1200	0.07	0.90	1680	08

Note: *r-value is computed based on normal anisotropy i.e. r_{mean}-value.

Applying the power law to DP and TRIP steels results in over-predictions of the uniform elongation (strain at maximum load) for the DP steel, and an under-prediction of the uniform elongation (strain at maximum load) for the TRIP steel. These observations can be pictorially explained in Figure 2.30, which shows the work hardening rate (WHR) intersection with the stress-strain curves for DP, TRIP, and HSLA steels. These variations are the result of the way the n-value is extracted from the stress-strain diagrams, i.e. as the slope of the stress vs. strain in the logarithmic scale. So a different mathematical formula should be adopted to describe the HSS and AHSS deformation behavior.

The approach described in following text uses the work hardening potential (WHP) instead of the WHR. The WHP is the WHR but normalized to the flow stress, as in

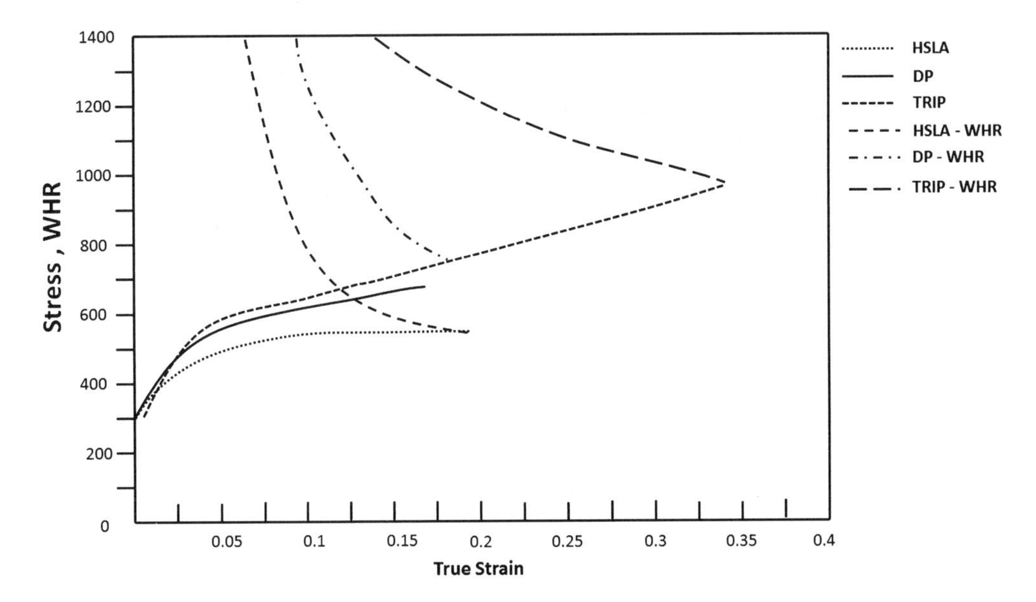

Figure 2.30 The work-hardening rate behavior for different steel grades; the intersection is the *n*-value

Equation 2.14. If one plots the WHP behavior with the true strain, two observations can be made:

- The work hardening potential WHP decreases monotonically with strain.
- The WHP appears to approach a terminal value at higher strains (asymptotic behavior).

From these observations, one can propose that WHR can be in the form (in terms of true strain) as in Equation 2.15, which can be further integrated to yield Equation 2.16:

$$WHP = \ln\left(\frac{\frac{d\sigma}{d\varepsilon}}{\sigma}\right) \tag{2.15}$$

Noting that when $\frac{d\sigma}{d\varepsilon} = \sigma$, the WHP = 0.

$$(\mathrm{d}\sigma/\mathrm{d}\varepsilon) = \mathrm{A}(1/\varepsilon) + \mathrm{B}$$

$$\sigma = \mathrm{A}\ \ln(\varepsilon) + \mathrm{B}(\varepsilon) + \mathrm{C} \tag{2.16}$$

Equation 2.16 contains three coefficients, A, B, and C, typically termed as A: Shape Coefficient, B: Tilt Coefficient, and C: the Strength Coefficient. These names imply the effect of each of these terms on the stress-strain behavior. Equation 2.16 is also termed the Shape Tilt Strength or the STS equation.

The effect of having a lower value of the Shape Coefficient, or A, means that the steel grade has lower initial (i.e. initial strains) WHR with concentrated strain distribution; however, the higher A value implies a higher initial WHR with broader strain distribution. For the Tilt or B factor, the lower value means lower WHR at large strains with earlier necking. Meanwhile, the higher value means sustained WHR at large strains with delayed necking. The Strength or C coefficient values correspond to the steel grade strength, i.e. the higher the C, the higher the steel strength.

In summary, the shape coefficient describes the steel WHR at initial strain, the tilt describes the WHR behavior at larger strain, while the strength coefficient describes the steel strength. When the true strain is equal to one, the power law and the STS equation become equal.

Example 2.1 illustrates the STS equation parameters for the DP, TRIP and HSLA steels:

$$\text{HSLA: } A = 66,\ B = 223,\ \text{and}\ C = 673.$$

$$\text{DP: } A = 117,\ B = 54,\ \text{and}\ C = 932.$$

$$\text{TRIP: } A = 98,\ B = 666,\ \text{and}\ C = 802.$$

Stamping the HSS and AHSS steels requires careful modifications to the stamping process variables because such grades possess reduced elongations (less formability window) due to their micro-structure. Additionally, they have increased resistance to compression due to their high tensile strengths, this means an increased sensitivity to buckling and splitting. The spring-back in such steels is also more severe, and the residual stresses (upon unloading) are higher, leading to instabilities in the final geometry and dimensions. Typical adjustments are done to reduce the die bend radius to minimize such shape changes, in addition, post-stretching might be applied after unloading the panel to reduce the residual stresses. Furthermore, the following paragraphs will discuss the standard stamping adjustments done when dealing with the HSS and AHSS grades.

The stamping press force capacity defines the maximum force that it can apply, however, this force is not the actual or available force applied on the panel (all the time, and at different distances from the bottom dead-center), thus the force on the panel should be evaluated to make sure that it is enough to exert plastic deformation for high strength steel. Additionally, the higher work-hardening behavior of a DP grade, for example, requires higher press loads when compared to the other conventional HSS and traditional grades with same panel thickness. However, the AHSS panels are typically lower in thickness (due to their higher strength), thus the required press load should be decreased or compensated for. Also, the high *n*-value for AHSS leads to spreading the strain gradients and reducing the maximum strain values. Figure 2.31 illustrates an example of a drawing and an embossing process for a mild, HSLA and a DP 350/600 grades. This means that for DP 350/600, designers can utilize it for all stamping areas that are formed in pure stretch, such as embossments, character lines, and other design features with localized strain gradients, because a high *n*-value implies a better material ability to distribute the strain without localizing it. However, at higher strains, DP will have a lower *n*-value (high shape but low tilt coefficient value) which means that it will behave similarly to the HSLA at high strains. On the other hand, TRIP grades will have a high *n*-value throughout their entire strain range (high values of tilt and shape coefficients). Thus, the sustained high *n*-value behavior results in a better suppression of strain localization due to certain design features or shapes. Additionally, when stretch-formed, TRIP will still distribute the strain because the deformation will cause additional transformations of the retained (meta-stable) austenite to the martensite phase, thus, adding more strength to the deformation zones (around the stamping features) in addition to the redistribution of deformation to areas of less strain.

For the deep drawing deformation modes, one can inspect the limiting draw ratio (LDR) test results to measure the HSS and AHSS grades performance. For most HSS and AHSS, the LDR averages are around the value of 2 (LDR is the maximum ratio of blank diameter to punch diameter possible in a drawing mode). The DP grades will have an LDR value similar to that of most HSS. On the other hand, the TRIP

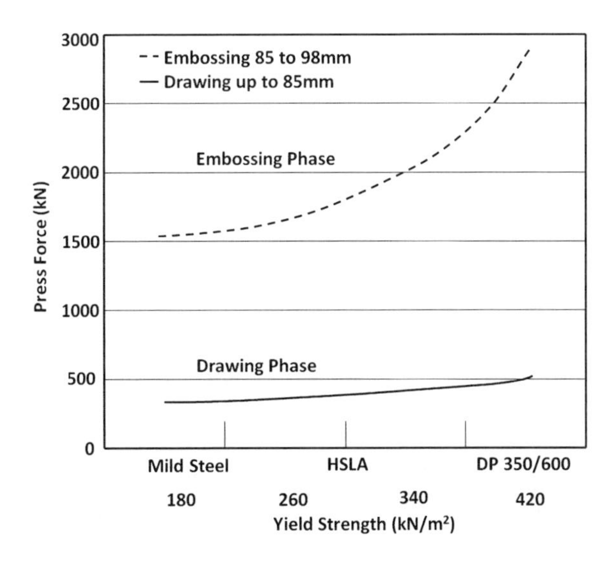

Figure 2.31 An example of a drawing and an embossing process for a mild, HSLA and a DP 350/600 grades

grades possess higher LDR values, which mean an improved deep drawability. The transformation of the retained austenite into martensite is dependent on the transformation mode so the behavior of TRIP grades is different and dependent on the deformation modes involved: drawing or stretching. This fact affects the strain distribution along the drawing features; the amount of transformed martensite generated by shrink flanging in the flange area would be less than that in plane strain deformation (the cup wall). This difference in transformation results in a stronger wall area than the flange.

For the bending deformation mode, a high total elongation is needed to provide enough material flow to outer fiber stretch of the bend before a split occurs; knowing that the material total elongation decreases as its strength increases, HSS and AHSS require designers to specify a larger bend radius for same panel thickness. Note that designers typically specify the bend radius in relation to the panel thickness, i.e. R_{bend}/t. However, for equal strengths, the AHSS grades will have higher total elongation due to its work-hardening characteristics which translates into a smaller minimum bend radius. At the same time, designers should inspect the λ-value (the hole-expansion test) to check for the edge quality because deformation might localize, hence create low local elongations especially at the edges. Additionally, the design considerations should include the formability of the tailor welded blanks (TWBs) (explained in detail in Section 2.4) because such blanks go through a laser welding process that affects the panels' formability at the weld region.

Another important issue that should be taken into account when dealing with AHSS is their spring-back trends. As discussed in the previous section, spring-back increases as the strength increases, which means that for AHSS spring-back is more severe than that of traditional or even HSS grades. Recommended design practices to reduce the spring-back in AHSS include configuring the structural components, mainly the rails and crossbars to be open-ended channels, also the punch radius should be as sharp ($<2t$) as formability and geometry allow, also the die clearance should be reduced to constrain the bending and the unbending process for the angular change spring-back. For draw dies, this clearance is kept at $1.1 \times t$ of the panel. Furthermore, it is recommended that the AHSS and HSS binder pressure (blank holding pressure) should be increased to at least twice that of conventional grades. Designers should design transitions in the stamping geometry to be gradual to avoid stress concentrations. Typical practices to solve unavoidable spring-back issues in AHSS are based on using multiple step forming or sequential stamping process.

So, the stamping of HSS and AHSS requires careful assessment of the stamping conditions and dedicated deformation analysis based on the HSS and AHSS material behavior. Stamping windows and practices designed for traditional grades should not be used readily for HSS and AHSS blanks. Furthermore, the blanking and trimming operations should account for the fact that the shearing process will lead to high stress and hardness concentrations at the edges, when cutting or trimming HSS or AHSS due to their work-hardening behavior. Also, the blanking and cutting clearances should be increased due to the higher material strength, clearances are about 0.5–06 × t for mild steel while for AHSS are around 0.1–0.12 × t. Additional challenges arise when trying to establish the forming limit diagrams for certain AHSS grades such as the TRIP. This is due to the fact that the TRIP grade n-value changes according to the transformation of the austenite into martensite, which in turn depends strongly on the deformation mode and the strain values. So, even different locations within the same panel can have different martensite levels due to the different strain paths (balanced biaxial, plane strain, uni-axial tension, compression) and the varying amounts of deformation. The Automotive Steel Partnership have published several technical reports to describe the formability of the HSS and AHSS grades and the recommended stamping practices, in terms of tool development, setting the press tonnage and configuration, and even the effect of the lubricants on the different deformation modes. Other good resources for such recommendations and analyses are the American Iron and Steel Institute publications, such as their report on formability characterization of a new generation of high strength steel [10]. Readers interested in the final performance characteristics of the HSS and AHSS, in terms of their crash worthiness, fatigue, and repair ability are referred to the report entitled *Characterization of Fatigue and Crash Worthiness of New Generation of High Strength Steel FOR Automotive Applications (Phase I and Phase II)* [11].

2.2.3 Stamping Aluminum Sheet Panels

Even though early adoption of the aluminum as a material for automobile bodies was in the form of extrusions, to constitute a space-frame design, current applications of aluminum are in the form of stamped panels. The lower formability of aluminum when compared to steel motivated its early usage in the form of extruded structures, more about Al extrusions can be found in *Aluminum Automotive Extrusion Manual* [12]. However, new advances in aluminum metallurgy enabled its application as stamped material, formed on the same presses as steel and using similar forming practices. The use of stamping presses to deform aluminum blanks facilitates the Al adoption in the automotive industry and enables the construction of complete hybridized body structures in one OEM stamping facility, thus, reducing the cost of subcontracts and the logistics involved. Aluminum is typically used for the body panels that do not require major deformation processes such as the flat or semi-flat panels, e.g. the vehicle hood. Designers have also selected aluminum for the vehicle hood because not only does it reduce the overall weight of the vehicle by a good percentage (due to its size), but also the design of the hood (inner and outer) can be configured to increase the aluminum hood stiffness without major thickening of the used panel. Even though aluminum has a density that is 1/3 that of steel, its Young's modulus is almost 1/3 that of steel. Young's modulus has a direct impact on the material stiffness, the higher the modulus, the better stiffness. It is worth mentioning also that some OEMs are still producing the space-frame based on 100% aluminum, such vehicles are Rolls-Royce, and the Audi A8 and A2. Other applications of aluminum panels in the automobile body can be found in the deck-lid and the roof (as in the Daihatsu Copen, and the Nissan Cima) panels [13].

Aluminum possesses following advantageous characteristics: light weight (due to its low density, 1/3 that of steel), good corrosion performance (due to the stability of the aluminum oxide layer), its relative strength (still weaker than steel), and its recyclability, which is still challenged by the sorting and collection efforts due to the lack of large aluminum structures (such as ships, buildings, or bridges). Aluminum comes in two different groups from a fabrication point of view: wrought Al and casting Al.

The designation of the aluminum alloys describes the major alloying element through the first number, e.g. 2XXX refers to copper as the main alloying element, while the 3XXX refers to manganese, the 4XXX to silicon, the 5XXX to magnesium, 6XXX to magnesium and silicon, 7XXX to zinc. The second number in the alloy designation refers to the modifications, if any, if no modifications are applied to the alloy, then it should read 0, while the third and fourth digits refer to the alloy number within that series. The main Al alloys used in automotive bodies are:

- The 2XXX: especially, the 2008, 2010, and 2036, which experience bake strengthening, thus adding to the final panel strength; with the 2036 having the highest strength value, however, it also has the worst formability.
- The 5XXX: especially, the 5182, 5454, and the 5754, which do not possess any strain hardening during the baking process but at the same time these alloys have the best formability. But 5XXX can be used for exterior panels due to the formation of Lueder bands on its surfaces when deformed.
- The 6XXX: especially, the 6022, 6111, and 6009. These alloys go through a bake hardening process, with the 6111 having the highest strength, the 6009 has a good formability while in bending mode, and the 6111 shows good elongation while in the stretching mode.
- The 7XXX series is used mainly for Al extrusions.

The temper treatment designation for the Al alloys comes immediately after the alloy designation and is defined according to Kauffman (in Kutz, 2002):

- *F*: *As-Fabricated,* which applies to wrought or cast Al products that are made through a shaping process without specific or special thermal control/monitoring or without a strain-hardening process being used.
- *O: Annealed*, which applies to the wrought Al alloys that are annealed to achieve a lower strength temper, typically applied to increase workability, and also to cast products that are annealed to improve ductility and dimensional stability.

- *H: Strain hardened*, which describes alloys with their strength improved through a strain hardening process. Additional thermal treatments might also be applied.
- *W: Solution heat treated*, which defines alloys that age naturally upon a solution heat treating process.
- *T: Thermally treated* alloys used to achieve stable tempers (other than a designation F). More about the Al heat treatment and temper designation can be found in [14].

Further designation of the thermally treated alloys, i.e. *T* designation, follows the following sequence: the first digit defines the specific sequences of basic treatments with the numbers 1–10 used to identify the following treatment conditions [15]:

- *T1:* Cooled from an elevated temperature shaping process and naturally aged to a stable condition.
- *T2:* Cooled from an elevated temperature shaping process, cold-worked, and then naturally aged to a stable condition.
- *T3:* Solution heat-treated, cold-worked, then naturally aged to a stable condition.
- *T4:* Solution heat-treated, then naturally aged to a stable condition.
- *T5:* Cooled from an elevated temperature shaping process then artificially aged.
- *T6:* Solution heat-treated then artificially aged.
- *T7:* Solution heat-treated and over-aged/stabilized.
- *T8:* Solution heat-treated, cold-worked and then artificially aged.
- *T9:* Solution heat-treated, artificially aged, and then cold-worked.
- *T10:* Cooled from an elevated temperature shaping process, cold-worked, and then artificially aged.

The main alloying elements in some of these Al alloys are shown in Table 2.6, which shows the percentages by weight, while Table 2.7 displays the assessment of some of the Al alloys with respect to the processes involved in the automotive

Table 2.6 Aluminum grades' main alloying elements

Alloy designation	Fe	Si	Cu	Mn	Cr	Zn	Mg
2008	0.40	0.50	0.70	0.30	0.10	0.25	0.35
2010	0.50	0.50	0.90	0.20	0.15	0.30	0.60
5182	0.35	0.20	0.15	0.35	0.10	0.25	0.50
5754	0.40	0.40	0.10	0.50	0.30	0.20	3.00
6009	0.50	0.70	0.35	0.45	0.10	0.25	0.65
6111	0.40	0.85	0.70	0.35	0.10	0.15	0.80

Table 2.7 The main formability characteristics of the aluminum grades

Aluminum alloy Series	Formability (include spring-back)	Weld-ability (arc)	Weld-ability (resistances spot)
2XXX (2008 & 2010 series)	Medium Formability	Medium Weld-ability	Medium Weld-ability
6XXX (6009, 6022, 6111 series)	Medium Formability	Medium Weld-ability	Excellent Weld-ability
5XXX (5740 & 51820 series)	Excellent Formability	Excellent Weld-ability	Low Weld-ability

Note: Al alloys' behavior may change upon heat treatment applied.
Source: [12]

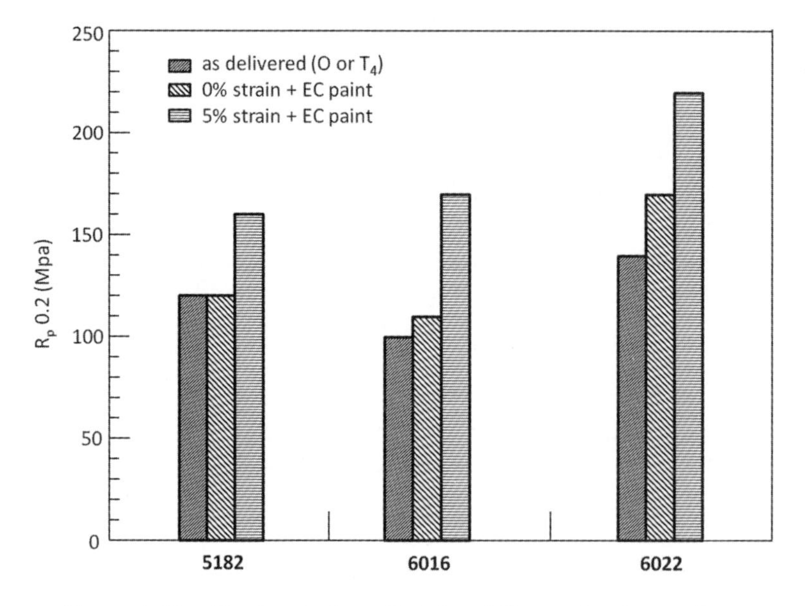

Figure 2.32 The hardening effect for different aluminum alloys

manufacturing . Also, Figure 2.32 shows the strain and bake hardening that Al alloys undergo.

The body panels' designers study and analyze the functional requirements from each panel and select the suitable aluminum grade, heat treatment and thickness, so that it can replace the current steel design. For example, if designers need to replace a panel based on the dent resistance criteria, then the direct replacement, i.e. based on manipulating the thickness will be according to

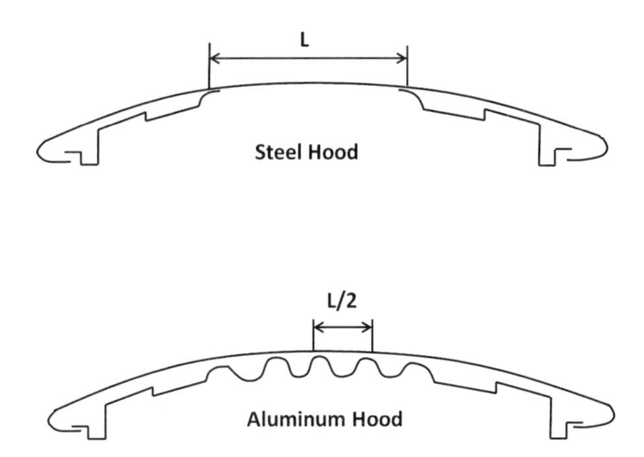

Figure 2.33 A schematic showing a steel hood and an aluminum hood

$\frac{t_{Steel}}{t_{Al}} = \frac{YS_{Al}}{YS_{Steel}}$. However, increasing the thickness of the Al panels will negate the net weight savings gained from using Al. Better replacement will analyze the dent resistance to be a function of the panel thickness, its yield strength, shape, and finally the unsupported area. Thus manipulating the shape and reducing the unsupported area will have the same effect as increasing the thickness without adding extra weight.

For example, replacing a steel hood with an aluminum hood can be done by reducing the span (the unsupported area from the hood inner) by half which will result in increasing the hood stiffness by almost 2.5 times without the need to increase the thickness. This can be done by increasing the number of contact points between the hood inner and outer panels. Figures 2.33 and 2.34 explain this example pictorially and mathematically, respectively. Al panels' stiffness can be increased by increasing its moment of inertia by a factor of 3 (easy if inner panels exist) to match that of a steel panel (from a stiffness point of view), or by increasing the panel thickness (for panels without inners, such as fenders and quarter panels). If the replacement is done directly based on increasing the panels' thickness, then the cost added (due to the higher Al raw material cost) relative to the weight saved (due to the Al lower density) can be seen in Figure 2.35 for each of the automobile body panels. It is worth mentioning that the added cost from using Al is not dependent on the production volume as is the case with polymers and composites, because with polymers the main problem is the processing and tooling costs, while for Al it is the cost of the raw material (Table 2.8). Several cost analyses have been done for Al automotive panels and are best summarized by Diffenbach [in 9, 12].

However, steel and aluminum have almost the same natural frequency. The natural frequency of a panel is dependent on its stiffness (linearly proportional) which is controlled by the panel geometry, and Young's modulus. In addition, the natural

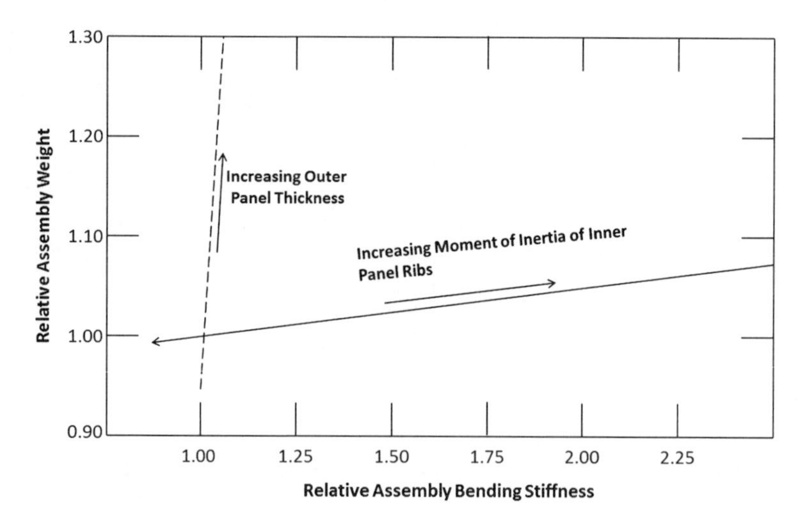

Figure 2.34 The effect of increasing the aluminum panel thickness versus that of increasing the moment of inertia, showing the weight implications

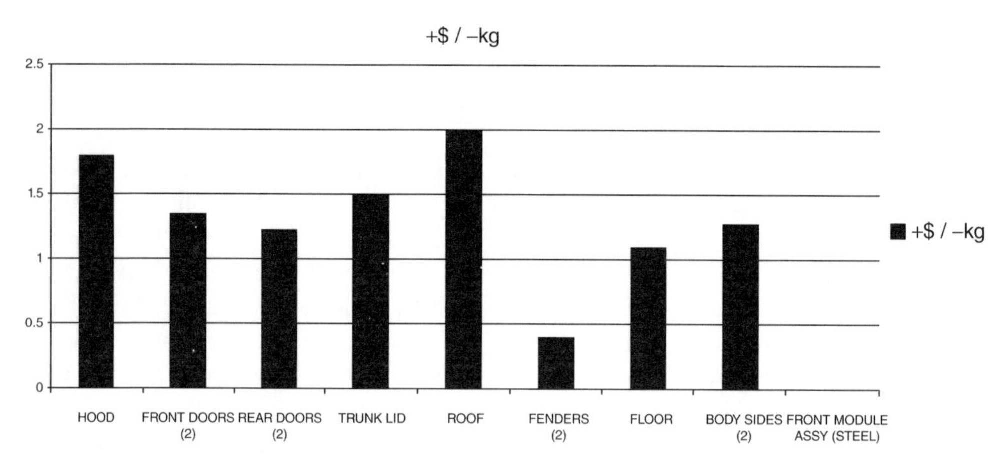

Figure 2.35 The cost increase per unit weight saved for automotive panels, when steel panels are replaced with aluminum

frequency is also affected (inversely proportional) by the panel mass, i.e. density. Steel has a Young's modulus and a density that are three times that of Aluminum. So Al panel vibration response will be similar to that made of steel, given that both panels have the same geometry and thickness.

From formability point of view, aluminum is less sensitive to the strain rate, which affects its crash-worthiness characteristics. However, recent advances in Al panel and crush zones' design have resulted in better crash test results. Also, the formability of a typical Al alloy is evaluated at about 2/3 that of a DQ steel grade, with more severe

Table 2.8 The cost multiplier for each of the different steel grades, including aluminum

Material type	Relative cost (%)
Hot rolled steel coils	80
Hot dipped galvanized rolled steel sheets	112
Electro-galvanized rolled steel sheet	135
High strength low alloy	115
Bake Hardenable	125
Transformation induced plasticity steel	150
Aluminum (5XXX Alloy Series, automotive formable grades)	500
Dual phase	145
Cold rolled steel is considered the basis for comparison, i.e. cold rolled is 100%	

spring-back due to the low Young's modulus value. Other unique Al stamping characteristics are:

- The need for lubrication for both sides of the blank.
- Al performs better in the drawing than the stretching deformation mode.
- The Al *r*-value is about 60% of that of steel.
- Al alloys are more sensitive to the orientation from the rolling direction, i.e. higher planar anisotropy than that of steel grades.
- Al forming limit diagrams are not yet established related to the panel thickness.

The standard tensile test for Al is shown in Figure 2.36 relative to that of Steel. Also, Figure 2.37 displays the FLD for that of Al relative to that of steel. Table 2.9 shows the general guidelines for stamping aluminum in regard to the draw die clearances and the tool radius. Other recommendations with regard to the part design are: avoid abrupt changes in the panels' cross-section; shallow draw beads should be used for hoods and deck-lids applications while full beads should be applied for deeper draws as the case in door inners; and lock beads should be used for flat or semi-flat panels such as the roof.

2.3 Automotive Stamping Presses and Dies

This section is intended to describe the main types of the stamping presses and the different configurations of the dies used within the automotive stamping process. Additionally, the accessory tooling required to facilitate the automation of the sequential stamping will be briefly explained.

The automotive presses are typically of two types: the mechanical presses and the hydraulic ones. Mechanical presses use the energy stored in their flywheel which is

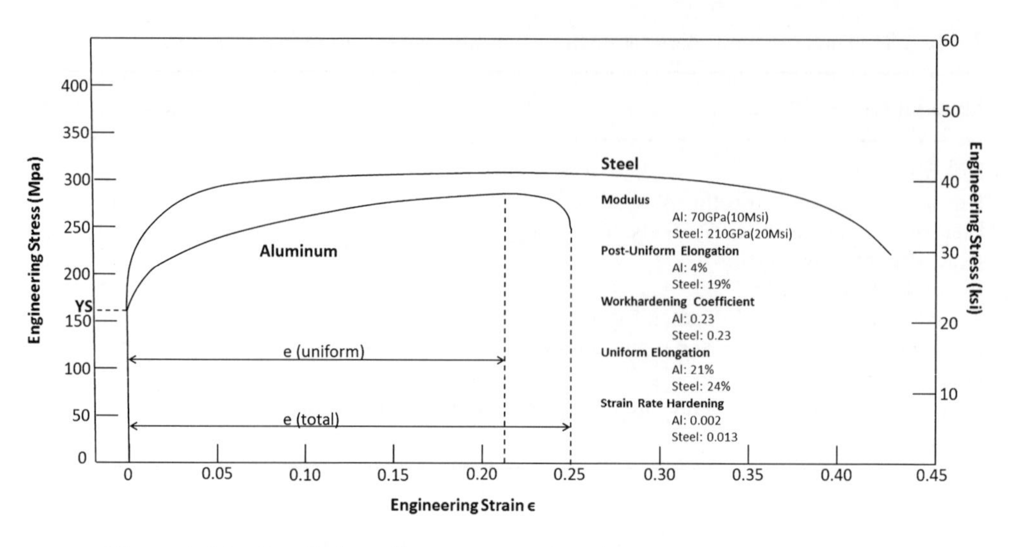

Figure 2.36 Detailed tensile test comparison between steel and aluminum

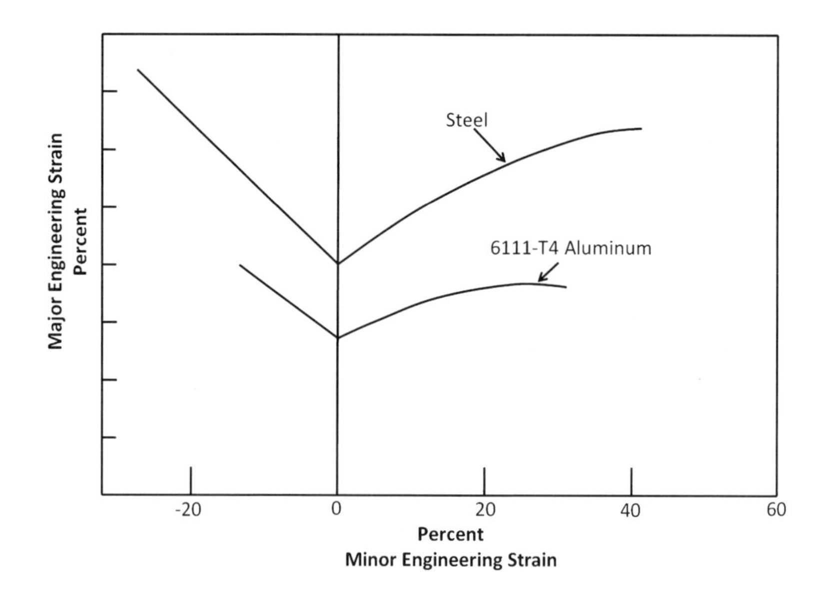

Figure 2.37 The formability comparison between aluminum and steel shown using FLD

then translated into a reciprocating motion that forces the press slide through a crankshaft, eccentric shaft or eccentric gear. Such presses come in two different configurations: presses with a gap frame and with a straight side. The gap frame presses come in a wide range of tonnage capacities from about 20 to 600 tons, with speeds ranging from about 20 to 800 strokes per minute (SPM).

Table 2.9 The main design guidelines for tooling for forming aluminum

Drawing tooling and dies	Drawing die characteristics; clearances and blank design
Draw Tool Radius	Punch Radius: $4 \times t_{Panel} - 12 \times t_{Panel}$ Die Radius: $4 \times t_{Panel} - 12 \times t_{Panel}$ Draw Bead Depth: $6 \times t_{Panel} - 8 \times t_{Panel}$ Draw Bead Radius: $5 \times t_{Panel} - 7 \times t_{Panel}$
Draw Die Clearance	First Draw: $1.1 \times t_{Panel}$ Second Draw: $1.1 \times t_{Panel} - 1.5 \times t_{Panel}$ Third &/or Sub-sequent Draw $1 \times t_{Panel} - 1.2 \times t_{Panel}$

Source: [12]

On the other hand, the hydraulic presses control their force by configuring their hydraulic pressure which then moves one or more rams in a predetermined sequence. The hydraulic presses cover a wide range of applications because of their large bed sizes and their ability to provide their full tonnage at any point in the stroke, which is not the case for the mechanical presses. Hydraulic presses are widely used for deep drawing and for the short runs with frequent die changes.

The automotive presses are typically designated based on the following system: *SU4-1000-144-96* where: 1000 stands for the tonnage, 144 is the bed size, i.e. 144 inches from left to right, and 96 is the bed size from front to back, i.e. 96 inches. The letters describe the following possibilities: SS is a straight side press, G is a gap press, SE is straight side eccentric, SC is straight side crank, and DU is the Danly under-drive press, S is a single action press, and T is a transfer press. So SU translates into single action press with under-drive. Other automotive presses include the hemming presses used to join the doors' inner and outer panels; such presses rely on single or multi-stage flanging dies.

The automotive presses are arranged in different line configurations according to their action on the part, and the specific part design; these configurations include:

- *Tandem press line:* where the different presses are arranged without any intermediate material storage in between. For example, a double action press can be accompanied by three or four other single action presses to achieve the deep drawn shape and then the trimming, piercing, flanging and restrikes actions by the single action presses. The panel flow between these presses in a tandem line can be manual or automated. The tandem press arrangement is typically used for low to medium volume panels, with around six strokes per minute work sequence. An automated arrangement of a tandem press line is shown in Figure 2.38.

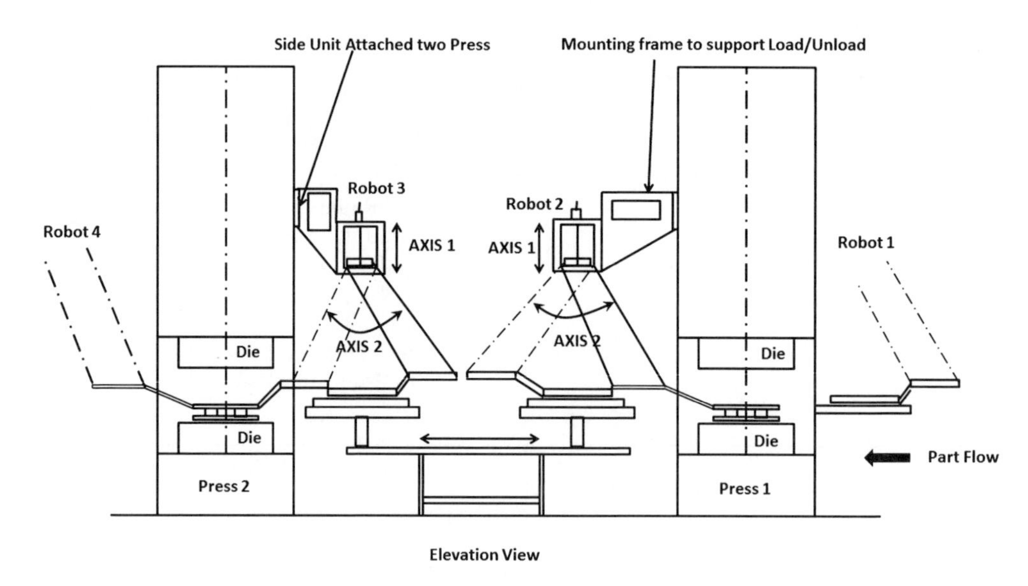

Figure 2.38 A tandem press line configuration

- *Transfer press line:* where several forming steps or tooling are combined in one press, with all in press panel handling using automation fixtures (tooling) that utilizes suction cups to transport the panel from one forming station to the next. The transfer press line consolidates the stamping presses, however, it works in serial mode, where any stoppage in one station leads to stoppages in all other stations within the same transfer line. The transfer press line can use a tri-axis or cross-bar transportation system to transfer the panels from one station to the next.

2.3.1 Automotive Dies

The stamping die functions in four different ways:

1. The die position and locate the blank correctly within its cavity according to datum point or line.
2. The die clamps the blank using its binder to hold the blank in place while stamping.
3. The die deforms the blank in three different actions: it cuts (trims, pierces, and shears), it folds (flange), and it stretches and draws the metal.
4. The die releases the formed panel using its mechanisms of cams and rejection pins.

Stamping dies are classified according to different categories:

1. *Manufacturing processes:* blanking, punching, bending, deep drawing, etc. Forming dies in automotive stamping, are of two main types: (a) Toggle die: runs

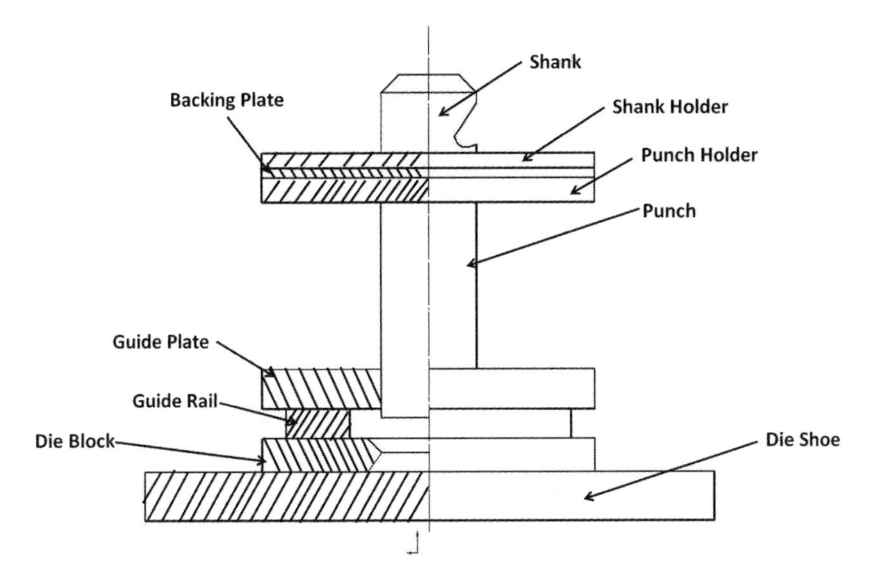

Figure 2.39 A typical automotive die and its basic elements

in double action, with a binder on outer ram (slide) to hold the blank, and a punch mounted on inner ram to form the blank; and (b)Lower cushion die: nitrogen, air pressurized cushions to control force. Typical die schematics are shown in Figure 2.39.

2. *Number of operations:* Single-operation (simple), and multi-operation (or combination) dies.
3. *Number of stations:* Single station and progressive dies. *Single station dies* may be: (a) *combination:* a die in which both cutting and non-cutting operations are done at one press stroke); or (b) *compound*: a die in which two or more cutting operations are done at every press stroke.

   Progressive dies are made in two or more stations. Each station performs an operation on the part or provides an idle station so that the work piece is completed when the last operation has been done. After the first part has traveled through all the stations, each subsequent stroke of the press produces another finished part. A progressive die is displayed in Figure 2.40. Progressive dies are typically used in small parts, fast production, and used only for cut, fold. However, progressive dies do not allow for good nesting and tend to increase the scrap rate. Typical designs of the progressive dies should have enough allowances for idle stations (does not perform any work) to provide some spacing for certain mechanisms (cams), and to accommodate any future changes in the product (e.g. change punch locations).

   Another type of stamping dies is the transfer die which differs from the progressive die in that the individual parts move from one die station to the other (between

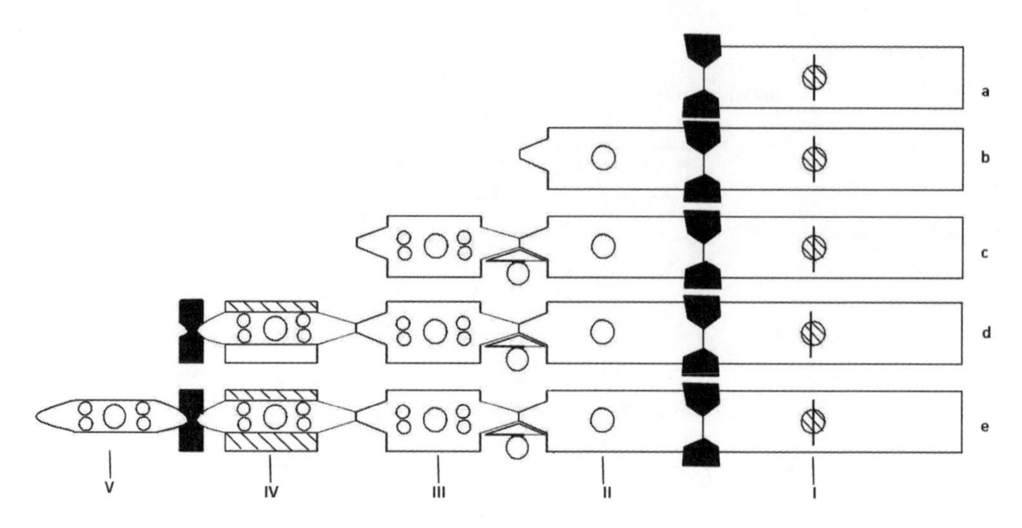

Figure 2.40 A panel development as it goes through a progressive die. Reproduced by permission of Industrial Press © 2010

each press stroke) by the use of mechanical cams, which are part of the die tooling or sometimes built into or mounted on the press.

4. *Production quantities:* high, medium, and low, which can be further categorized into:
   i. *Class A:* these dies are used for high production only. The best of materials are used, and all easily worn items or delicate sections are carefully designed for easy replacement. A combination of long die life, constant required accuracy throughout the life, and ease in maintenance are prime considerations, regardless of tool cost.
   ii. *Class B:* these dies are applicable to medium production quantities and are designed to produce designated quantities only. Die cost relative to total production volume becomes an important consideration. Cheaper materials may be used, provided they are capable of producing the full quantity, and less consideration is given to ease of maintenance.
   iii. *Class C:* these dies represent the cheapest usable tools that can be built and are suitable for low-volume production.
   iv. *Temporary dies:* these dies are used for small production and represent the lowest cost tools that will produce the part.

2.3.2 Die Operation and Tooling

This section explains the different functionality of the stamping dies, as stated in the previous section: locate, hold, affect, and release, followed by a discussion of the die development process. The die operation will be explained through its different mechanisms and components.

2.3.2.1 The Blank Holder

The blank holder is one of the main important parts that stamping engineers rely on not only to hold the panel in place but also as a variable to manipulate or to fine tune the metal flow within the die cavity. Also, the blank holder's function prevents the appearance of wrinkles in the panel upper flange area. The most important parameter is the blank holder pressure. If the blank holder exerts too little pressure, or if the punch or draw-ring radii are too large, wrinkles will appear. Wrinkles are sometimes caused by too much metal trying to flow over the draw ring, which usually happens when the clearance between the punch and the die is not adequately designed.

The blank holder pressure can be evaluated through its force, which can be mathematically described in Equation 2.17:

$$P_d = 0.3 \times \left(\left[\frac{D}{d} - 1 \right]^3 + \frac{d}{200t} \right) \times UTS \tag{2.17}$$

where D is the blank diameter, t is the panel thickness, d is the inside cup diameter, and UTS is the Ultimate Tensile Strength of the blank.

2.3.2.2 Draw Beads

Draw beads are often used to control the flow of the blank into the die ring. Beads facilitate or constrain the material flow by bending or unbending it during drawing. The draw beads that help the flow of the material are located on the die ring, while these beads that are used for restricting the flow are in the blank holder and in corresponding places on the die ring. Figure 2.41 shows a draw-bead configuration for facilitating material flow while Figure 2.42 shows the bead for restricting the material flow.

Some correlations exist to calculate the draw bead radius to facilitate material flow, such as the one in Equation 2.18:

$$R = 0.05 \times d\sqrt{t} \tag{2.18}$$

2.3.2.3 Blanking and Shearing Dies

The blanking dies are configured based on the shape of finished blank and the material thickness. The thickness decides on the die clearance needed.

The shearing process consists of three phases: (1) the elastic deformation phase where the force applied by the die is less than the material yield strength; (2) the

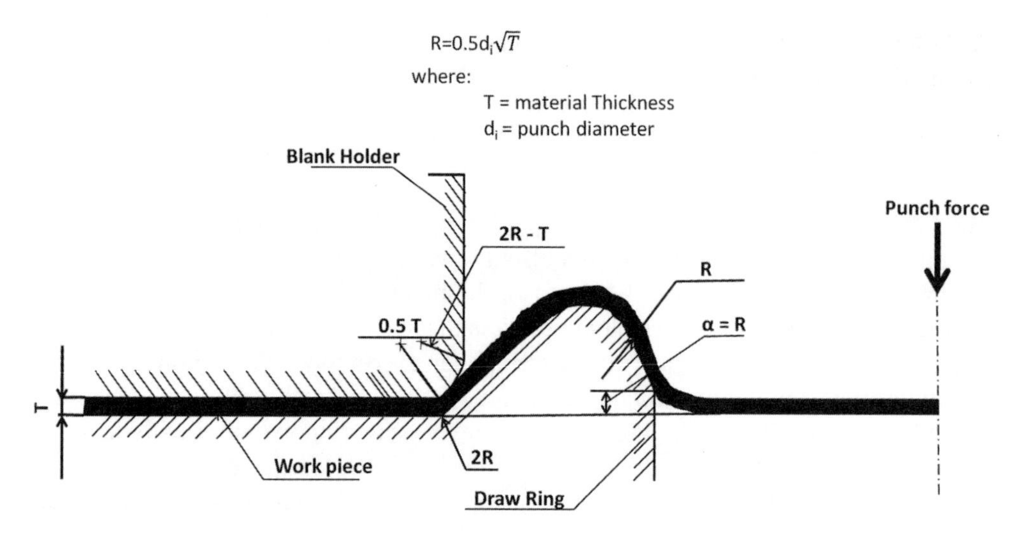

Figure 2.41 A draw bead used to facilitate the metal flow. Reproduced by permission of Industrial Press © 2004

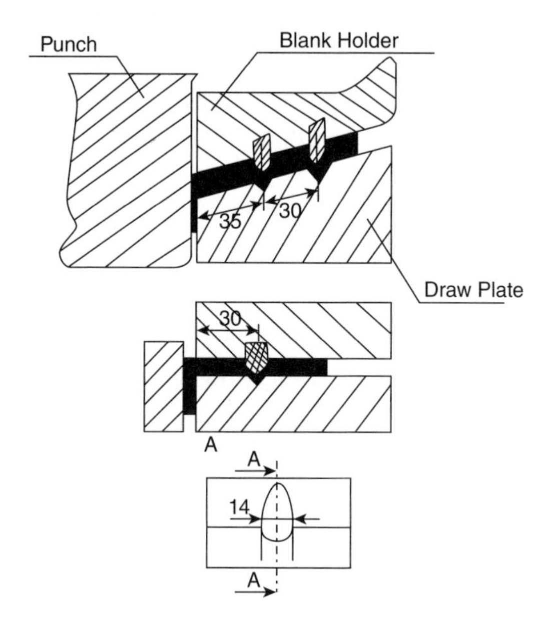

Figure 2.42 A draw bead configuration used to restrict the metal flow. Reproduced by permission of Industrial Press © 2004

plastic deformation phase where the force is larger than the YS but less than the UTS; and, finally, (3) the cutting or shearing phase when the stress (shearing force per unit area) applied is equal or larger than the shearing stress. The cutting force can be manipulated using the cutting angle, i.e. the cutter's angle. The shearing force can be computed using Equation 2.19 and Table 2.10 [16]:

Table 2.10 The relative amount of penetration of the upper blade into the material based on material grade and thickness

Panel thickness	Low carbon steel	Medium carbon steel	high alloy steel	Aluminum panels
$1.00 > t_{panel}$	0.75	0.65	0.50	0.80
$2.00 > t_{panel} > 1.00$	0.70	0.60	0.45	0.75
$4.00 > t_{panel} > 2.00$	0.65	0.55	0.40	0.35
$t_{panel} > 4.00$	0.50	0.45	0.35	0.60

Note: All dimensions are in mm.

$$F = n \times k \times UTS \times \varepsilon \times t^2 / \tan\varphi \tag{2.19}$$

where $n = 0.75$ to 0.85 for most materials, $k = 07$ to 0.8 that is the UTS/τ of the material, ε is the relative amount of penetration of the upper blade into the material (from Table 2.10), and φ is the inclination angle of the upper cutter.

The blanking and punching dies use the punch force to cut the metal into precise shapes; controlling the punch force and speed are vital to ensure the accurate blanking and punching process. Additionally, the interface conditions between the punch and the die in terms of lubricant selection, and the blade edge conditions and clearances must be controlled. Similar to shearing, the blanking process consists of three different phases: the elastic, plastic and the fracture phases where the applied stress by the punch exceeds the blank shearing stress. Additional requirements on the edge quality (micro-cracks, burrs) are also important during the blanking process. Figure 2.43 displays the blanking/punching process schematics, showing the different zones.

The blanking clearance determines the quality of the process (i.e. the blank edge quality) and it also determines the punch force requirements. The clearance can be easily estimated once the die and punch diameters are determined to be as in Equation 2.20:

$$C = \frac{D - d}{2} \tag{2.20}$$

where D is the die diameter and d is the punch diameter, additionally Table 2.11 includes some of the clearance values for different materials.

Other means of evaluating the clearance are through correlations developed for different thicknesses, and for different die surface finishes. For example, Equations 2.21 and 2.22 describe the required clearance for material thicknesses above and below 3 mm:

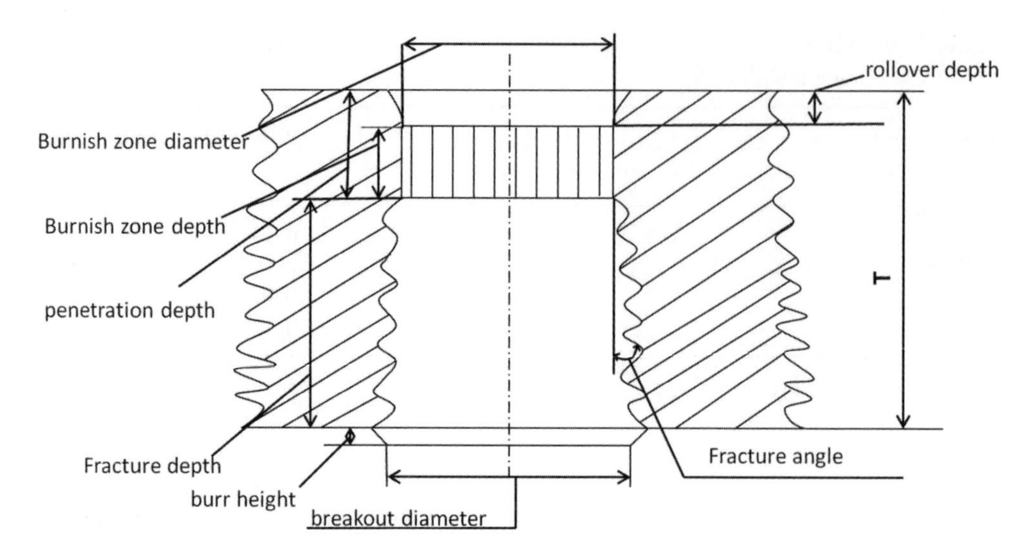

Figure 2.43 The blanked cross-section and its different regions

Table 2.11 The clearance values for several material types and thicknesses

Panel thickness	Low carbon steel	Medium carbon steel	High alloy steel	Aluminum panels
0.50	0.025	0.03	0.035	0.05
1.00	0.05	0.06	0.07	0.10
2.00	0.10	0.12	0.14	0.20
3.00	0.15	0.18	0.21	0.28
4.00	0.20	0.24	0.28	0.40
5.00	0.25	0.30	0.36	0.50

Note: All dimensions are in mm.

$$C = \frac{k \times t \times \sqrt{0.7UTS}}{2} \quad \text{For } t < 3\text{ mm} \tag{2.21}$$

$$C = \frac{(1.5k \times t - 0.015)\sqrt{0.7UTS}}{2} \quad \text{For } t > 3\text{ mm} \tag{2.22}$$

with t being the material thickness, k a coefficient dependent on the die type, ranges = 0.005–0.035 and typically $k = 0.01$ for most die types [16].

2.3.2.4 Bending

The bending die is one of the most used stamping dies due to its ability to form different channel structures and other shapes, but also because such shapes increase the panel stiffness by increasing its moment of inertia.

The bending process is achieved through applying a bending moment on the blank; the terminology of the bending process is shown in Figure 2.44, and include the neutral line where no deformation (strain = 0) takes place, and the bend radius, angle and allowance.

Different correlations exist to compute the bending forces required to produce the different bend shapes, such as the U and V shapes, however, the most important thing to compute is the bend radius or the minimum bend radius to the blank thickness r_{bend}/t. To evaluate the minimum bending radius, one can compute the strain as in Equation 2.23 for any location within the blank thickness, i.e. at a distance of R.

$$\varepsilon = \frac{R - R_{neut}}{R_{neut}} \tag{2.23}$$

where R_{neut} is the radius measured from the neutral axis (or line); also knowing that the maximum strain happens at the outer side of the blank, i.e. $R = R_{inner} + t$; then when the neutral line is right in the middle of the blank thickness, i.e. $R_{neut} = R_{inner} + t/2$. Then the tensile strain can be described according to Equation 2.24; also after inspecting Equation 2.24, one can conclude that as R_{inner}/t decreases, the bend radius becomes smaller, thus the tensile strain on the outer blank side increases.

$$\varepsilon = \frac{1}{\left(\dfrac{2R_{inner}}{t}\right) + 1} \tag{2.24}$$

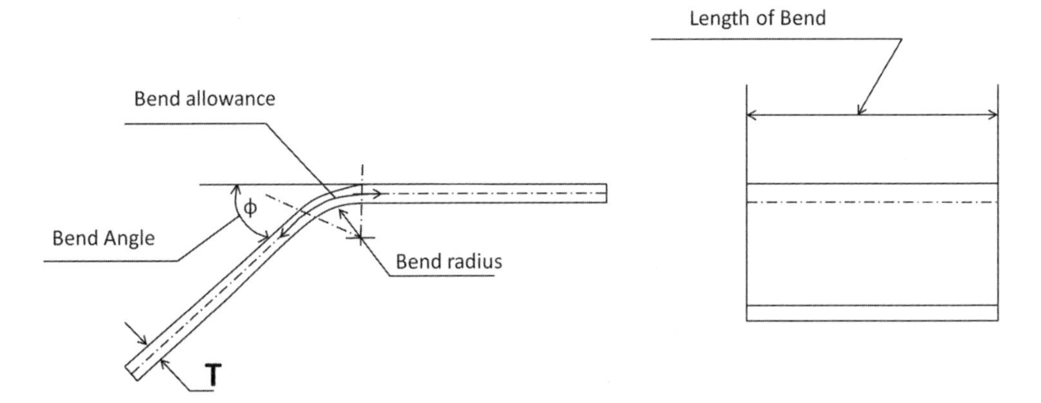

Figure 2.44 The bending radius and its different bending radii

Then the minimum bend radius can be further extracted from Equation 2.24 as in Equation 2.25, by substituting with ε_{split} which is the strain at which the outer blank splits:

$$R_{min} = \frac{t}{2}\left(\frac{1}{\varepsilon_{split}} - 1\right) \tag{2.25}$$

2.3.2.5 Deep Drawing

In deep drawing, the die punch forces the metal to flow between the die cavity and punch surfaces, typically the deep drawing process can generate final shapes that do not require subsequent steps, in addition, it generates small scrap amounts when compared with other stamping processes. Panels produced through deep drawing include the trunk pan and the oil pan. Deep drawing requires careful adjustment of the process variables (lubrication, blank holder forces, etc.) to achieve the final shape without any splits or fractures. The blank holder force should be calculated according to the equations in previous sub-sections, to ensure that the force is not too high to cause splits in the cup wall or too low to cause wrinkles. Deep drawability is often expressed as the blank diameter to the punch diameter, and the number of needed successive drawing processes is described by the ratio of the part (cup, pan) height to its diameter. Additionally, one can compute the required blank size to produce a certain deep drawn shape through the formula in Equation 2.26, where A_{shape} is the area of the final produced shape. Other calculations are needed when the final shape is not symmetrical.

$$D = 1.13 \times \sqrt{A_{Shape}} \tag{2.26}$$

Draw beads might be utilized to increase the metal flow into the die cavity, thus facilitating the deep drawing process especially in the case of non-symmetrical shapes.

2.3.2.6 Coatings and Lubrications

The blank facial conditions affect its performance in the different formability modes, some more than others. This part of the text discusses the effects of the coatings (galvanized, etc.) applied in the steel mill and the lubrications applied in stamping facilities, on the formability of materials. It is important to understand the lubrication and coating process for steel coils going through a stamping process because of the friction impact on formability. Friction affects the interaction between the stamping tooling and the blank surface, hence it changes its deformation and metal flow tendencies. Friction is like formability, it is a system property, not a material or process property but a combination of the two, in addition to the surrounding conditions such as temperature and humidity. Even though the Coulombic friction model describes friction coefficient μ as independent of the contact pressure, this model may not be appropriate to include in formability simulation to describe the facial interactions. For example, the frictional behavior in the vicinity of draw beads is complicated in nature and does not follow the Coulombic model, which motivated the development of several tests to measure the frictional forces at these locations; an example of such tests is the draw bead simulator (DBS) test. This test measures the bending-under-tension (BUT) stresses to describe the formability trend when the blank is stretched and bent over a die radius.

In stamping, researchers have recognized two frictional regimes or modes in the metal deformation process. The first is the boundary regime, which features close or intimate contact between the panel and the die surfaces; such a regime is dominant at low speeds, at low lubricant viscosities, and at high die pressures. The second regime is termed the hydrodynamic regime, which describes the conditions when the blank is completely separated from the die by the lubrication film. The hydrodynamic regime can be observed at high speeds, high lubricant viscosities, and low forming pressures. Most researchers believe that in sheet metal forming, the most dominant frictional regime is a mixed one of both the boundary and the hydrodynamic ones. The Stribeck curve describes the interrelationship between the applied lubricant viscosity, sliding velocity, and the contact pressures in terms of the Coulombic friction coefficient μ.

In lubrications, the main functionality of an applied lubricant is to improve the following: the separation between the blank and the tooling (i.e. die), the cleanliness of the blank, its corrosion resistance, and the blank surface profile (fill surface roughness deviations). Additional benefits from lubrications are better formability (especially for deep drawn parts and for Al coils), in addition to longer die service life.

The lubrication (typically called mill-oil) is typically applied through an electrostatic oiler at the steel mill to increase the corrosion resistance of the blanks. This oil layer is then removed by the blank-washer at the stamping facility, followed by applying another lubricant film. The mill lubricant is typically classified into two types: the mill-oil and the prelube. The mill-oil is cheaper than prelube and is easy to apply, however, its performance in increasing the corrosion resistance is less than that of the prelube. The prelube is typically green or red in color. The lubrication applied by the steel mill is typically applied at the rate of 80–200 mg/sq. ft. The exact value is computed based on the coil roughness profile, the oil film should be about 50% of the *Ra* value, i.e. if the *Ra* is equal to ~1.6 micron, then the oil should be applied at ~75 mg/ sq ft (or 0.8 g/m^2).

With regard to the lubrication applied within a stamping facility, the blank-washer applies a thin layer of blank-wash compounds on the coils that might mix with the steel-mill oil residuals to form the final lubricant film that will go through the press. There is another lubrication (similar to the blank-wash compound) applied inside the press by sprayers. The lubrication compounds are also different depending on the blank material, for example, when lubricating low alloy steel, mineral oils with medium to heavy viscosity might be used, while for stamping stainless steel blanks, corn oil or powdered graphite particles might be used. However, for stamping aluminum coils, mineral oil or sulphurized fatty oil blends can be used.

The pre-coatings applied at the steel mill also affect the overall formability of the metal sheets at the automotive stamping facilities. Steel mills provide pre-coated sheets in the following coatings: galvanized sheets, electro-galvanized and zinc-enriched coated sheets. The hot-dip galvanized coat has a matt finish, while the electro-galvanized films have a shiny finish (so an electro-galvanized sheet is typically used for outer *Class A* panels). Such coatings are applied to improve the corrosion resistance of sheet metals. In regard to pre-coated films, automotive OEMs are typically interested in: (a) evaluating the coating layer adhesion to sheet metal, which is done through a ball-impact stretching test; (b) the coating film's thickness and uniformity, which is quantified using the applied coat-film weight per unit area, e.g. 60 g/m^2. Additionally, OEMs are interested in the effect of the coating type and thickness on the metal formability and the formation of certain defects such as galling. Researchers [17] have summarized the effect of the pre-coat layers on formability as follows: pre-coated steels have a different lubrication behavior from cold-rolled sheet metal under the same lubricant conditions; the adhesion of the zinc layer to the metal surface should be evaluated at different deformation schemes; and the performance of the coatings in terms of corrosion resistance can vary depending on the deformation mode that the panel underwent.

Additionally, chrome coating is typically applied on the die surfaces to enable lower grade die material to be used, hence reducing the total cost, but at the same time improve the die surface resistance to scoring and wear.

The automotive die development is a very time- and computation-intensive process, hence developing a die for a major panel can cost the OEM up to $ 5 million per die. Advances in computer-aided design and manufacturing (CAD/CAM) have led to a better die development process, a die development process using CAD/CAM is depicted in Figure 2.45.

2.4 Tailor Welded Blanks and their Stamping

The tailor-welded blanks, coils, and tubes (TWB/C/T) open the door to customizing the sheet metal thickness and grade at different locations within the same sheet or coil, and to varying the wall thicknesses along tube shapes. The main objective behind the TW technology is to allow for mass savings in the automobile body, by customizing the thickness or grade at specific locations. For example, the door inner panel has to satisfy a stiffness requirement only at the hinge area, so earlier applications used a thicker sheet to stamp the inner, while a TWB allows the OEM to have

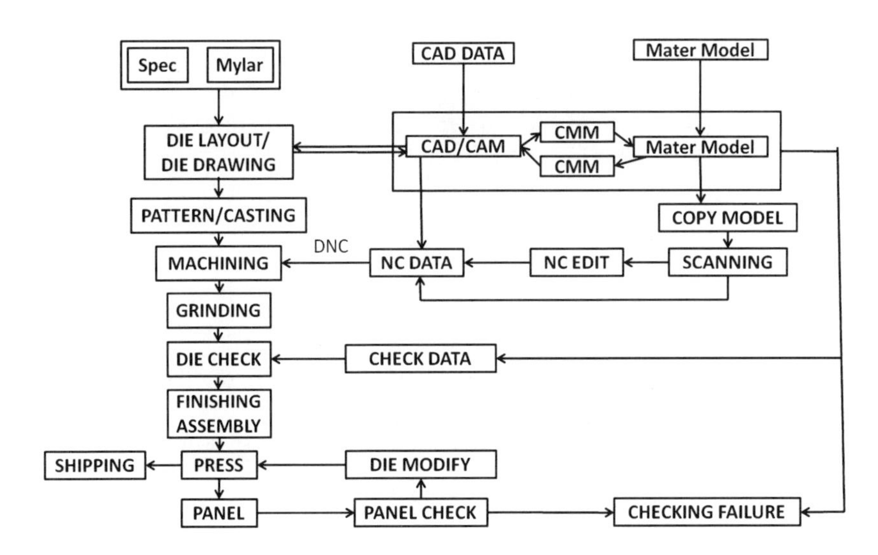

Figure 2.45 The die development process using computer-aided design

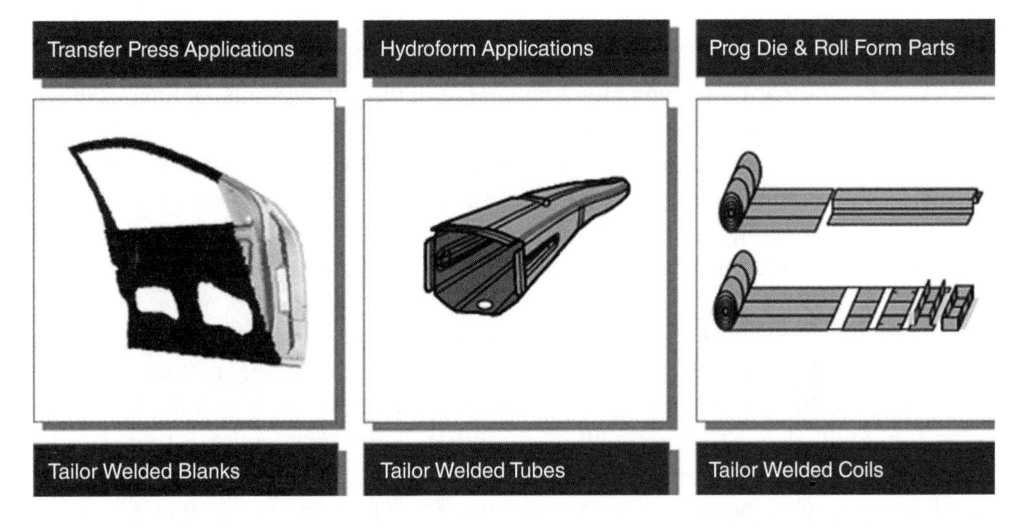

Figure 2.46 The different configurations of the tailor welding technology

a thicker portion in the hinge area and the rest of the sheet can be made from a thinner gauge to reduce the overall weight of the panel. Also, TW technology enables improvements in the vehicle structure performance and might lead to cost reduction. Figure 2.46 displays the different TW technology implementations.

Figure 2.47 An example of a door inner tailor-welded blank. Reproduced by permission of American Iron and Steel Institute and Auto/Steel Partnership © 2010

The tailor welded technology cuts different (in thickness and/or grade) metal sheets or coils, then joins them (welds). To achieve this, edge preparations are done to control the final coil or blank properties and prepare it for the subsequent welding process. Edge preparations typically include; blank die preparations, precision shear, laser cutting, and cold working [18].

The main joining techniques used in TW technology are: laser welding (the most recent), resistance mash seam welding, and spot welding (the cheapest, and easiest). Each of these welding schemes allows for different weld seam configuration, such as the single straight line seams, multiple and angular straight seams, or curvilinear seams.

The straight line seam configuration allows for a simple straight line weld application, with a seam length up to 1.3 meters. Most common applications for this type of welding are: front door inners; rear door inners, with shorter welds than the front door; longitudinal rails; pillars, with the B-pillar as the most common; cross-rails; reinforcements; in addition to other applications such as floor pans, rocker panels, shock towers, wheelhouse inners, sill panels, etc. The multiple straight line weld seams apply two or three welds in the same axis, the most prominent TW panels achieved through this weld are the body side outer and the engine rails. Figure 2.47 shows this type of weld configuration when applied to a door inner panel.

Another important type of TW application is the patch type, where it places one blank of sheet metal on top of another to strengthen the final panel. The two blanks are then spot welded then formed. The patch does not require any edge preparations because linear welding is not applied (i.e. no butt weld joint design or preparations are needed). Thus the patch is more flexible and costs less due to the use of spot welding not laser welding. However, due to the spot welding surface effects and its

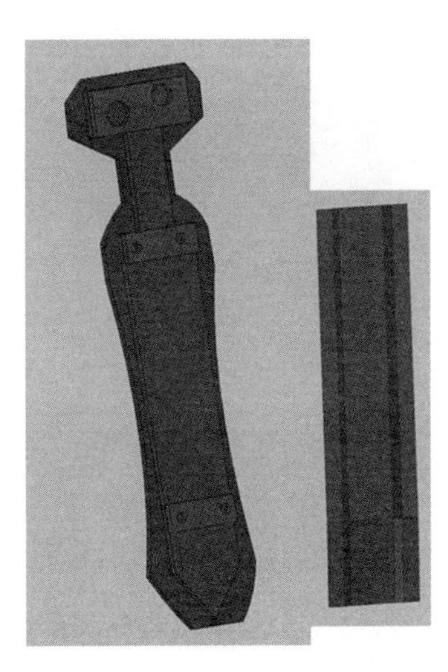

Figure 2.48 A patch tailor-welded blank arrangement. Reproduced by permission of American Iron and Steel Institute and Auto/Steel Partnership © 2010

weld bead issues, it might affect the sheet formability and fatigue performances. Figure 2.48 displays an example of the patch TW for a floor pan bar.

For the Tailor Welded Coils (TWCs), different coils of different thickness or grade are welded to form a new coil with varying thickness and/or sheet metal grade. The TWCs provide major advantages for the automotive stamping activities that include: enabling roll forming and progressive die applications; major mass savings; a reduction in the overall dimensional variations within the body shell; and improving the shell overall fit and finish (better control of gaps and flush setting by controlling metal rigidity). The TWC can join two or three different coils, hence it offers good customization to the finished coil and consequently to the final panel. Also, it can join different steel grades ranging from mild steel, to HSLA and DP grades, with thicknesses ranging from 0.6–3.0 mm and widths from 50–1200 mm. Table 2.12 describes the joining compatibility between the different steel grades in addition to stainless steel, in regard to the TWC application. Figure 2.49 displays some implementations of TWC, using different thickness and widths configurations.

The Tailor Welded Tube (TWT) technology aims to produce variations in tubes' longitudinal cross-sections and wall thicknesses. This allows for design customization in space-frame-based shells. These advantages come to offset some of the inherent limitations in the current hydro-formed, conventional (not TW) tubes such as their limited aspect ratio, i.e. their diameter to wall thickness ratio in addition to the reduced formability window.

From a formability point of view, the TW technologies B/C/T affect the material formability through its impact on material strength, the applied weld ductility, and the weld seam orientation relative to the metal flow patterns. The metal flow histories and patterns are mainly controlled by the grade selection, the weld seam location,

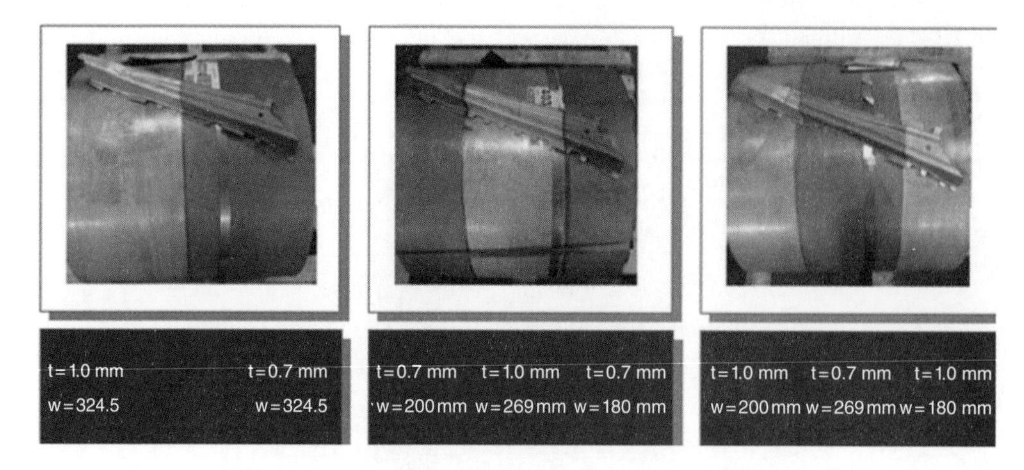

Figure 2.49 Different example of tailor welded coils

Table 2.12 Compatibility of the different steel grades for tailor-welded applications

Steel grade	HSLA	IF	DP	Stainless Steel
HSLA	●	●		●
IF	●	●		●
DP			●	
Stainless Steel	●	●		●

and finally the distribution of the contact pressures and forming stresses. Both laser and mash welds reduce the ductility of parent mild steels by 50% to 75% for parallel strains. Additional considerations relate to the weld hardness profile, in other words, the weld peak hardness relative to the weld-seam width. So the selection of the joining technology is vital in determining not only the cost and edge preparations, but also to decide on the final blank formability characteristics. Experimental results from [19] indicated that weld failures in laser welded blanks are dependent on the weld line orientation relative to the major strain direction, in addition to the relative strength of joined blanks. Further experimental results showed that the blank failure was parallel to the weld seam in the weaker material when the major strain was perpendicular to the weld orientation. The size of the welded blank or test sample affected how close the failure was to the weld. Wider material samples showed the failure closer to the weld. The weld-seam formability can be controlled through designing the welding process according to the ratio of hardness to the area of the weld heat affected zone (HAZ), which can be done by reducing the weld-seam width, and reducing the weld peak-hardness.

Table 2.13 The different possible welding schemes and the required preparations and formability behavior for tailor-welded applications

	Mash	Induction	CO_2	Nd-YAG	Electron beam
Pre-weld preparations	Low	Low to Medium	Extensive/ high Precision	Precise	Precise
Post-weld preparations	Not needed	Required	Not needed	Not needed	Not needed
Seam formability	Good	Low	Good	Good	Very good
Steel grades	Possible	Low applicability	Possible	High applicability	Possible
Curvilinear welds	Not applicable	Not applicable	Possible	High applicability	Possible

Table 2.13 shows the different possible welding schemes and the required preparations and formability behavior. The formability issue in the TW technology is one of the reasons that slow the spread of the HSS and AHSS in the TW, in addition to the weldability for such steel grades.

The main implementations for TW technology in the automotive body panels are in: the engine rails, the door inners, the pillars, the body-side inners, the lower compartment rails, and for general reinforcements such as the floor pan reinforcement.

TW technology offers many advantages to the automotive stamping industry; however, careful costing calculations and production planning studies should be done before an OEM decides to adopt such technology (Table 2.14). The costing calculation for TW is shown in Section 2.8. The TW technology tends to add several additional processing steps that consequently add cost, processing time, and effort. If the welding process is sub-contracted due to the lack of expertise in the welding technology (such as laser welding) or the lack of equipment, then the following processing steps should be considered in the production plan:

- the stacking and de-stacking process of the blanks;
- the cutting and edge preparations;
- the welding process and the required inspection and control functions;
- the dimpling step, to facilitate the stacking of blanks with different thicknesses;
- the cleaning or oiling of the weld-seam;
- in addition to any transportation steps.

Table 2.14 Some of the automotive OEMs' main welding schemes for their tailor-welded blanks

Automotive OEM	Tailor welding technology	Weld capability	Weld system characteristics
Toyota Motor Company	CO_2 and YAG Laser	Multi-axis, with curvilinear welds	Blank-die edge preparation strategy
Nissan	CO_2 Laser welding	Straight line welds and curvilinear welds	Blank-die edge preparation strategy
Honda	Nd-YAG Laser welding	Single and multi-axis welds	Precision shear is used for blanks

At the same time, all the above added steps should be considered from the perspective of the added advantages, for example, TW tends to reduce the number of parts and consequently the number of dies and the number of forming steps. Also TW reduces the scrap rate and allows for better utilization of the offal and nesting practices.

2.5 Advances in Metal Forming

The previous sections have focused on the current practices in automotive sheet metal forming, while this section addresses some of the more recent advances in the sheet/tube metal forming technologies that might not be widely used but still show future potential.

2.5.1 Hydro-forming and Extrusions

The hydro-forming process is a process where tubular sections are produced by an internal fluid pressure that expands the tube walls conforming to a die shape; other definitions of hydro-forming include: "Hydro-forming processes are metal-forming processes based on the application of pressurized liquid media to generate a three-dimensional work-piece shape" [20]. Hydro-forming is done using a multi-forming sequence. The sequence starts by placing the tube in a press, and then the press cavity is closed while the tube ends are capped. Then a fluid is pumped into the tubes under high pressure, forcing the tube wall to expand, following the press cavity shape; further cutting and welding processes follow to complete the expanded tube into its final shape. The hydro-forming press acts in two folds: first, to hold the die halves against the fluid pressure and then to separate (open) them once the tube is completely

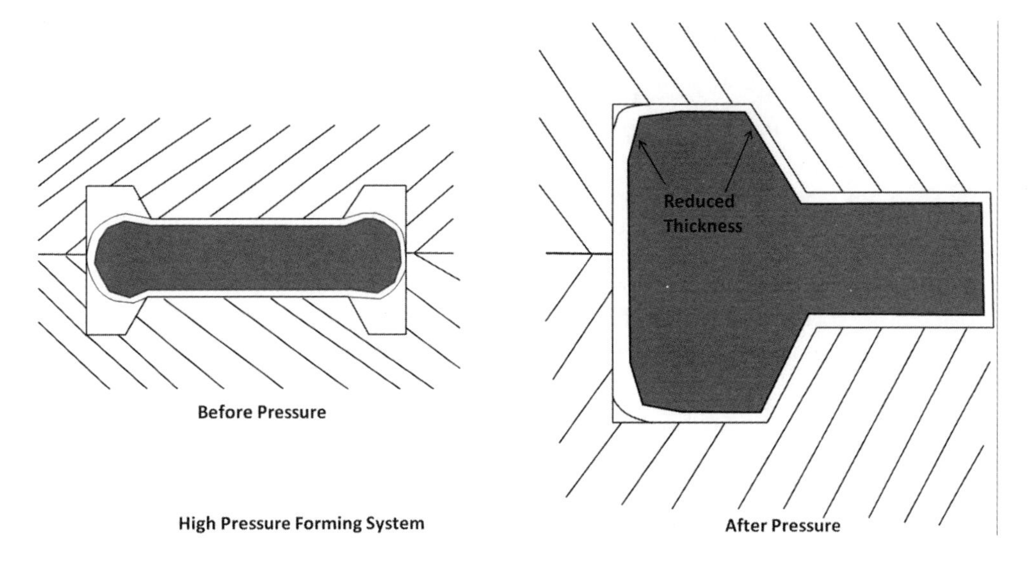

Figure 2.50 The different stages of the hydro-forming process

expanded (i.e. eject the finished product). The most common hydro-forming press is the hydraulic press with different frame configurations such as the four-column frame. Hydro-forming presses are selected based on the required clamping force needed to control the die halves; this force is dependent on the maximum forming pressure and work-piece area. Additionally, the presses include a variety of control modules and pressure intensification units, in addition to the hydraulic valve systems.

The main difference between the hydro-forming and other stamping-based, stretch-forming processes is that hydro-forming provides uniform pressure to the different tube areas. This not only allows lower formability material to be used in hydro-forming but also makes the process easier to control. Additionally, the hydro-forming technology allows for weight-reduction and cost saving because it reduces the number of parts through part consolidation and further reduces the number of production steps. However, the main and most prominent advantage for hydro-forming is that it enables the forming of lightweight materials such as aluminum, magnesium, and metal matrix composites (MMCs) for automotive applications. Other types of hydro-forming are known as sheet hydro-forming, which rely on high pressure hydro-forming or hydro-mechanical deep drawing.

However, the hydro-formed components will have larger thinning rates than stamped components. This effect is more apparent in the wall corners, as shown in Figure 2.50. The basic calculation for hydro-forming internal (fluid) pressure should be such that it provides enough stress to exceed the yield strength of the material, at the corners. Thus, the walls are stretched over the die cavity. The formula for the hoop stresses of a thin-walled pressure vessel can be used as an approximation to

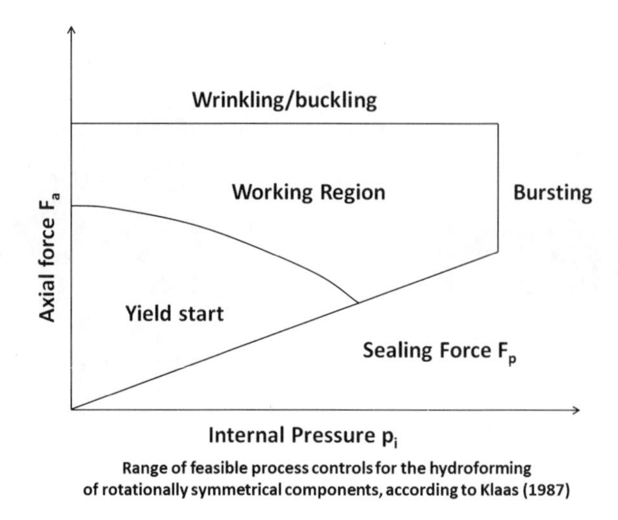

Figure 2.51 Feasible process control regions for hydro-forming

the relationship between the tube yield stress (YS) and the fluid internal pressure, given a determined internal diameter and the tube wall thickness as in Equation 2.27:

$$YS = \frac{D_{tube} P_{fluid}}{2t} \tag{2.27}$$

where *YS* is the tube yield strength, *t* is its thickness and *D* is its internal diameter, while P_{fluid} is the fluid internal pressure.

From Equation 2.27 one can conclude that as the material stretches over the corner, the requirement on internal pressure increases due to the reduction in the radius and to the work-hardening effect (i.e. the increase in the flow stress). This issue becomes more important if OEMs consider the use of HSS and/or AHSS in hydro-forming presses. To solve the dramatic increase on the internal pressure requirements, a multi-stage pressurizing process is proposed in recent applications. The first stage of internal fluid pressure is applied while the press is being closed, and a second high pressure stage is introduced after the cavity is fully closed. This two-step forming pressure forces the tube wall to flow into the corners without major stretching (wall thinning) thus filling the die cavity. [21] indicated that the formability limits for hydro-forming are dependent on the hydro-forming variables' control, mainly the forming load and the internal fluid pressure, according to the feasible process control window as in Figure 2.51. However, the best way to evaluate the formability of the formed tubes is still based on conducting a tensile test, with specimens taken from the tube in the longitudinal direction as in Figures 2.51 and 2.52. Other tube formabil-

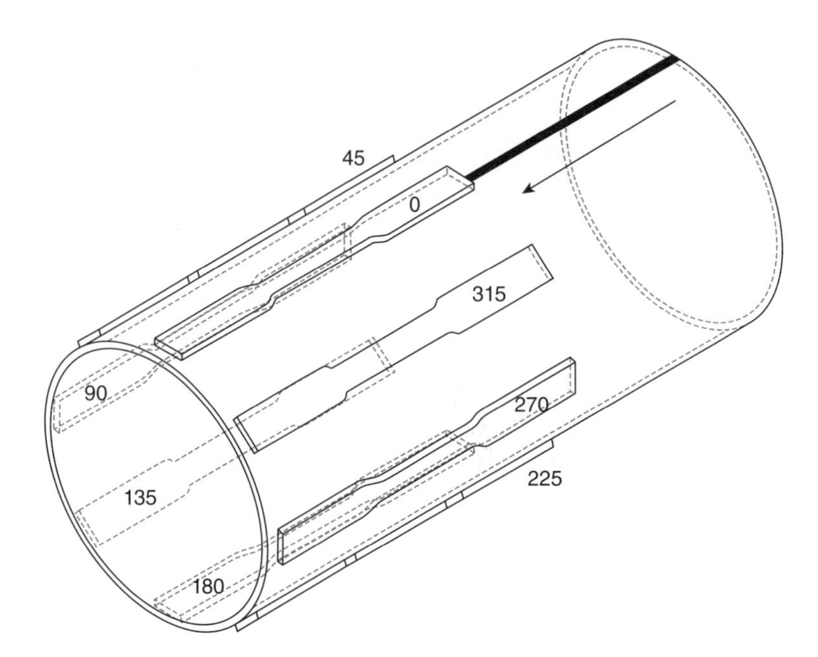

Figure 2.52 Sample SAE tensile test for hydro-formed tubes. Reproduced by permission of American Iron and Steel Institute and Auto/Steel Partnership © 2010

ity tests include the ring hoop tension (RHT) test, and the tube free-expansion (burst) (TFE) tests.

The requirement on the blanks' surface conditions in hydro-forming are even more severe than these in stamping, because of the high contact stresses (due to the high internal pressure) at the interface stresses, in addition to the plastic deformations across the tube. Hence lubrication is used to reduce the sliding frictional pressure between the die and the tube surfaces.

2.5.2 Industrial Origami: Metal Folding-Based Forming

This sub-section discusses a new approach to forming sheet metal (mainly aluminum and steel) into automotive body panels, using the principles of Origami. Origami is the art of folding paper, however, the Industrial Origami Incorporated (IOI) have adopted Origami principles to produce automotive grade body panels by creating fold-defining geometries, i.e. smiles/slots, through laser cutting or water jet machines or even a punch machine, to enable the folding of such panels into useful shapes. An example shape created through this technology is shown in Figure 2.53, with a zoom view of the smiles in Figure 2.54.

Figure 2.53 An example of the Industrial Origami dash panel

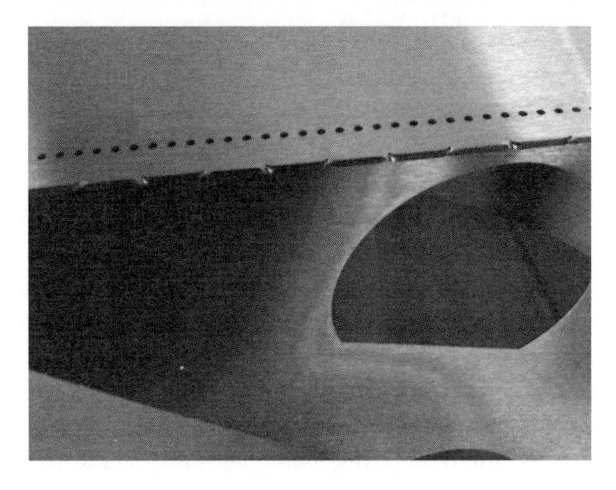

Figure 2.54 A zoomed view of the folds, showing the smiles

The main benefits from this technology for the manufacturing of automotive body panels can be summarized as:

- It helps in consolidating parts by integrating different folds to create different shapes from one sheet.
- It can be applied to wide range of thicknesses and materials without major equipment or tooling change-over.
- It reduces the stacking, de-stacking, and other logistical (non-value-added) steps in the stamping facilities.
- It reduces the overall production cost and, most importantly, it reduces the time and cost of launching new vehicles or vehicle models.

- It reduces the investment cost when developing/designing new vehicles, because such technology reduces the number of dies which translate into major investment (around \$5 million for each major body-panel die).

2.5.3 *Super-plastic Forming*

Due to the difficulties associated with forming aluminum, magnesium and other lightweight (low density) materials, using the standard press-based metal forming practices, the super-plastic forming technology is undergoing extensive research to explore its potential in automotive manufacturing. Super-plasticity is typically defined as the material ability to be deformed into large uniform elongations (in some cases almost Tot. Elong. = 800% of original length) and to yield rupture elongations under pure tension loading.

The super-plastic forming process is typically done at elevated temperatures, using low force. The main mechanism behind the super-plastic forming is based on grain boundary sliding, so the super-plastic forming is typically done on materials with very fine grain size, less than 10 micron, with an optimum strain rate also described in terms of the metal grain size as in Equation 2.28, where the P depends on the material type, for example, P is equal to 3 for Magnesium:

$$\varepsilon_{sp} \propto d^{-p} \qquad (2.28)$$

Inspecting Equation 2.28 shows the extremely slow strain rates applied in the super-plastic forming process, which translates into longer (in hours) forming process, when compared with the conventional press-based done in seconds. Additionally the super-plastic forming requires higher forming temperatures, typically around 50% of the material melting point. However, the super-plastic process offers a large straining capacity during the stretch forming mode achieved with low forming forces and low tool stresses due to the low material flow stress during the forming process, typically in the order of 60 kN/m^2.

Other material requirements for the super-plastic forming include high resistance to grain growth, and formation of pores.

2.5.4 *Flexible Stamping Procedures*

Several OEMs are trying to develop better practices in their stamping lines and technologies to consolidate the number of processes/steps involved in forming the body panels. An example of such procedures is a flexible press line developed by Toyota which is a three-step system: (1) a drawing step achieved using liquid pressure to achieve the required press-forming patterns for the specific panel; (2) a trimming step utilizing a high speed 3-axis laser cutting technology to generate holes and

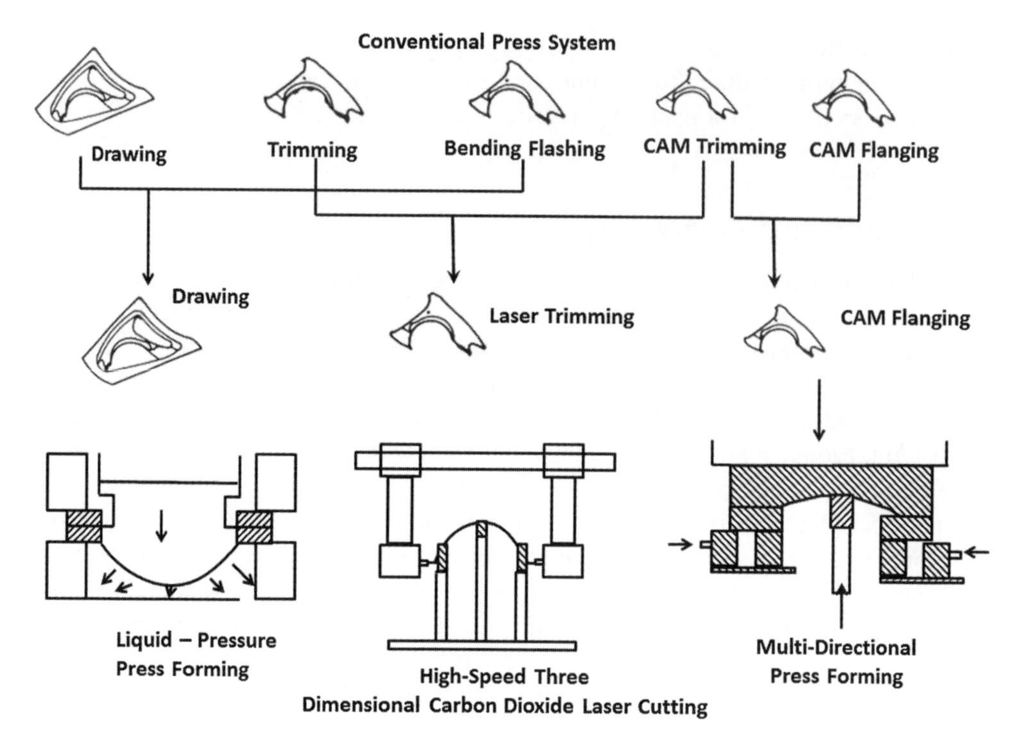

Figure 2.55 An example of the flexible stamping procedure when applied to forming a fender

trim lines; and, finally, (3) a cam flanging step done through a multi-directional pressure applied from six different directions to achieve multi-directional forming profile. An example of forming a fender is presented in Figure 2.55, to highlight the difference between the typical five-step process and the Toyota-proposed process which consists of the three steps.

2.6 Stampings Dimensional Approval Process

The produced stampings, around 400–500 pieces in total, vary in size and thicknesses; in addition, their material is made from different steel grades or aluminum alloys. So automotive OEMs have developed three different strategies for stampings and dies' dimensional approval, because the quality of a stamping in terms of its dimensions, shape (strain) and appearance, once integrated into a vehicle shell will affect the customers' perception of the final vehicle quality. OEMs quantify the customer perception of dimensional conformance through quantifying: (a) the wind noise; (b) water leak; and (c) gaps and door fit.

During the die approval process, OEMs will apply one of the following dimensional validation strategies:

- *The net build or sequential validation process*, which validates each individual (major) panel separately. In this strategy, all dimensions on each individual stamped part (major panels only) should achieve a process capability value C_p and a process capability index C_{pk} greater than a set threshold such as1.33,or 1.67. Both the C_p and C_{pk} indices assess the ability of a process to produce outputs within their specification limits. C_p is determined by dividing the total tolerance by six times the standard deviation; while C_{pk} index differs from C_p because it includes the deviation of the mean from its nominal. Most manufacturers achieve C_p requirements, but often fail to achieve C_{pk} requirements due to mean deviations from the nominal. In other words, their processes have sufficiently low variation but are off target. C_p and C_{pk} are mathematically described through Equations 2.28 and 2.29.

$$C_p = \frac{USL - LSL}{6\sigma_{sub}} \tag{2.28}$$

$$C_{pk} = \min\left(\frac{USL - \bar{X}}{3\sigma_{sub}}, \frac{\bar{X} - LSL}{3\sigma_{sub}}\right) \tag{2.29}$$

  Following the net-build approach results in great confidence in each panel's dimensional conformance, however, following this approach results in longer die validation times; in addition, it does not describe the dimensional conformance of the completed BiW. The net–functional build approach is the oldest dimensional validation strategy and was popular among the North American OEMs (the big three).
- *The functional build approval process*: approves the dies from a complete assembly perspective. The assembly perspective is achieved through an assembled vehicle shell from the different panels using screws, also known as the "Screw Body." Manufacturers would screw mating panels together to check for assembly interference. Once dimensionally stable, defect-free panels are achieved, where the process is repeatable, the dies are approved [22].

  The evaluation of this screw-body vehicle then drives changes to those subassemblies affecting the final BiW conformance: The use of the functional build process appears most applicable in the assembly of either two non-rigid components or a non-rigid to a rigid component. In general, a component may be considered non-rigid if it has a blank thickness less than 1.5 mm [23].The functional build strategy was popular among the Japanese OEMs. This strategy provided a quick and efficient validation process that reduced the overall die development time and cost.

The approval of the body-side panel cannot be evaluated without considering the center pillar reinforcement panel because this panel is a structural component and thus will have greater influence on the final assembly. If the body side panel is 1 mm outboard from the centerline of the car, but the center pillar is at nominal, then the overall assembly will probably shift toward nominal. This shift occurs because the mating surfaces are parallel. Thus, the less rigid body side panel will conform to the rigid inner structure.

The example above highlights the main advantage behind using the functional build approval process. However, there some important issues to consider when applying this strategy to body panels, such as:

- Defining an acceptable assembly based on potential to produce sub-assembly dimensions to nominal involves risk, because this will require that all dies are at the same level of development, which might cause some delays.
- Delayed dimensional decisions until complete evaluation of weld tools, which might eventually require die rework.
- OEMs should not delay die approval decisions based on potential to compensate in assembly because of resource limitations to make changes in time.

These issues produced the development of the "Event Based Functional Build" (Figure 2.56):

- The event-based functional build process is a combination of the above two strategies. This strategy requires individual panels' loose conformance, followed by the approval for a screw-body. "Event-based functional build is characterized as a process utilizing objective criteria that are relatively loose in dimensional con-

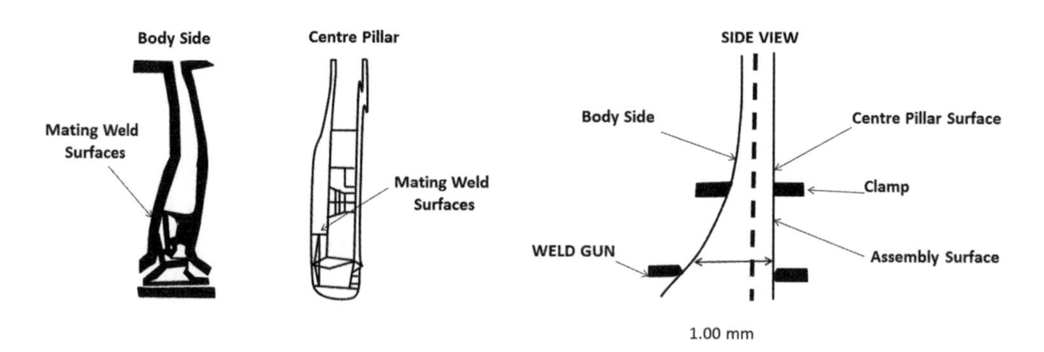

Figure 2.56 An example of the event-based functional build process. Reproduced by permission of American Iron and Steel Institute and Auto/Steel Partnership © 2010

formance and stringent in terms of timing" [22]. Some component criteria are necessary to insure part dimensions are relatively stable without excessive mean deviations. This strategy applies the functional build scheme for selected panels based on their rigid or non-rigid classification within the BiW structure. The die approval flow based on the event-based functional build process can be pictorially shown in Figure 2.57.

To apply any of the dimensional conformance strategies, OEMs utilize different measurement tools. The main tools include hard gages (using checking fixtures) and coordinate measuring machines (CMM) and optical coordinate measuring machines (OCMM). The choice between these measurement tools is dependent on the application, cost, repeatability, the tool mobility, and the tool flexibility. The tool flexibility means the possibility of adding new points of measurements so new panels or dimensions can be evaluated. A body-side fixture is shown in Figure 2.58, while a CMM is displayed in Figure 2.59.

The dimensional conformance measurement relies on a coordinate system to describe the location of each panel or feature relative to a reference point. All part measurements are relative to a part datum scheme described on geometric dimensioning and tolerancing (GD&T) drawings. This system employs the following coordinate system: fore/aft (X), in/out (Y), and up/down or high/low (Z). The 0,0,0 point

Non-Functional Build Conformance Strategy	**Rigid (panel thickness > 1.5 x dim.), Non-complex components and panels**	
Functional Build Conformance Strategy	**Event-based Functional Conformance; Major Outer Panel [body-side outer, fenders, etc)**	
	Rigid and Non-Rigid inner, Complex panels	**Non-Rigid inner; under-body**
		Rigid, Complex; windshield frame

Figure 2.57 The event-based functional build process flow showing the recommended validation process for each panel

Figure 2.58 A hard fixture used for dimensional approval. Reproduced by permission of American Iron and Steel Institute and Auto/Steel Partnership © 2010

Figure 2.59 Dimensional approval using a CMM. Reproduced by permission of American Iron and Steel Institute and Auto/Steel Partnership © 2010

of the car is the front, lower, center position. A schematic of such coordinate system is shown in Figure 2.60.

The holding fixtures (hard gages), follow an *n-2.1* locating scheme to position parts: n locators position a part in a primary plane or direction, two locators then position the part in a secondary direction, leaving one locator for the tertiary direction. Thus, fixing the part in 3-dimensional space satisfies the six degrees of freedom constraint. Some OEMs will replace the three locators for the secondary and tertiary directions by using two round pins, one fitting a circular hole and the

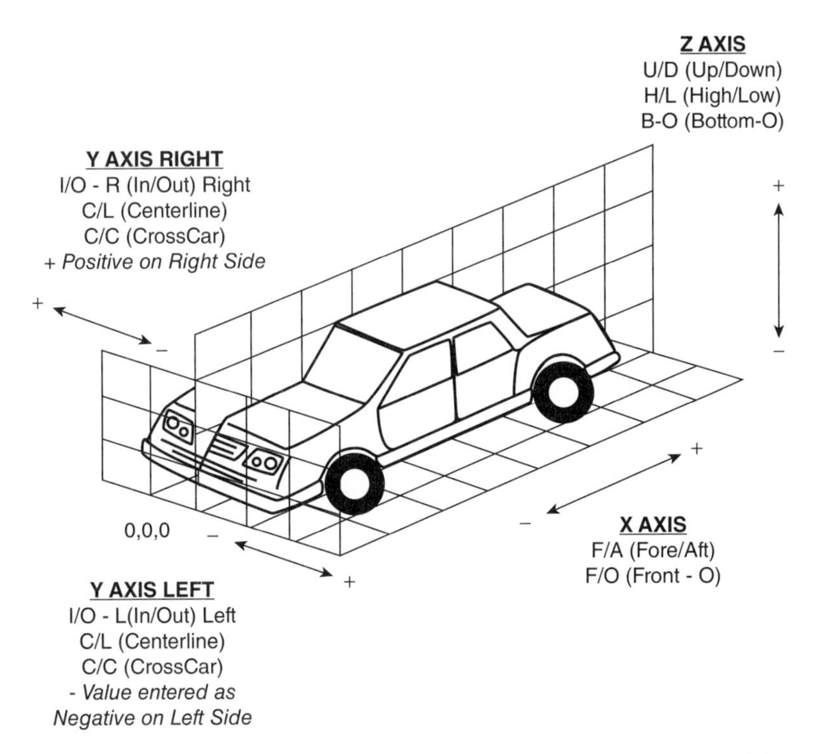

Figure 2.60 The coordinate system for the automotive dimensional approval. Reproduced by permission of American Iron and Steel Institute and Auto/Steel Partnership © 2010

other a slot, for certain parts. The pin locates the part in two directions, in/out and fore/aft. The slot then becomes the other locator for the secondary direction. Figure 2.61 shows such a measurement system.

The criteria to re-work a die are defined based on the cost implications and the actual deviation in stamping dimensions, as shown in Figure 2.62, noting that the major factors affecting the dimensional quality include: stamping variation, assembly process and automation, product design, and the dimensional engineering strategies of part measurement systems.

2.7 Stamping Process Costing

This section discusses the costing process for automotive stampings; the cost analysis will aid the designers not only to select the material in terms of type, grade, and thickness; but also to help in deciding on the different design options such as the tailor-welded technology.

The stamping costing process can be done using the major panels within the body-shell, around eight panels instead of computing every single part (around 400 parts).

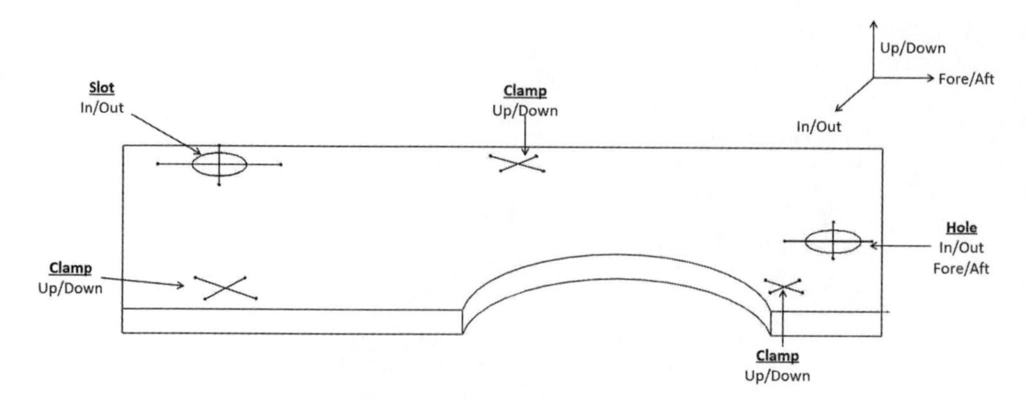

Figure 2.61 An example of a locator system

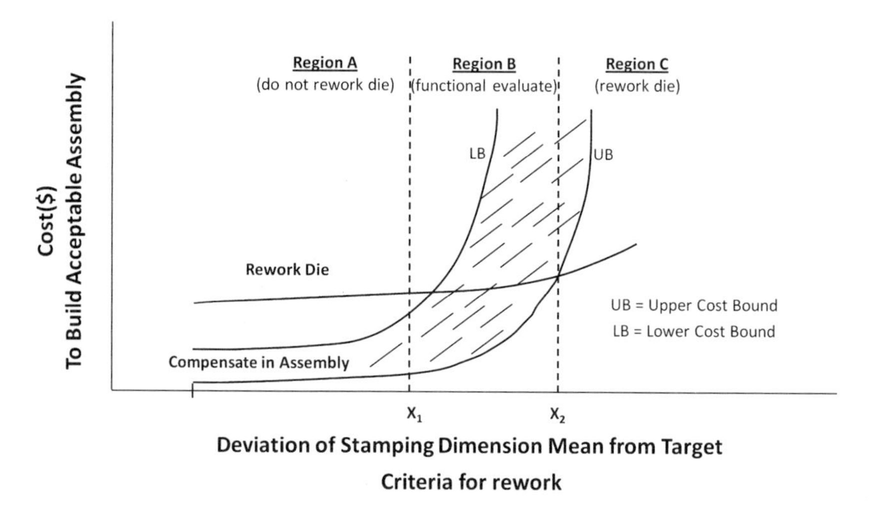

Figure 2.62 The die re-work criteria. Reproduced by permission of American Iron and Steel Institute and Auto/Steel Partnership © 2010

These eight major panels control the overall cost of the automobile body structure. The costing should include following costs and variables:

- *the fixed cost:* includes the facility investment in terms of die development cost and capital equipment, in addition to insurance and other maintenance costs.
- *the variable cost:* includes the raw material cost (i.e. coils), the operating cost in terms of the power usage and the consumables;
- *the sub-contract cost (if any):* which covers costs such as laser welding if done by a supplier.

The following text explains the costing process in two different cases: the first is for the stamping process as a whole, while the second is for the selection of a tailor-welded door inner design.

2.7.1 *Case I: The Stamping Process*

Computing the overall cost of the stamping process helps in knowing the stamping step cost contribution to the overall vehicle manufacturing cost; additionally, it helps in showing any potentials for cost savings.

The costing for the stamping process can be done from two different perspectives: the detailed perspective where each cost (from the above) is computed to result in the overall cost, while the second perspective is based on modeling the process to result in a cost function dependent on the process variables, such as materials and power cost. This section will discuss the detailed perspective because it provides the process details and the different inputs, outputs and wastes. Also, for this calculation each panel's functional performance criteria (dent resistance, bending stiffness) should be recognized so that initial material selection (i.e. steel grade) is done properly because the cost difference between different grades can reach up to 2 or 4 times.

2.7.1.1 Detailed Cost Analysis

For the detailed cost analysis; one should compute: the material cost, which is done through multiplying the coil surface area by its thickness to result in its volume. Then the volume is multiplied by the material density to yield its weight. Finally, the weight is multiplied by the material cost/ton or Lb according to the steel mill pricing. To do this, the blank surface area for each of the major body panels should be known, to simplify this process one can rely on the CAD file for each panel assuming a margin of error. Additionally, the scrap (offal) should be factored in. The major panels to consider are: the four doors with their inner and outer panels, the roof, the hood (inner and outer), the under-body, the fenders, and body-sides (left and right), and the quarter panels. The calculations presented in this section are based on a BiW that was scanned using a CMM so that the blanks' surface areas are known. The retrieved BiW surface is shown in Figure 2.63.

To evaluate the operating cost, the stamping press information were assumed to be: a press rate of 600 $/hr with 2 strokes/minute, blanking press rate 400 $/hr, total production hours per year = 5000 hrs, vehicle production volume = 200,000 vehicles/yr, labor rate of $30/hr, fringe benefits may be added to bring the total of labor rate to $45/hr, laser welding cost (for TWB) if done through subcontract is ~$2.5/meter of weld, Building maintenance, insurance and utilities cost around $55/ year per sq. ft. Additionally, for any robots used in the stamping process, one can assume a general robot rate of $6/hr with robot efficiency of around 70–80 %. Additional costs

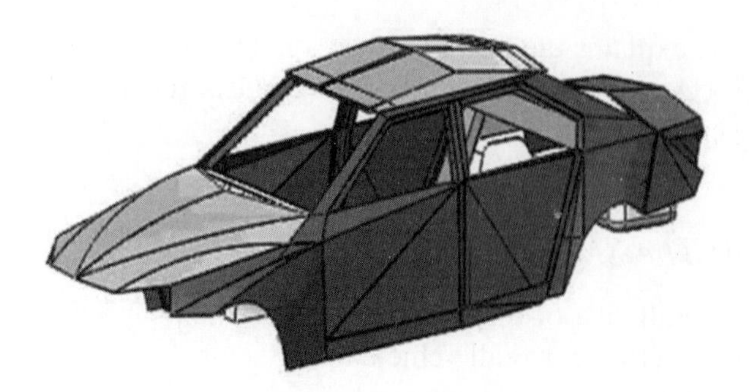

Figure 2.63 The BiW used for the cost example

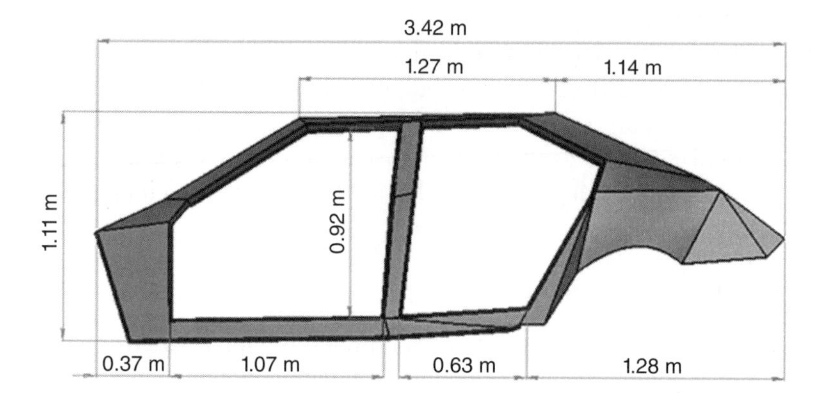

Figure 2.64 Dimensions of the BiW

come from the tooling investment; each major body panel die costs around \$3–5 million in design and development efforts that should be spread over the vehicle model life, which is typically around 4–5 years. Table 2.15 shows the inputs in terms of the blanks and expected scrap percentage. Also Table 2.16 explains the detailed calculations for the body-side outer panel, shown in Figure 2.64.

The calculations in Table 2.16 enable the engineers to evaluate the different contributions (material, operation, etc.) to the final cost for each panel, then evaluate their material selection, their nesting practices (to reduce scrap), and other available technologies (tailor welding) cost implications. Additionally they can perform a sensitivity analysis to see the effect of the material pricing fluctuation, and other variables such as electricity costs, etc.

The second costing example presented in this section relates to the TWBs design selection for door inner tailor-welded panels with the goal of achieving an acceptable added cost to the weight saved, i.e. +\$/-kg.

Table 2.15 The input table describing the blanks and the expected scrap percentage

Assumptions				FORMING			
Part Description	# Parts/Car	Material Characteristics	Sub-Assemblies	BLANKING			STAMPING
				Length *[mm]*	Width *[mm]*	Area *[m^2]*	Used Surface Area *[m^2]*
Front Door	2	Impact	Outer Surface	1200	650	0.780	0.698
			Inner Structure	1200	1200	1.440	0.761
Rear Door	2	Impact	Outer Surface	1100	700	0.770	0.490
			Inner Structure	1200	1100	1.320	0.549
Hood	1	Impact	Outer Surface	1778	1270	2.258	1.957
			Inner Structure	1778	1270	2.258	0.979
Trunk	1	Impact	Outer Surface	1500	850	1.275	1.062
			Inner Structure	1500	850	1.275	0.531
Floor Pan	1	Impact	Bottom Pan	2350	2000	4.700	3.410
Rear Wheel Well	2	Torsion	Structure	1100	700	0.770	0.578
Trunk Pan	1	Impact	Structure	2000	1750	3.500	1.990
Firewall	1	Impact	Structure	800	1700	1.360	1.050
Rear Bulkhead	1	Impact	Structure	1550	1800	2.790	1.780
A Pillar	2	Bending	Inner Structure	800	200	0.160	0.115
B Pillar	2	Bending	Inner Structure	250	1000	0.250	0.199
C Pillar	2	Bending	Inner Structure	350	1000	0.350	0.123
Roof	1	Impact	Outer Surface	1600	1300	2.080	1.753
Body Side Outer	2	Impact	Outer Surface	3500	1250	4.375	1.667
Rockers	2	Torsion	Inner Structure	1800	210	0.378	0.362
Front Fender	2	Impact	Outer Surface	1350	800	1.080	0.439
Front Module	1	Torsion	Assembly			0.000	
Rear Module	1	Torsion	Assembly			0.000	
Roof Cross Members	2	Bending	Inner Structure	1200	200	0.240	0.196

Table 2.16 Detailed calculations for the different costs involved

Body Side Outer		Independent Variables		
		Variable	Units	Value
		# Components		1
Forming	Blanking	Final Part Area (Top)	m^2	1.667
		Length of Blank	m	3.500
		Width of Blank	m	1.250
		parts blanked/min	part/min	8
		# of times blanked		1
	Stamping	# of times stamped		1
		parts stamped/min	part/min	8
Joining	Spot Welds	# of spot welds		500
	Arc Welds	Length of arc welds	m	0.000
		Electrode diameter	mm	2
		Wire feed rate	m/s	0.148
		Gas flow rate	cfh	25.00
		Travel speed	m/s	0.072
		Current	Amps	300
		Voltage	V	26
	Laser Welds	Length of laser welds	m	0.000
		Travel speed	m/s	0.072
	Adhesive Joints	Length of Joint	m	9.00
		Price of Adhesive	$/gall	300.000
		Price of Sealant	$/gall	301.000
		Cross sect. area of Sealant	mm^2	1.886
		Cross sect. area of Adhesive	mm^2	5.715
		Travel speed	m/s	0.072

Material	CR Carbon
Ref Thickness [mm]	0.70
Relative t for Impact	0.70
Part Mass [kg]	9.18
Total Length of Weld	0
Total Cost	$69.10

Dependent Variables			Costs		
Variable	Units	Values	Costs	Unit	Value
% Scrap	%	62%	Blanking Cost	$/part	$0.83
Mass of Scrap	kg	14.918	Scrap Revenue	$/part	$3.29
Blank Area (Top)	m^2	4.375	Material Cost	$/part	$14.58
			Stamping Cost	$/part	$1.25
			Forming Cost	$/part	$13.38
			Spot Weld Cost	$/part	$50.00
Energy Cost	$/part	$0.00	Arc Weld Cost	$/part	$0.00
Welding Time	sec	0.00			
Robot Time	sec	0.00			
# of Robots		1			
Electrode Cost	$/part	$0.00			
Gas Cost	$/part	$0.00			
Gas Volume	ft^3	0.000			
# of bottles		0.000%			
Welding Time	sec	0.000	Laser Cost	$/part	$0.00
Robot Time	sec	0.000			
# of Robots		1.000			
Robot Time	sec	178.654	Adhesive Cost	$/part	$5.72
Application Time	sec	125.058			
# of Robots		1.000			
Cost of Sealant	$/part	$1.35			
Cost of Adhesive	$/part	$4.08			
			Joining Cost	$/part	$55.72

Inputs Data
Calculated Data
Referenced Data

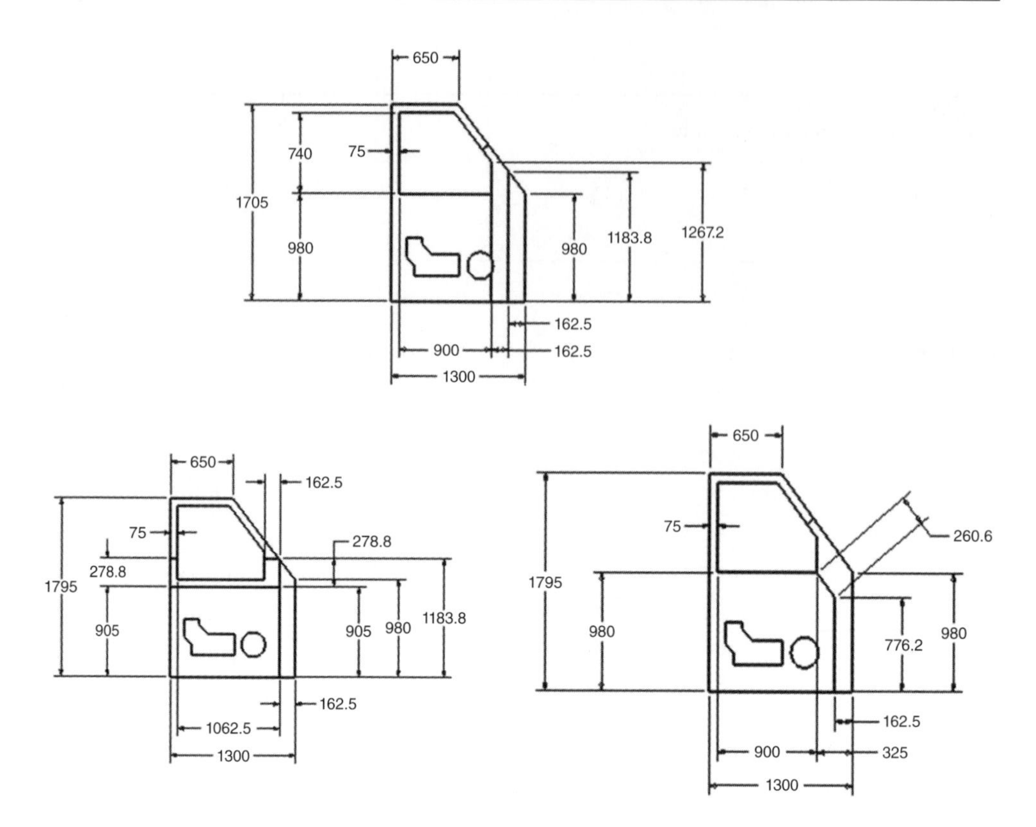

Figure 2.65 Different designs for the tailor-welded door inner

2.7.2 Case II: Tailor-Welded Door Inner Cost

The door targeted for improvement (weight reduction) is a 20 Lbs (9 kg) with main dimensions of the blank of 980 × 1300 mm. The door's main dimensions will help in selecting the right coil width from the available widths that steel mills provide. In this costing exercise three TWB designs will be considered, as shown in Figure 2.65, the designs are based on using two/three different blanks to form the door inner hinge area (with requirement on stiffness), noting that the panel dimensions are selected from available coil widths. The designs are called Alpha, Bravo, and Tango respectively.

The design analyses include, weight computation based on each blank surface area (dependent on the design), thickness (selected based on the yield strength of each grade), and density (the same because all are steel grades). Additionally, the laser welding cost for each design should be computed because each design features dif-

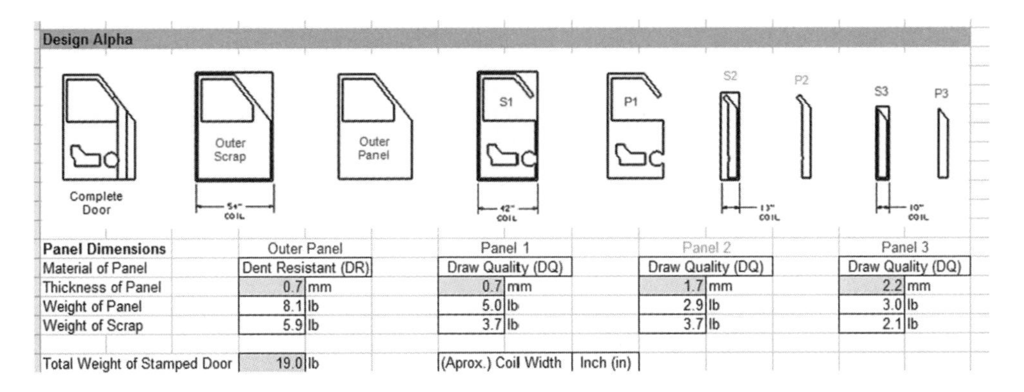

Panel Dimensions	Outer Panel		Panel 1		Panel 2		Panel 3	
Material of Panel	Dent Resistant (DR)		Draw Quality (DQ)		Draw Quality (DQ)		Draw Quality (DQ)	
Thickness of Panel	0.7	mm	0.7	mm	1.7	mm	2.2	mm
Weight of Panel	8.1	lb	5.0	lb	2.9	lb	3.0	lb
Weight of Scrap	5.9	lb	3.7	lb	3.7	lb	2.1	lb
Total Weight of Stamped Door	19.0	lb	(Aprox.) Coil Width	Inch (in)				

Figure 2.66 The different weight calculations for the first scheme of TWB door inner

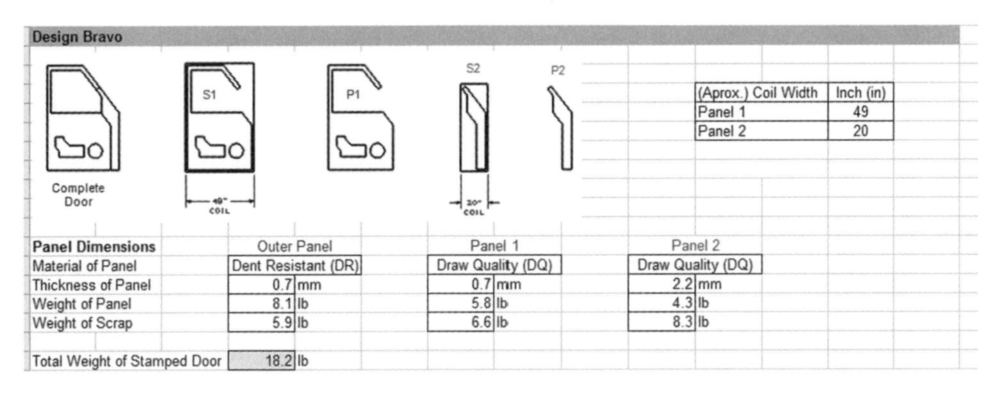

(Aprox.) Coil Width	Inch (in)
Panel 1	49
Panel 2	20

Panel Dimensions	Outer Panel		Panel 1		Panel 2	
Material of Panel	Dent Resistant (DR)		Draw Quality (DQ)		Draw Quality (DQ)	
Thickness of Panel	0.7	mm	0.7	mm	2.2	mm
Weight of Panel	8.1	lb	5.8	lb	4.3	lb
Weight of Scrap	5.9	lb	6.6	lb	8.3	lb
Total Weight of Stamped Door	18.2	lb				

Figure 2.67 The different weight calculations for the second scheme of TWB door inner

ferent weld configuration and length. The analyses for each of these designs are shown in Figures 2.66–2.68.

The analyses in these figures have also utilized the stamping process inputs from case I. The design that is selected is design Bravo, because it meets the functional requirement for the door inner panel in addition to +$/–kg. The panels used to form the hinge area require a nesting scheme so that the engineering scrap is reduced and the material utilization is improved.

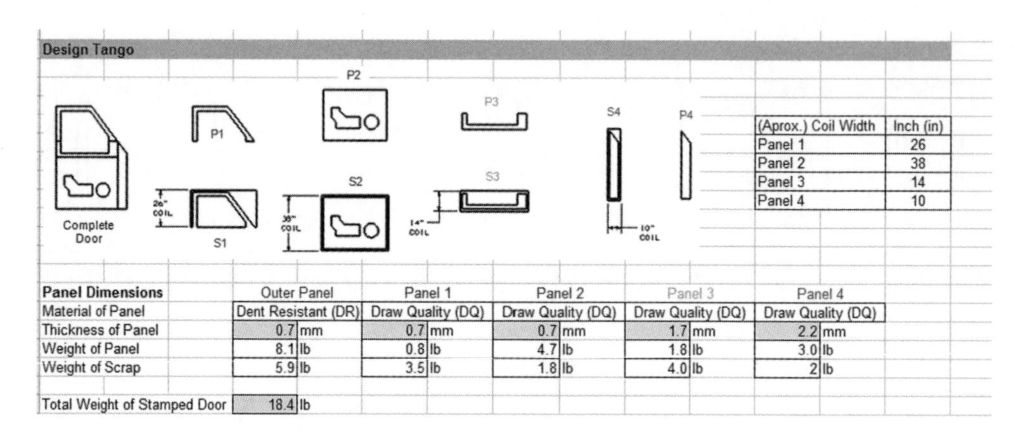

(Aprox.) Coil Width	Inch (in)
Panel 1	26
Panel 2	38
Panel 3	14
Panel 4	10

Panel Dimensions	Outer Panel	Panel 1	Panel 2	Panel 3	Panel 4
Material of Panel	Dent Resistant (DR)	Draw Quality (DQ)	Draw Quality (DQ)	Draw Quality (DQ)	Draw Quality (DQ)
Thickness of Panel	0.7 mm	0.7 mm	0.7 mm	1.7 mm	2.2 mm
Weight of Panel	8.1 lb	0.8 lb	4.7 lb	1.8 lb	3.0 lb
Weight of Scrap	5.9 lb	3.5 lb	1.8 lb	4.0 lb	2 lb
Total Weight of Stamped Door	18.4 lb				

Figure 2.68 The different weight calculations for the third scheme of TWB door inner

Exercises

Problem 1

In your own words, explain thefollowing statements about sheet metal forming:

- Most automotive stampings pass the C_p but fail the C_{pk}.
- HSLA is used for automotive stampings to reduce weight.
- Scrap utilization in automotive stampings does not have a major impact on overall cost.
- Progressive dies might have idle stations.
- High strength steel (DP, TRIP, etc.) requires special stamping consideration.
- The formability of TWBs is a function of the weld hardness H to the area of the Heat Affected Zone.
- Tailor-welded blanks made from HSLA requires higher welding speeds.
- Tailor-welding blanking requires a dimpling station.
- Sequential validation for stampings is a time-consuming process.
- Automotive stampings are evaluated based on six formability indices.

Problem 2

A circle grid analysis of a vehicle front fender panel resulted in the strain map (%) shown below. The fender is made of a HSLA 0.50 mm steel panel.

- Draw the forming limit diagram (FLD) and indicate the location of the five strain measurements shown in the fender strain map, *using the steel FLD sketch.*
- Based on the FLD, indicate if any of the strain-measurement locations might split.
- Describe the deformation mechanisms for each strain measurement location.
- Based on your FLD and the deformation map, comment on this formability practice, and suggest improvements (if any), to solve any possible problems.
- *Note: any improvements suggested should utilize the same steel grade.*
- On your FLD, draw or indicate the line that represents zero change in the steel panel thickness (i.e. zero strain in the panel thickness).

- Based on your FLD, indicate whether the metal has more deformation capacity in "Draw" deformation than the "Stretch" mode.

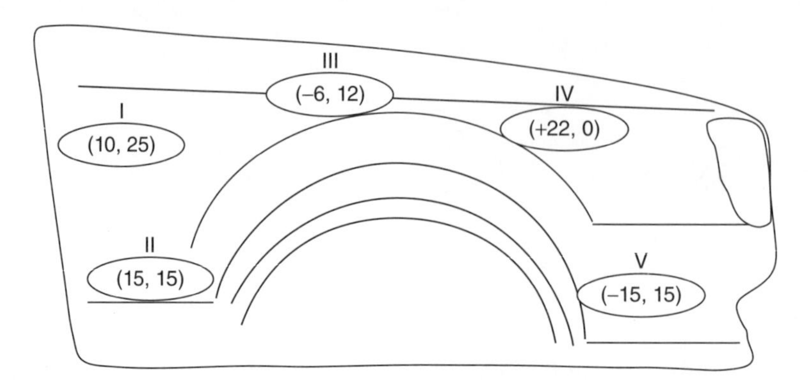

Problem 3

Using the FLD and LDH graphs shown below which correspond to Al and to an unknown steel grade ($Y_s = 25$ ksi), and the fender strain % map, answer the following:

1. Based on the strain map, is it possible to form the fender from Aluminum, if not, suggest improvements for *each specific* failure strain point.
2. What would be the *maximum stress* for steel in plane strain-mode, if you know that the thickness of Al panel is 1.4 mm, and steel strength coefficient is 900 MPa?
3. Calculate the remaining deformation capacity for location 1 and its final thickness.

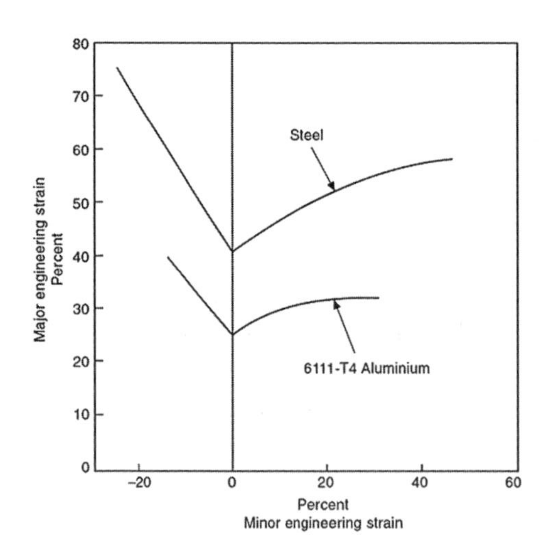

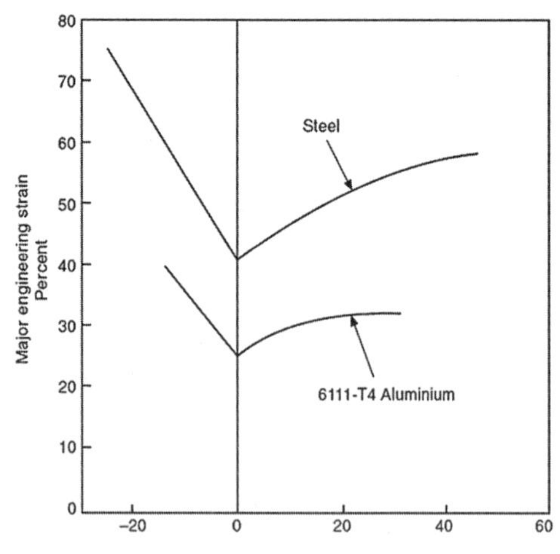

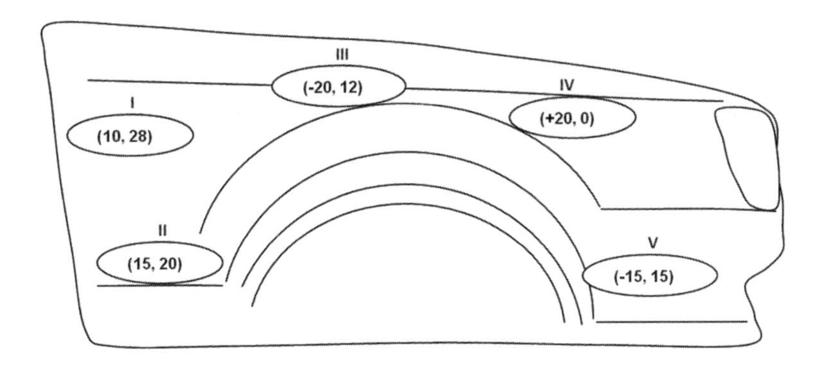

Problem 4

The stress-strain diagram shown below represents different metals' tensile test results after going through paint curing oven. After carefully inspecting this curve, fill in the blanks:

Note: A, B, C, D, might be any of the following: BH, EDDQ, DQSK, Al, HSLA, IF.

1. Grade D will have the highest ______________________ at same load as other grades.
2. Grade C might be ______________________.
3. Grade A might be ______________________ or ______________________.
4. Grade D has the same modulus of elasticity as ______________________.
5. The grade with the highest stiffness is ______________ while the one with the lowest stiffness is ______________________.
6. Grade ______________ is the strongest, while Grade ______________ has the highest formability.

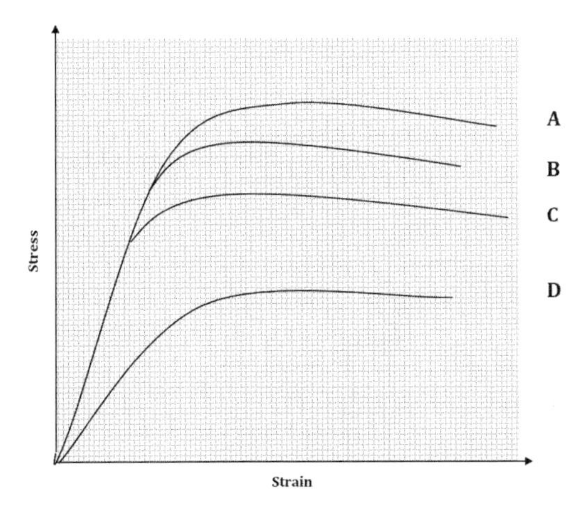

Problem 5

Compute the ratio of maximum bending radii for a 0.7 mm BH panel, before and after it goes through the paint oven.

Problem 6

Compute the ratio of expected spring-back between IF and Al panels, given that the modulus of elasticity for Al = 1/3 that of steel.

Problem 7

For a stamping conformance study, C_p is found to be = 2.0, the mean is 2.0 mm, and the UL = 2.3, LL = 1.8, what's the C_{pk} value?

Problem 8

Modify a current design, provide a sketch of new design, taking into consideration the number of operations needed to make the part.

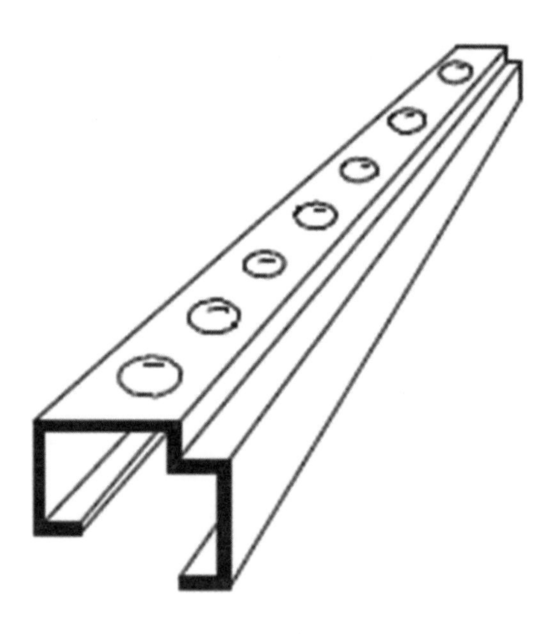

Problem 9

Modify the following designs from a formability point of view, and indicate the main motivation behind modifying the current design. (Hint: consider the nesting of current design.)

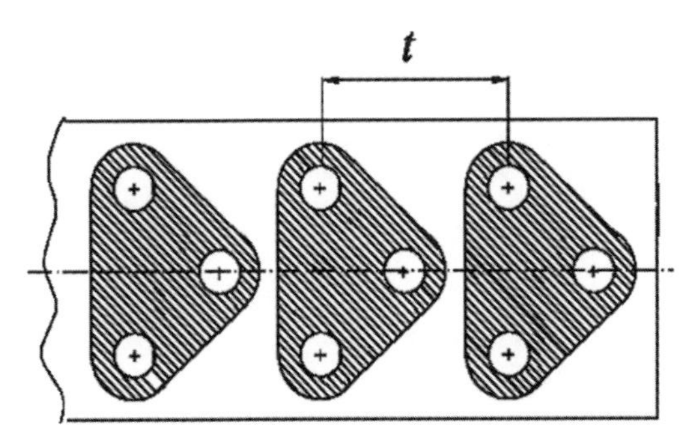

Problem 10

Based on the STS parameter table shown below: Part A: Contrast the formability of HSLA and DP steel grades from following attributes: Work-hardening rate at initial strains, Work-hardening rate at larger strains, Strength, Necking (early or delayed). Part B: Compute the stress for HSLA, for a true strain of 0.4 and 1, using the (a) STS equation and (b) the power law.

Parameter	HSLA	DP	TRIP
A (Shape)	66.3	117	98
B (Tilt)	223	54.8	666
C (Strength)	673	932	802

3

Automotive Joining

3.1 Introduction

This chapter focuses on the joining activities involved in the automotive body-weld or body-shop area. These activities are mainly: (a) the fusion-based joining steps achieved through the metal inert gas (MIG), tungsten inert gas (TIG) welding and the resistance seam, projection, and spot welding schemes. The fusion welding is done in automated, semi-automated, and non-automated fashions. However, this chapter will focus mainly on the arc and resistance welding practices. (b) Adhesive bonding is also applied to join the door inners and outers through the hemming process in addition to many other applications. Recent advances in the automotive joining technologies such as friction stir welding (FSW) will also be discussed later in this chapter.

Also, this chapter will discuss the robotic welding activities and some of the activities involved in regard to its automation. The text will also address the variety of fixtures and fixture transportation systems which are used to hold and fix the assembled BiW panels during the welding process.

3.2 Fusion Welding Operations

The fusion welding processes feature the joining of two mating substrates by creating a melting pool in their joint location, i.e. the localized coalescence location. So the main characteristic of a fusion-based welding process is that the welding is done at or above the melting temperature of the substrates. Thus, knowing the substrate thermal properties (thermal conductivity, specific heat, thermal expansion coefficient, and density) is important to evaluate the thermal effects on the weld quality and on the final structure geometry and dimensions. Typically the melting is achieved through different heating sources: oxy-fuel, electric arc, and high energy beam

---

The Automotive Body Manufacturing Systems and Processes, First Edition. Mohammed A. Omar.

Table 3.1 The different weld designations and types as classified by the AWS

Welding group	Welding process details	ASWE designation
Arc welding	Flux Cored Arc	FCAW
	Gas Metal Arc	GMAW
	Gas Tungsten Arc	GTAW
	Plasma Arc	PAW
	Submerged Arc	SAW
	Shielded Metal	SMAW
Resistance welding	Projection	RPW
	Resistance Seam	RSEW
	Resistance Spot	RSW
	Upset Welding	UW

(laser), shielded from the environment using an inert gas. The fusion-based welding activities provide great design flexibility to automotive designers because this enables them to join and place metal reinforcements and shapes into the specific locations where they are functionally needed. However, due to the high temperature involved in fusion welding, the area surrounding the weld pool will go through a metallurgical change in its grain size, leading to a heat affected zone (HAZ) with long grains that are susceptible to failure due to their low strength, and low ductility. The HAZ shape and characteristics depend on the weld seam width, and the base metal thermal properties, the rate of heat deposition, the rate of cooling after the weld is made, and the geometry of the weld bead and the presence of cracks, residual stresses, inclusions, and oxide films. Pre-heating the weld area might improve the weld metal properties, especially for metals with high thermal conductivity, such as aluminum and copper.

Table 3.1 shows the different types of fusion welding. The automotive welding practices focus on the utilization of resistance welding and arc welding techniques. Additionally, some brazing and soldering processes are used. The brazing joins the materials through their coalescence by raising the joint temperature to brazing level in the presence of a filler metal, which is distributed through the capillary action into the joint. The difference between soldering and brazing is in their filler metal melting temperature. However, both soldering and brazing find limited applications in automotive production.

3.2.1 Basics of Arc Fusion Welding and its Types

The arc welding application depends on the successful delivery of the following main functionalities from the welding machine, welding material, and the welding motion control (manual or automated):

- Starting the arc and maintaining its distance and configuration.
- The correct feed of the filler (sometimes the electrode) into the arc in terms of the electrode type and rate.
- Controlling the arc welding speed to achieve the designed penetration.
- Controlling the arc location, i.e. welding locations and bead placement.
- Controlling the arc for adjustments and corrections.

The arc welding components, responsible for delivering the above-mentioned functionalities are:

- The arc welding power sources which range from 5–40 kW, with voltages from 10–50 V, and amperage settings of 5–1000 A. These sources are of three types: Constant Power (CP): typically used for stick welding. Constant Voltage (CV): used for MIG welding; and Constant Current (CC): used for gas tungsten and plasma arc welding.
- The arc welding wire feeders, which are of two main types: the electrode wire feeder for consumable electrodes; and the cold wire feeder, for non-consumable electrodes.
- The welding torches, which can be air-cooled, or water-cooled, and can have a concentric or a side delivery of the wire through a straight or a bent configuration.

Additionally, the arc welding process relies on specific preparations of the substrates (to be joined) in terms of their surface cleanliness and their relative orientation. The substrate's weld location or joint location can be configured in one of five main different types: (a) the Butt joint where parts are aligned in the same plane; (b) the T-shaped joint, where the parts are aligned at right angles; (c) the Edge joint, where the weld is done for the edges of parts parallel or nearly parallel; (d) the Lap joint which is done for overlapping parts in parallel planes; and, finally, (e) the corner joint where the parts are at right angles to each other and at the parts' edges.

Also, within these five joint configurations, different weld shapes (described by their cross-section shape) exist, such as the fillet weld, the single V, the double V and the square shape, etc. A fillet weld cross-section in shown in Figure 3.1, and Figure 3.2 displays the shape of a groove weld. Figures 3.1 and 3.2 show the different weld shape parameters that are typically used to assess the weld quality through visual inspection. Furthermore, the American Welding Society further defines the welds according to their positions: flat, horizontal, overhead and vertical.

Within the different types of arc welding; the automotive industry focuses on the MIG or gas metal arc welding (GMAW) and TIG welding. These welding techniques join metals by heating the panels to their melting point with an electric arc. The arc is between a continuous, consumable electrode wire and the metal being welded. The arc is shielded from contaminants in the atmosphere by a shielding gas. The arc

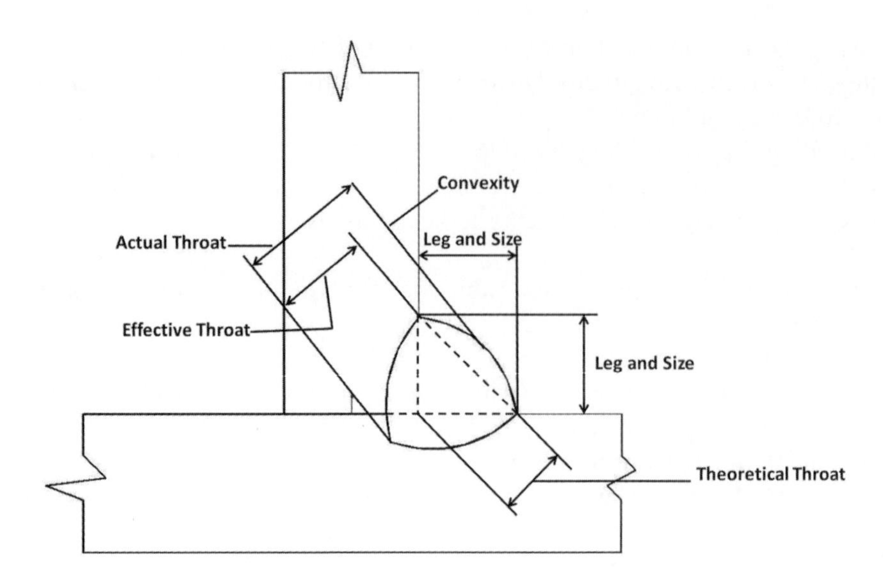

Figure 3.1 A fillet cross-section

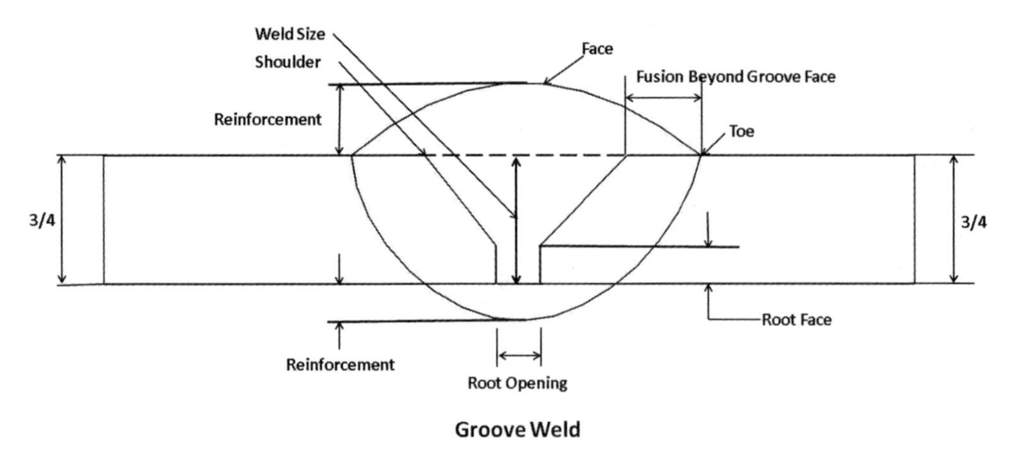

Figure 3.2 A groove weld cross-section

power in MIG and TIG welding is a product of the welding machine's current and voltage. The arc shape is usually classified into two concentric zones: an inner core which is the plasma that carries the current, and an outer flame region. The arc can be made from Direct Current (DC) and/or an Alternating Current (AC); recent applications of arc welding utilize sinusoidal pulsing direct current or square wave alternating current. Arc welding typically utilizes either a non-consumable electrode, through generating enough heat to melt the base material and the filler metal but without melting the electrode itself. This non-consumable electrode is used in TIG welding. Consumable electrodes, where the generated heat melts the base metal and the electrode, i.e. the molten electrode becomes the filler metal, as in the MIG welding scheme.

3.2.2 Metal Inert Gas MIG Welding Processes

MIG welding is typically applied in different locations within the body-weld and through different implementations such as: (a) the semi-automatic welding application, where the equipment controls only the electrode wire feeding. Movement of the welding gun is controlled manually. (b) The machine-based welding application, where a machine controls the electrode wire feeding and the movement of welding gun; constant adjustment is needed to set the manipulator path. (c) The automatic welding uses equipment which welds without the constant adjusting of controls by a welder or operator. On some equipment, automatic sensing devices control the correct gun alignment in a weld joint, and, finally (d) the robotic welding application which will be discussed in this chapter in detail.

MIG welding is used mainly to weld steels but can be extended to other materials. The MIG welding electrode should match the substrate material and the shielding gas should also be selected based on the base metal characteristics; mainly the substrate affinity to react with oxygen. MIG welding can be used to join panels with minimum thicknesses of 0.04 mm. The MIG welding process can be controlled based on three different levels of weld variables, as defined by [24]:

- Pre-defined variables, which cannot be changed during the welding process, such as the electrode size and type, the welding current type and the shielding gas composition. These variables are selected, based on the joint geometry and the joined metals.
- Primary adjustable variables, which are used to change or adjust the weld characteristics in terms of penetration depth, and bead/seam width. These variables include the variations in the weld-travel speed, the arc voltage, and the welding current amperage.
- Secondary variables, which are related to application process, i.e. the manipulation of the welding gun in terms of its tip distance to the welding-pool and its nozzle angle. Figure 3.3 shows the effect of tip distance on the weld bead width.

In MIG welding, the correct adjustment and setting of the above variables result in a controlled welding process, depositing the right amount of filler, applying the correct bead width and profile, and the right weld penetration. The interactions between these variables are displayed in Figures 3.4, 3.5, and 3.6, showing the impact in terms of weld penetration, bead width, and bead profile (height), respectively.

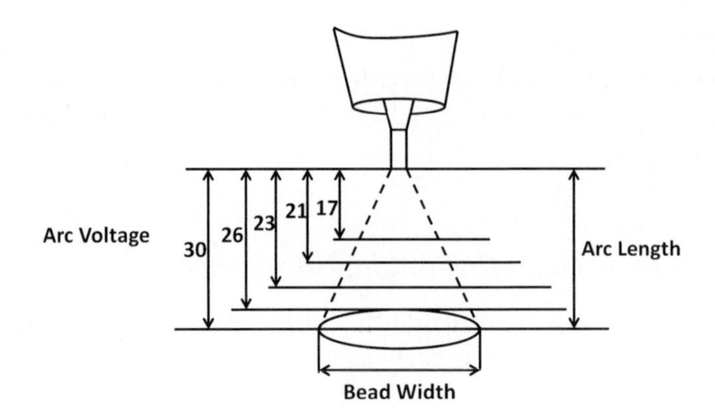

Figure 3.3 The effect of tip distance on the weld bead width

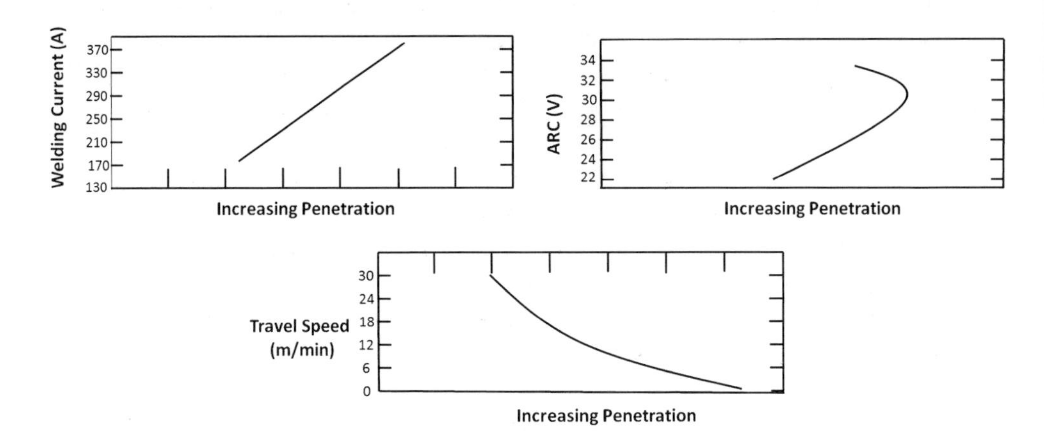

Figure 3.4 Weld control variables and their effect on the penetration. Reproduced by permission of Taylor Francis © 1995

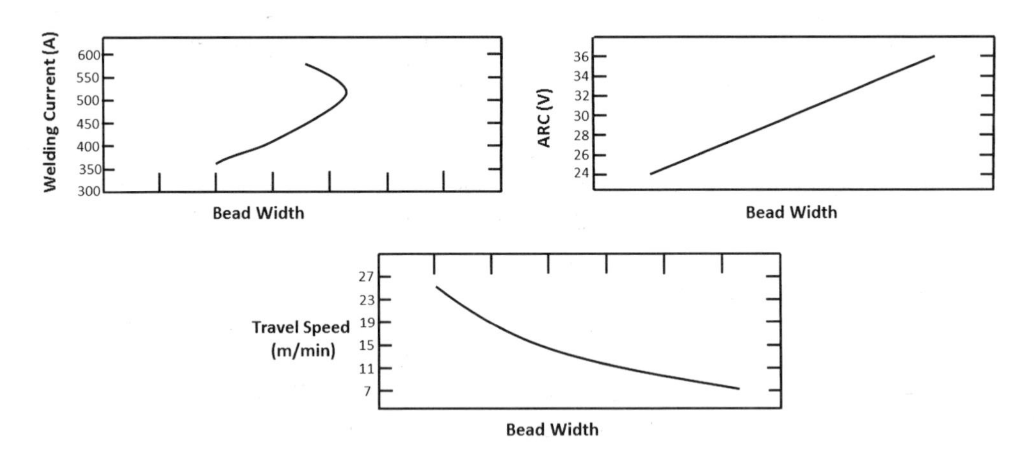

Figure 3.5 Weld control variables and their effect on the bead width. Reproduced by permission of Taylor Francis © 1995

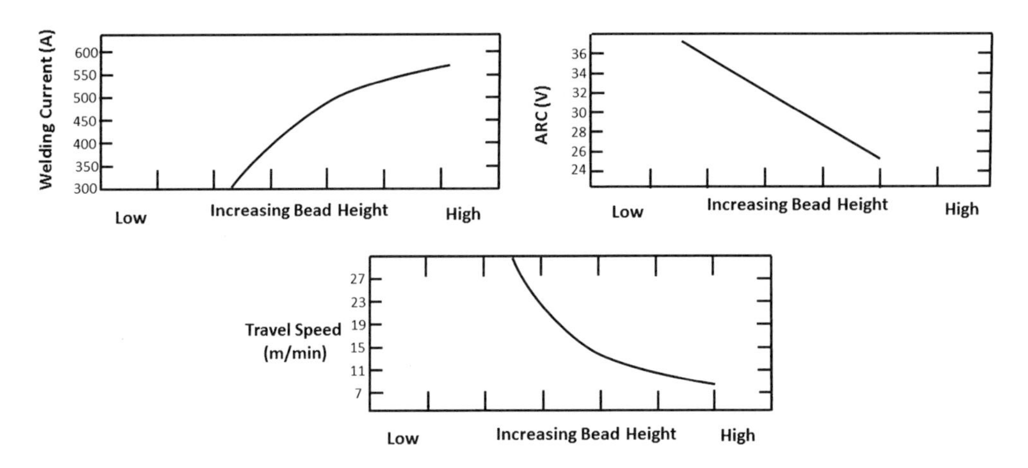

Figure 3.6 Weld control variables and their effect on the bead profile. Reproduced by permission of Taylor Francis © 1995

The transfer mechanism of the electrode material across the arc into the welding pool is very important to the MIG welding process. There are four transfer modes:

1. The *spray transfer mode* which is distinguished based on: the metal droplets' size, frequency and characteristics across the arc. Spray transfer takes place at high current density, using a small diameter electrode, resulting in droplet sizes smaller than the electrode diameter. However, this droplet size leads to low deposition efficiency, with little spatter, typically found in arc environments shielded with Argon as the shielding gas. Also, the spray transfer's high current density leads to deep penetration, hence this transfer mode is not recommended for thin applications or vertical weld positions.
2. *Globular transfer*, with droplets with similar size to the used electrode, hence it produces spatter while welding. This transfer mode is typically found in CO_2-dominated shielding environments.
3. *Short-circuiting*: the short circuit happens due to the high feed rate of welding wire which leads to it contacting the weld pool. Thus the weld pool surface tension draws the molten wire material. This transfer mechanism is limited to the welding of steel because of CO_2 gas shielding.
4. *Pulsed spray*: requires a special power supply to pulse the metal deposition into the weld pool. This transfer is similar to the spray transfer but with better control over the deposition rate. Additionally the discrete drops are transferred in a regular manner with little spatter. The droplets' size is similar to that of the globular transfer, thus the pulsed spray mechanism has both the advantages of the globular mode (in terms of deposition rate) and the spray mode (in terms of the spatter). These types are displayed in Figure 3.7.

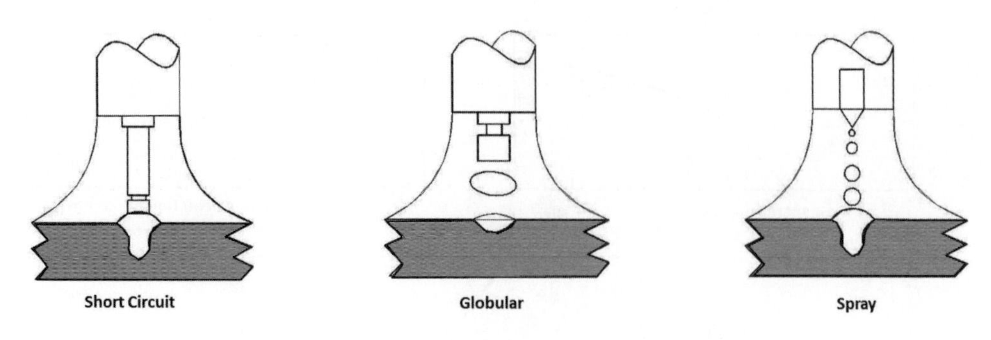

Figure 3.7 The electrode transfer mechanisms

Table 3.2 The different shielding gas types, composition, and applications

Shielding gas type	Composition	Used for
Argon + Oxygen mixture	96% Ar + 4% O_2	Mild Steel, Low alloy Steel
Argon + Carbon Dioxide	75% Ar + 25% CO_2	Mild Steel, Low alloy Steel, Stainless Steel
Argon + Helium	50% Ar + 50% Helium	Aluminum, Magnesium, Copper alloys
Argon	100% Argon	Non-ferrous metals
Carbon Dioxide	100% CO_2	Mild Steel, Low alloy Steel
Argon + Helium + Carbon Dioxide	7.5% Ar + 90% He + 2.5% CO_2	Stainless Steel

The design of the MIG welding process relies on the selection of the shielding gas environment (type and flow rate), the power source (CC, CV, or CP), the welding-wire (feed rate and material), the weld speed and configuration. This selection process should take into consideration the base metal characteristics such as its affinity to oxygen and thermal properties.

To select the suitable shielding environment for MIG welding, it is important to understand its functionality; the shielding gas prevents the contact between the atmospheric air and the weld-pool, hence protecting it from any reactions or contaminations. Such gases are typically classified as *Weld-Grade* or *Dry* with dew point—40 °F because any moisture will cause porosity pockets in final weld joint. Also, when welding aluminum, inert gases such as Argon should be used because of the aluminum reactivity or reaction rate with oxygen. However, when welding steel alloys, a CO_2 environment is typically selected due to the lower price of CO_2. Table 3.2 shows the different shielding gases and their typical recommended applications [24].

Material Thickness				Electrode Diameter		Welding Current	Arc. Voltage	Wire Feed	Travel Speed	CO2 Gas
Gauge	in	mm	Type of Weld	in	mm	(A, dc)	Elec. Pos.	(in/min)	(in/min)	Flow (cfh)
18	0.050	1.3	Fillet	0.045	1.1	280	26	350	190	20-25
			Square Groove	0.045	1.1	270	25	340	180	20-25
16	0.063	1.6	Fillet	0.045	1.1	325	26	360	150	30-35
			Square Groove	0.045	1.1	300	28	350	140	30-35
14	0.078	2	Fillet	0.045	1.1	325	27	360	110	30-35
			Square Groove	0.045	1.1	325	29	360	110	30-35
			Square Groove	0.045	1.1	330	29	350	105	30-35
11	0.125	3.2	Fillet	1/16	1.6	380	28	210	85	30-35
			Square Groove	0.045	1.1	350	29	380	100	30-35
3/16	0.188	4.8	Fillet	1/16	1.6	425	31	260	75	30-35
			Square Groove	1/16	1.6	425	30	320	75	30-35
			Square Groove	1/16	1.6	375	31	260	70	30-35
1/4	0.250	6.4	Fillet	5/64	2	500	32	185	40	30-35
			Square Groove	1/16	1.6	475	32	340	55	30-35
3/8	0.375	9.5	Fillet	3/32	2.4	550	34	200	25	30-35
			Square Groove	3/32	2.4	575	34	160	40	30-35
1/2	0.500	12.7	Fillet	3/32	2.4	625	36	160	23	30-35
			Square Groove	3/32	2.4	625	35	200	33	30-35

Figure 3.8 Metal inert gas welding schedule for globular transfer

Additionally, the overall MIG welding process can be configured using a *welding schedule*. The welding schedule describes the suitable welding conditions in terms of the base material type and thickness in addition to the weld configuration. The welding schedule describes the recommended electrode diameter, welding current, arc voltage, wire feed, the weld travel speed, and the shielding gas flow rate. The automotive OEMs rely on such schedules to control and configure their MIG welding operations. The MIG welding schedule for a mild carbon and low alloy steels is shown in Figure 3.8.

Also, different welding schedules exist for the different metal transfer modes across the arc, for example, the schedule in Figure 3.8 is for a globular transfer mode, while the schedule in Figure 3.9 is for a short circuit transfer mode.

The potential defects in MIG welded joints are typically related to the deviations in the welding variables such as the weld tip distance or nozzle orientation, or to surroundings effects such as inclusions or porosity. Other defects are caused by the joint design and shape.

One can define the main types of the MIG defects to include:

- *Porosity*, which is the formation of air/gas pockets within the weld; these can be found as a cluster of small pockets or as continuous features along the weld cross-section. The main causes of weld porosity are mainly related to: (a) the deviations in the shielding gas flow rate; (b) contaminated or wet shielding gas; or (c) exces-

Material Thickness			Diameter		Welding Current	Arc. Voltage	Wire Feed	Travel Speed	CO2 Gas
Gauge	in	mm	in	mm	(A, dc)	Elec. Pos.	(in/min)	(in/min)	Flow (cfh)
24	0.025	0.6	0.03	0.8	30-50	15-17	85-100	20-Dec	15-20
22	0.031	0.8	0.03	0.8	40-60	15-17	90-130	18-22	15-20
20	0.037	0.9	0.025	0.9	55-85	15-17	70-120	35-40	15-20
18	0.050	1.3	0.035	0.9	70-100	16-19	100-160	35-40	15-20

Figure 3.9 Metal inert gas welding schedule for spray transfer

sive welding voltage or current. Additionally, high weld travel speed, leading to quick solidification of the pool, before any gases can escape from the joint, other causes include contaminants on the base metal surface and/or the welding wire.

- *Inclusions*, which are typically classified as slag and oxide inclusions. Both types weaken the weld because it serves as a crack initiator in the weld structure. The oxide inclusions are more severe when welding aluminum or stainless steel due to the thick oxide layers on their surfaces. Also, inclusions are related to high welding speeds.
- *Incomplete fusion of the weld*: this defect is more apparent when using the short-circuit metal transfer mode due to its low penetration depth. Also, the high welding speed, and low welding current can also lead to shallow penetration, hence causing the incomplete fusion.
- *Excessive weld spatter*: the amount of spatter from a MIG welding process depends on the metal transfer mode, however, excessive spatter is a spatter that might hinder the shielding gas flow to the arc; in addition, it affects the visual appearance of the welded joints. Spatter is typically controlled through managing the welding current voltage. Other issues with welding spatter are related to welding fires. Each welding station should be provided with an exhaust system to deal with the welding fumes; such exhaust system features a duct system connected to a filter bank. If the spatter possesses a high enough temperature while its travel through the duct system, it will ignite the cellulose-based filters, causing a lot of smoke in the weld area. Spatter and sparks arrestors are typically installed to capture or cool such sparks. Other scenarios of welding spatter are caused by oil contaminants on the substrate or the melting of the zinc coating on steel panels. The zinc melting temperature is around 415 °C, which is well below that of the joined substrates (steel panels). A complete analysis of the welding sparks and spatter in welding duct systems can be found in [25].

Other MIG welding defects include: melt-through or burn-through, which describes the case when the arc melts or burns through the bottom of the welded joint.

Table 3.3 TIG welding schedule for steel

Thickness (inch)	Electrode diameter	Shielding gas flow rate c. ft/hr	Welding current A	Filler rod diameter
1/16	1/16	10–12	80–120	1/16
3/32	3/32	12–14	100–140	3/32
3/16	3/32	12–18	100–140	1/8
1//2	1/8	18–22	200–250	1/4

Additionally, the cracking of the MIG weld can be caused by excessive constraining of a joint or a high cooling rate. The MIG weld cracks are typically classified into hot and cold cracks. The hot cracks take place while welding at elevated temperatures and can also be related to the sulfur and the phosphorus contents in the steel base metal, while the cold crack happen when the weld joint is completely solid and is mainly related to hydrogen embrittlement.

3.2.3 Automotive TIG Welding Processes

The TIG welding process is similar to the MIG process; however, the TIG welding utilizes a non-consumable tungsten electrode to conduct the current to the arc. The TIG welding arc is shielded by the use of shielding gas made of Helium or Argon, also the TIG welding features the use of a filler metal that is introduced from the outside of the arc manually or through an automatic filler feeder mechanism. TIG creates very precise, localized heating patterns, making it suitable for applications that require localized, not global, heating of the base metal with precise control over the welding pool size.

The TIG welding equipment features the power supply equipment, which typically applies an AC/DC current source, with high frequency that might be generated through a high frequency generator. The high frequency feature is required to sustain a stable arc during the “zero” voltage conditions in the alternating current cycle.

Also the TIG torch, which is designed to deliver both the electric current and the shielding gas to the welding pool. Some torches are cooled through air or water cooling circuits. TIG torches are typically classified based on their current amperage.

Typical TIG welding schedules are shown in Table 3.3 for the case of steel and Table 3.4 for TIG welding aluminum.

3.2.4 Automotive Resistance Welding Processes

Resistance welding is accomplished when a current is caused to flow through the electrode tips and the separate metal panels (to be joined); the resistance of the base

Table 3.4 TIG welding schedule for Al

Thickness (inch)	Electrode diameter	Shielding gas flow rate c. ft/hr	Welding current A	Filler rod diameter
1/16	1/16	15	60–90	1/16
1/8	3/32	17	115–160	3/32
3/16	1/8	21	190–240	1/8
1/2	1/4	31	250–450	1/4

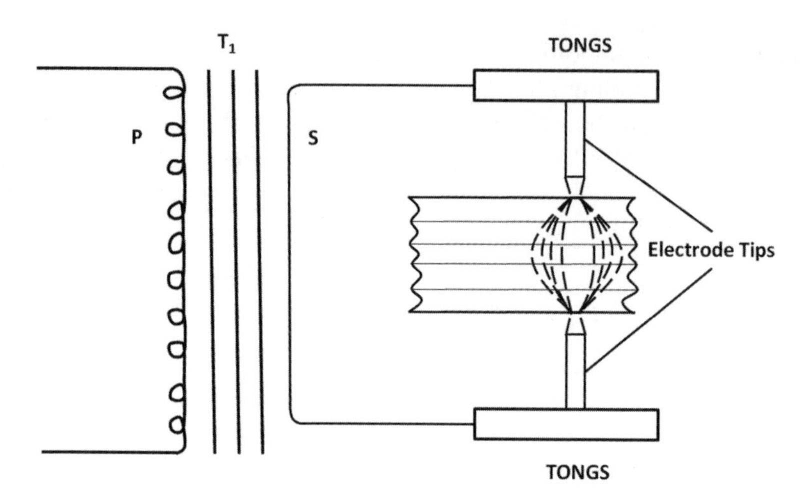

Figure 3.10 Spot welding circuit schematic

metal to the current flow causes localized heating in the joint, leading to the formation of the weld nugget at the interface between the two panels. The heat generated is simply described by Ohm's law which means that the heat is the product of the current value squared and the interfacial resistance (caused by the air gap between the two panels). The resistance welding process is achieved through two welding electrodes (upper and lower), while the joined panels are inserted between them, as displayed in Figure 3.10. Another important feature of the resistance welding process is that it forms internal joints (nuggets) within the joined panels; in MIG and TIG welding, the joint is external to the joined metals. However, the spot welding electrodes leave an impression on the panel surface; this impression is created due to the heating at the electrode tip caused by the small air gap between the tip and the panel surface. This impression limits the application of spot welding to interior panels only. At the same time spot welding is the most used welding practice in automotive manufacturing, a completed body shell features at least 4000–5000 spot welds.

The resistance welding passes through different phases or times; the first phase is the squeezing period, which is the time when mechanical pressure is applied to

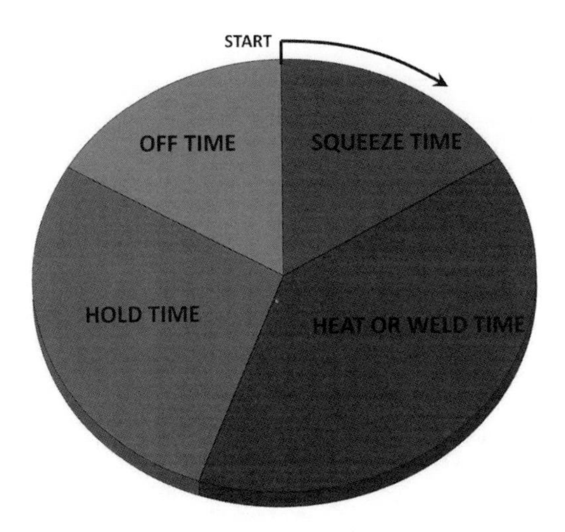

Figure 3.11 The spot weld cycle

restrain the joined panels. In this period no welding current is applied, however, this phase leads to good joint fit. The second phase or period is the actual welding time, which is the period in weld cycles (1 weld cycle = 1/60 second) when the current is flowing from the upper electrode through the joined panels to the lower electrode. The third phase is called the hold time, which is the time needed to keep the mechanical pressure on after the weld is made and the weld current is turned off. The final phase is the off-time period where the welding electrodes are separated, at this phase the weld join is permanent and the weld process is complete. These weld phases are shown in Figure 3.11. Typical values for these different phases are; the weld time ~3 cycles, with cooling time of 2 cycles.

The resistance welding process includes different variations depending on the shape of the electrode; the main types are: spot welding, projection welding, and seam welding. In seam welding, the electrodes are two copper wheels used to generate a continuous line of weld utilizing several discrete weld nuggets with a welding speed of around 2.6 meters/minute. Seam welding is typically used in automotive manufacturing to weld the upper and the lower portions of the automotive steel fuel tanks. Seam welding is done through static machinery, while spot welding can be applied in a robotic fashion (most common), portable equipment, and static online and offline welding machines. These three implementations are shown in Figure 3.12. Also, Figure 3.13 illustrates the basic welding circuit for a spot-welding machine. Understanding this schematic is important because the high frequency of using spot welding requires continuous monitoring and adjustment of the machinery involved.

The quality of the resistance welding process is described by the quality of the formed weld nugget, which can be quantified through the nugget size (diameter), the nugget location (depth), the nugget shape, and the spacing between the nuggets which

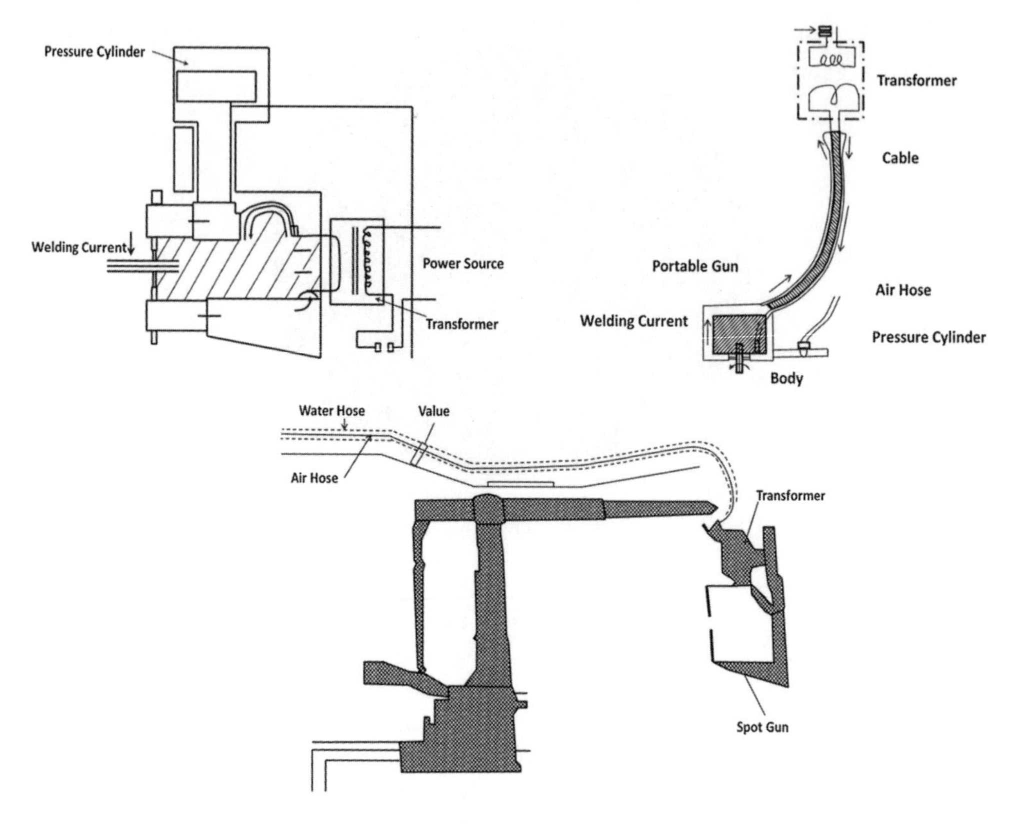

Figure 3.12 The different spot welding equipment

is also called the weld pitch. Figure 3.14 describes the different resistance weld characteristics. The main spot welding characteristics are the nugget and the spacing between these nuggets. The welding pitch is designed to prevent the welding current from shunting through the closest weld, because it represents the path of least resistance, thus reducing the current flow to the next nugget leading to an inadequate fusion. When welding aluminum and magnesium sheets, the minimum spot-welding pitch is higher than that of the mild steel and typically evaluated at 8 to 16 times that of the sheets' thickness. Management of the different resistances within the resistance welding process is vital to ensuring good nugget formation. These different resistances are:

1. the contact point between the electrode and top panel surface;
2. the resistance within the top panel;
3. the interface resistance of the top and bottom panels;
4. the resistance of the bottom panel;
5. the contact point between the bottom panel and the lower electrode;
6. resistance of the upper and lower electrode tips.

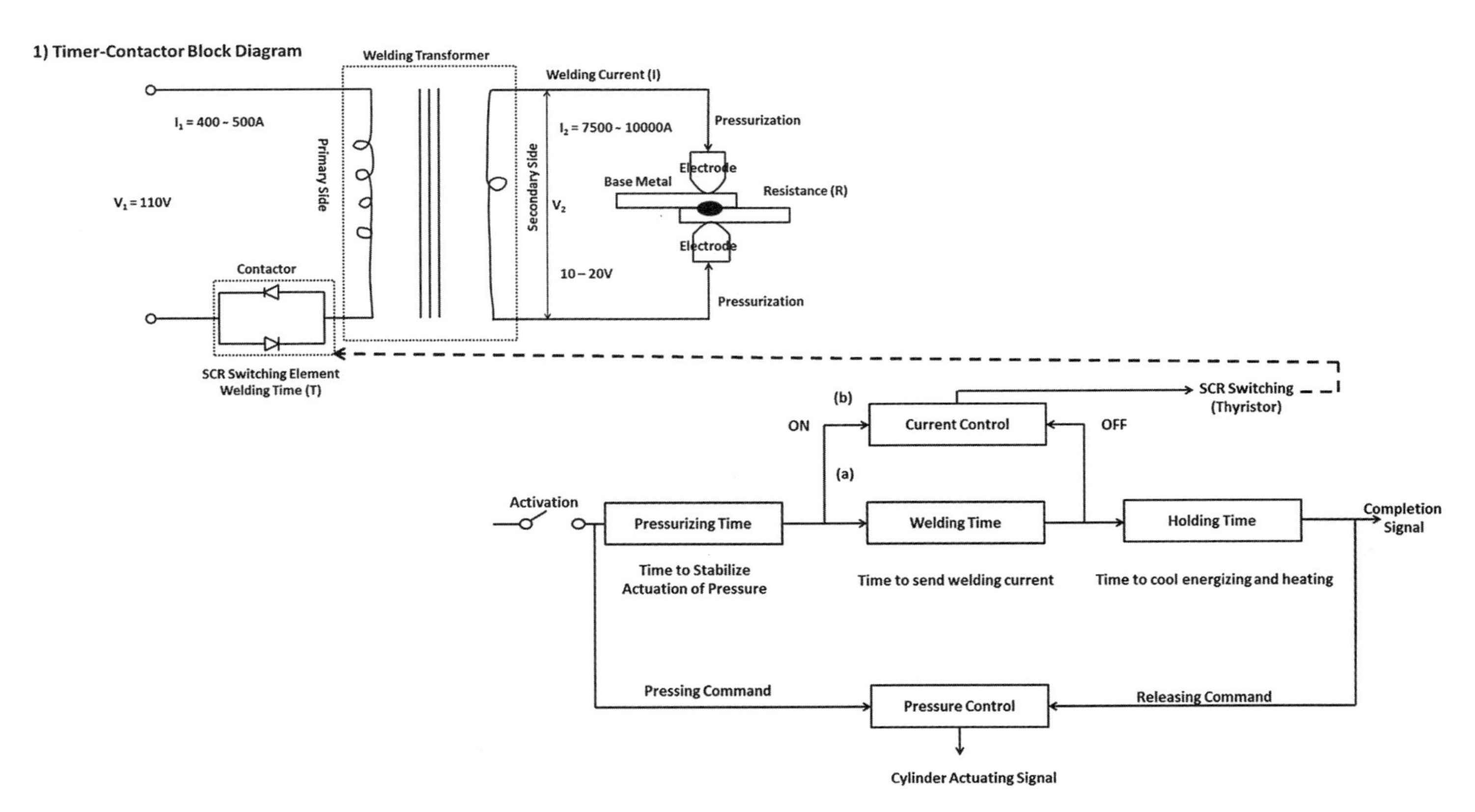

Figure 3.13 The details of the spot welding control

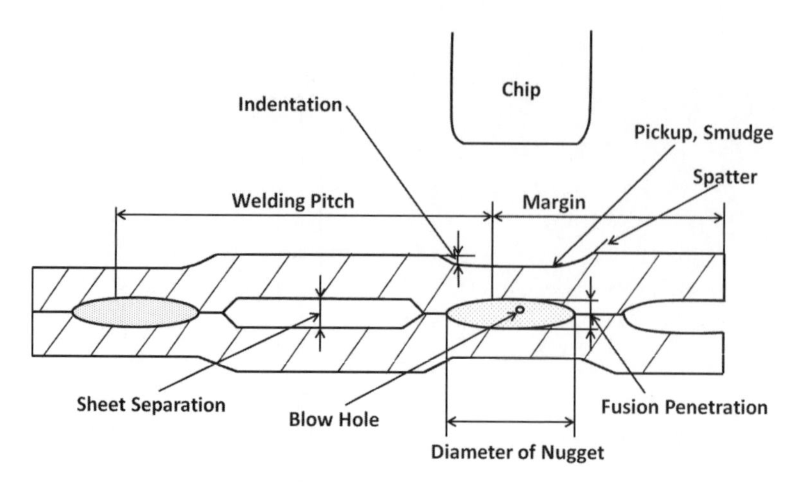

Figure 3.14 The spot welding joint's main characteristics

These resistances are affected by the surface roughness profile of the panels, any contaminants on the panel surfaces, the electrode tip conditions (age and material), and finally the amount of mechanical pressure applied to hold the two panels together. Additional factors are the panel material's internal resistance which is affected by the material type (aluminum versus steel) and the different alloying elements within the same type of material. For example, changing the steel grade for new car models leads to extensive trial and error efforts by production engineers to find the optimized welding conditions in terms of mechanical pressure and welding current and time. Typical welding currents for spot welding range from 9–16 kA, with mechanical pressure rated at 4–5 kN. Additionally, the selection of the electrode tip is vital in ensuring the right level of current flow through the electrode into the material and also to ensure the expected service life of the electrode. The electrode change and dressing frequency has a direct effect on the welding line up-time and down-time which affects the line productivity.

The spot welding tips are made of copper with alloying elements of chromium, cobalt and beryllium. The welding tongs and tips deliver the following functionalities:

- Conduct the welding current to the panels; hence the tip should have high thermal and electrical conductivities.
- Concentrate the pressure applied to the weld joint, thus the tips should have high strength values to avoid deformation.
- Conduct the heat from the work surface laterally,
- Maintain its shape and characteristics under working conditions; mainly its thermal and electrical conductivities.

These electrode tips are normally classified into two groups; Group A—Copper-based alloys, Group B—Refractory metal tips, the groups are further classified by number: Group A, Class I, II, III, IV, and V are made of copper alloys; while Group B, Class 10, 11, 12, 13, and 14 are the refractory alloys. Group A, Class I electrode tips are the closest in composition to pure copper.

As the class number increases, the hardness and annealing temperature values increase, while the thermal and electrical conductivity decreases.

Additionally, selecting the spot welding tip size is important, because it affects the current density (amperage per unit area) of the welding process, which has a direct impact on the weld nugget size. In other words, the size of the electrode tip point controls the size of the resistance spot weld. Typically the weld nugget diameter is slightly less than the diameter of the electrode tip point. If the electrode tip diameter is too small for the application, then the weld nugget will be smaller and weaker than its designed value. On the other hand, if the electrode tip diameter is too large, there is the risk of overheating the base metal and developing voids and gas pockets. In either case, the quality of the produced weld nugget would not be acceptable.

The tip diameter can be estimated through Equation 3.1, which is based on the panel thickness to be joined; Equation 3.1 estimates the tip diameter in inches. Other correlations suggest that the electrode tip diameter is equal to the square root of the panel thickness (for thin panels) and five times that for thick panels (>1 mm).

$$d_{tip} = 0.1 + 2 \times t_{panel} \tag{3.1}$$

If the weld is made between panels of different thermal characteristics, as in the case of spot welding copper and steel, an unacceptable weld will result for several reasons. The metals may not alloy properly at the interface of the joint. There may be a greater amount of localized heating in the steel than in the copper. To solve these issues, a heat equalization scheme is typically followed to ensure the delivery of the suitable amount of electric current to both of the panels. Some of the heat equalization techniques are listed below:

- the use of a smaller electrode tip area for the copper side of the joint to equalize the fusion characteristics by varying the current density in the dissimilar materials;
- the use of an electrode tip with high electrical resistance material, such as tungsten or molybdenum, at the contact point. The result is to create approximately the same fusion zone in the copper as in the steel.
- a combination of the two methods.

Figure 3.15 shows some of these heat equalization techniques.

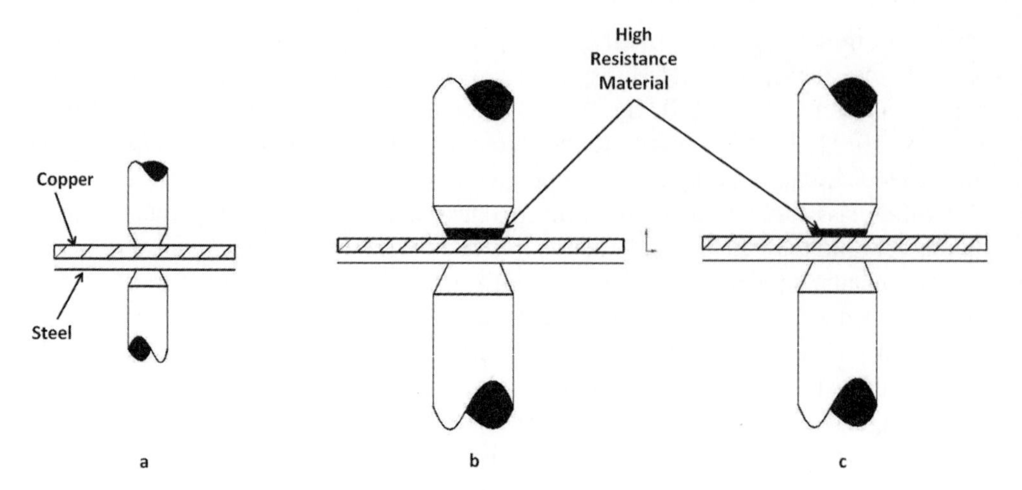

Figure 3.15 The different heat equalization schemes

Additional possible spot welding scenarios for multiple sheets can be achieved with multiple thicknesses up to eight layers and a 25 mm (1 inch) stack height. However, when spot welding aluminum panels, the maximum recommended stack height is around 8 mm only, and in the case of spot welding magnesium sheets, a 6.3 mm (1/4 in) height is possible in one spot-welding operation. For Al and Mg, a maximum of three layers is recommended. Also for steel when spot welding dissimilar sheet thicknesses, this reduces the number of panels that can be welded.

The typical daily maintenance and quality issues related to the electrode tip conditions, are mainly:

- abrasion of the electrode tip (decides on downtime);
- drop in electric current;
- non-effective shunt current.

Abrasion of the tip is the most common; as the tip abrades, the pressure area enlarges, thus the current density drops. The current density is the current flowing per unit area of welding tip; Figure 3.16 shows such effects with example numerical values. Additional variations in the weld appearance are related to what is called spot burrs, which are mainly caused by partially converged electric current, because the electrodes are applied in unbalanced planes; additionally, misaligned tips or deformed welding tips might cause spot burrs also. The spot burrs conditions are explained in Figure 3.17.

Sticking of the electrode tip can also occur when the welding spot gets overheated, then the tip and part of the panel melt and fuse, causing the tip to stick to the panel surface.

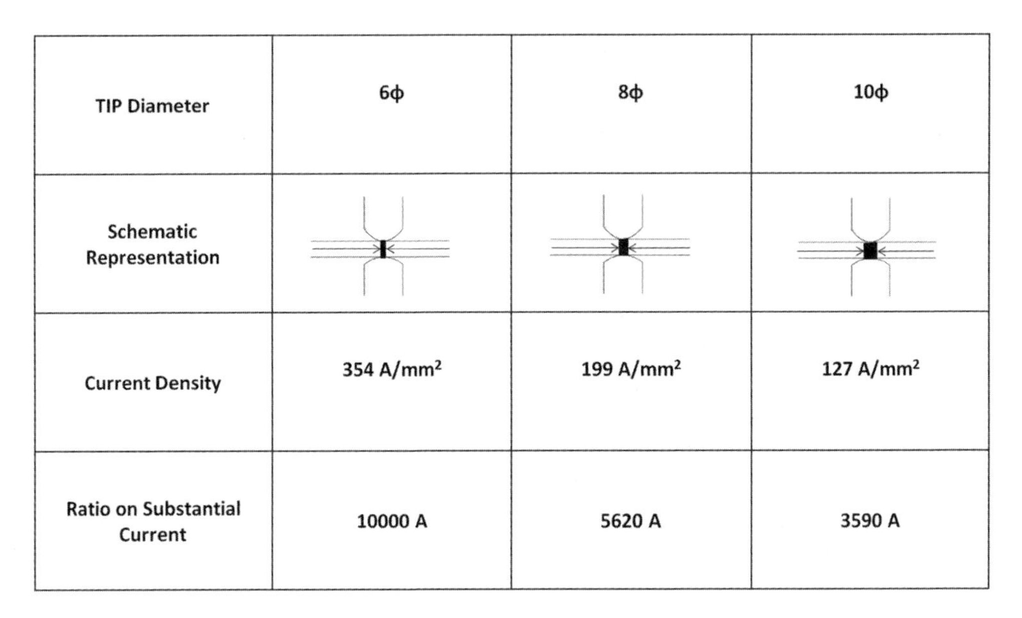

TIP Diameter	6ϕ	8ϕ	10ϕ
Schematic Representation			
Current Density	354 A/mm^2	199 A/mm^2	127 A/mm^2
Ratio on Substantial Current	10000 A	5620 A	3590 A

Figure 3.16 The current density effect

Factor	Schematic Drawing	Cause
A. Off-centered electrode tips. Improper electrode application	Off Centered	• Improper alignment of Tips. • Shank Deformation. • Point Holder misalignment.
B. Defective forming of electrode tips. Improper electrode application	Defective Forming	• Defective formation of Tips. • Defective Reforming of Tips.
C. Welding conditions improperly set. Proper settings are required for welding separation.	Excessive Calorific Value	• Current value Excessively set. • Improperly set welding pressure.

Figure 3.17 Main spot weld defects

3.2.4.1 Surface Conditions and Their Effect on Resistance Welding

All metals develop oxides on their surfaces that can be detrimental to resistance spot welding because it affects the current flow into the joined panels. In addition, the mill scale found on the hot-rolled steel surfaces tends to act as an insulator, thus reducing the amount of current and consequently weakening the weld nugget.

Mild or low carbon steel grades tend to develop hard, brittle welds as the carbon content increases, which typically require a post-heating treatment. Additionally, any quick cooling, i.e. quick quenching of the weld, leads to an increase in the probability of a hard nugget with a brittle micro-structure. Hot-rolled steels require proper cleaning of the surface to remove any remaining mill scale. On the other hand, the cold-rolled and hot-rolled steels with pickled and oiled surfaces are predictable in terms of their weld parameters and the final quality of the weld. However, if the oil concentration is excessive, then it might lead to the generation of carbon at the electrode tips, thus decreasing the weld tip service life. Degreasing or wiping is recommended for heavily oiled steel sheets.

When welding stainless steels, with their chrome-nickel rich alloying elements (austenitic), they tend to have high electrical resistance which makes them readily weldable with the resistance spot welding. However, the cooling rates for such welds should be quick to reduce the possibility of generating chromium carbide precipitations at the grain boundaries. In other words, the more time the weld is held at the critical temperatures (>800 °F), then the more the carbide precipitation.

For low alloyed and medium alloyed carbon steel grades, their electric resistance is typically higher than the case of low carbon or mild steels; therefore the spot welding current settings for such materials is lower. Additionally the weld nugget is more sensitive to the weld time, because it leads to metallurgical changes. There is certainly more possibility of weld embrittlement than there is with mild steel. Resistance spot welding pressures are normally higher with the low to medium alloyed steels due to the additional compressive strength inherent in their structure. So when welding these steel grades, longer weld time is needed to allow for ductile welds through slow cooling rates of the weld nugget.

For the case of spot welding aluminum and its alloys, the resistance spot welding machines with high KVA rating, >20 KVA are needed to produce good quality welds; this is mainly related to the aluminum's high thermal conductivity. On the other hand, the electrical conductivity of aluminum is also high, thus the welding transformers should be able to generate high currents to provide the amount of heat that is necessary to melt the aluminum and form the nugget.

Table 3.5 displays a resistance spot welding schedule [24], which describes the suitable settings in terms of the panels' characteristics.

The spot-welding process settings can also be computed through simple calculations based on the amount of heat (per material) needed to melt a volume equal to that of the nugget or to that of the product of weld length by its cross section (for

Table 3.5 A resistance spot welding schedule.

Thinnest Material		Recommended Electrode		Fused Zone		Recom.		Recom. Single		Expected shear strength in lbs/spot (kg/spot)					
										Aluminum				Stainless	
						Flange Overlap		Spot-Weld		Alloy		Carbon Steel		Steel	
Thickness		Contact Diameter		Diameter Expected		(Min)		Spacing		5052-H34		70 ksi		90 ksi	
in	mm	in	mm	in	mm	in	mm	in	mm	lbs	kg	lbs	kg	lbs	kg
0.010	0.25	0.13	3.0	0.10	2.5	0.38	10.0	0.75	19	154	70	286	130	374	170
0.021	0.53	0.19	5.0	0.13	3.0	0.44	11.0	0.75	19	264	120	704	320	1034	470
0.031	0.79	0.19	5.0	0.16	4.0	0.44	11.0	1.00	25	528	240	1254	570	1760	800
0.040	1.02	0.25	6.5	0.19	5.0	0.50	13.0	1.00	25	748	340	2024	920	2794	1270
0.050	1.27	0.25	6.5	0.22	5.5	0.56	14.0	1.00	25	1122	510	2970	1350	3740	1700
0.062	1.57	0.25	6.5	0.25	6.5	0.63	16.0	1.20	31	1584	720	4070	1850	5280	2400
0.078	1.98	0.31	8.0	0.29	7.5	0.69	17.5	1.50	38	2068	940	5940	2700	7480	3400
0.094	2.39	0.30	8.0	0.31	8.0	0.75	19.0	1.80	46	2596	1180	7590	3450	9240	4200
0.109	2.77	0.31	8.0	0.32	8.0	0.81	20.5	2.20	56	3036	1380	9130	4150	11000	5000
0.125	3.18	0.38	10.0	0.33	8.5	0.88	22.5	2.50	64	3564	1620	11000	5000	13200	6000
0.180	4.57	0.38	10.0	0.35	9.0	1.00	25.0	2.50	64	3784	1720	11660	5300	14520	6600

Table 3.6 The required heat to melt different materials per unit volume

Material type	J/mm^2	Btu/inch^3
Copper	6	85
Steel	10	135
Stainless Steel	9.5	135
Al	3	40
Magnesium	3	40

the case of arc welding). Such calculation is described in the sequence in Equations 3.2, 3.3 and 3.4.

$$\frac{H}{l} = e\frac{VI}{v} \tag{3.2}$$

$$H = u \times Volume = u \times A \times l \tag{3.3}$$

$$v = e\frac{VI}{uA} \tag{3.4}$$

where H is the heat input in J or BTU, l is the weld length, V is the voltage applied, and I is the current, and v is the welding speed, e is the welding efficiency.

The required heat to melt is summarized in Table 3.6 for some typically used materials.

Example (from [26]):

Two 1 mm thick steel sheets are being spot welded at a current of 5000 A, and current flow time = 0.1 sec. Using electrodes 5 mm in diameter, estimate the amount of heat generated and its distribution in the weld zone, Use an effective resistance of 200 micro-Ohm.

Heat generated due to a spot weld:

$$Vol = \frac{\pi d^2}{4} \times thickness = \frac{\pi (5)^2}{4} \times 2 = 39.3\ mm^3$$

The weld nugget volume:

$$H = I^2 R \times t$$
$$H = (500)^2 (0.0002)(0.1) = 500J$$

Specific heat required for diffusion: from tables 9.7 J/mm^2, and so the heat required to melt the nugget (volume):

$$H = u \times volume = 9.7 \times 39.3 = 381J$$

$$H_{dissipated} = 500 - 381 = 119$$

However, one should consider that such computations are only applicable if the substrates' alloying elements and surface conditions are constant and the only main resistance is the interfacial one (between the two panels), which is not the case for most of the welding operations due to the variations from the steel mill, contaminations, substrate roughness profile, and the electrode tip conditions. The following paragraph describes the resistance profile (with time) during the different spot-welding phases. Additionally, the concept of spot-welding lobes will be introduced, to illustrate the sheet metal conditions effect on the welding lobe.

3.2.4.2 Basics of Spot Welding, Lobes and Resistance Curves

A welding lobe is a graphical representation of the welding variables range over which acceptable spot welds are formed on a specific material. The welding lobe describes the welding window for the joined sheets under specific surface and mechanical pressure conditions. The lobe curve is determined by establishing spot welds using different weld-time, weld current combinations.

The welding lobes have a lower and upper bound; welds made with weld-current or weld-time exceeding the upper curve result in weld expulsion, on the other hand, welds made with weld-current or weld-time below the lower curve result in insufficient nugget size or yield a brittle nugget.

The welding lobes are established to translate the spot welding (electric) resistances behavior as the weld develops through the different phases; the squeeze-time, weld-time, etc. The electric resistances during a spot welding process go through five stages [27] of increase and decrease as the weld progresses; these stages are explained below.

At the first stage (squeeze-time), the mechanical forces applied on the sheet panels lead to an electrical contact at their surfaces in addition to the break-down of some of the surface features (roughness) due to the applied pressure. The drop in electrical resistance in this stage is drastic because of the quick reduction of the separating air gap.

In the welding phase (weld-time), the electric current is applied across the joint leading to heat generation at the panels' interface, which further leads to softening of the surface features (asperities). This softening leads to an increased area of contact between the two panels, hence the electric resistance decreases; however, the slope of this decrease is less than that of the first stage. Also as the heat increases, it flows laterally into the panels which leads to an increase in the panels' electric resistance. The continuous increase in heat leads to a further increase in the electric resistance, thus by the end of second stage, the electric resistance starts to increase. So, at the beginning of the second stage, the electric resistance drops but at the end the resistance increases.

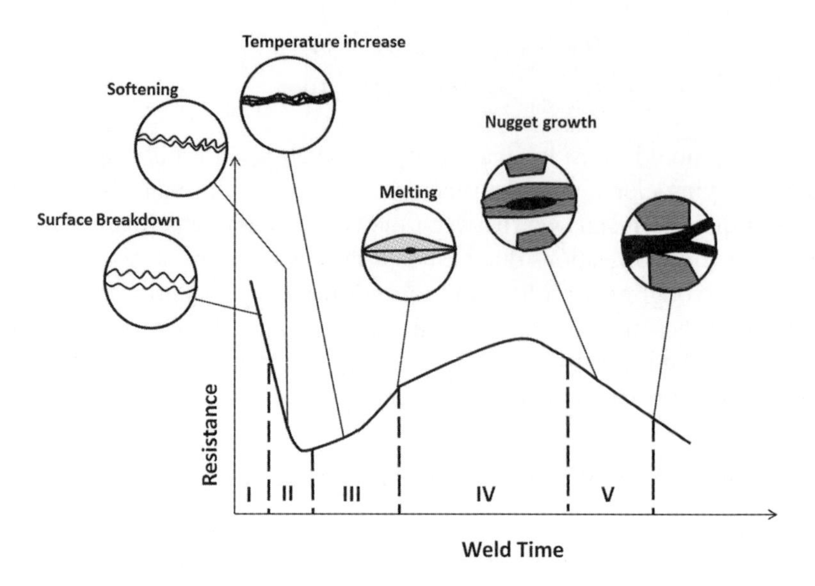

Figure 3.18 The interfacial resistance behavior with time

The electric resistance keeps increasing in the third stage due to the increase heating effect from the electric current which continues till the fourth stage, where the temperature increase reaches the material melting point. At melting, a phase change takes place, thus the added heat does not lead to temperature increase, additionally, the electric resistance reaches a peak and then drops as the material melting increases.

In the fifth and final stage, as the melting continues, the solid part is not able to support the mechanical holding pressure which leads to its collapse; additionally, the nugget continues to grow, leading to a drop in the electric resistance of the interface. The end of stage five is the upper bound of the spot welding process, exceeding this limit leads to weld expulsion. This resistance behavior with time is shown in Figure 3.18.

The spot welding lobe relies on the resistance behavior to set the welding parameters and establish the lower and upper bounds. In other words, the weld current and weld-time are set according to the electric resistance value to result in Ohm's heating effect needed to melt the material and form the nugget.

An example of the spot welding lobes is shown in Figure 3.19, which shows the weld time in weld cycles (1 cycle = 1/60 second) in the y-axis and the welding current in kilo amperage in the x-axis. The left-hand bound of the lobe is the minimum welding current and time combination that result in good welds, anything to the left of this bound will not result in a nugget. On the other hand, the right side represents the upper bound, anything above that limit will cause a weld expulsion.

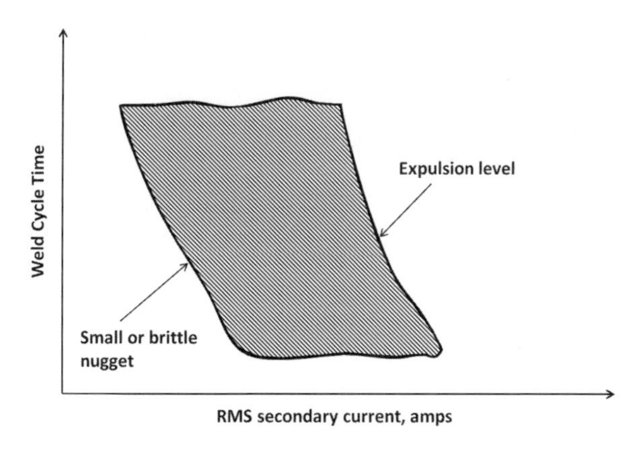

Figure 3.19 The spot welding lobe

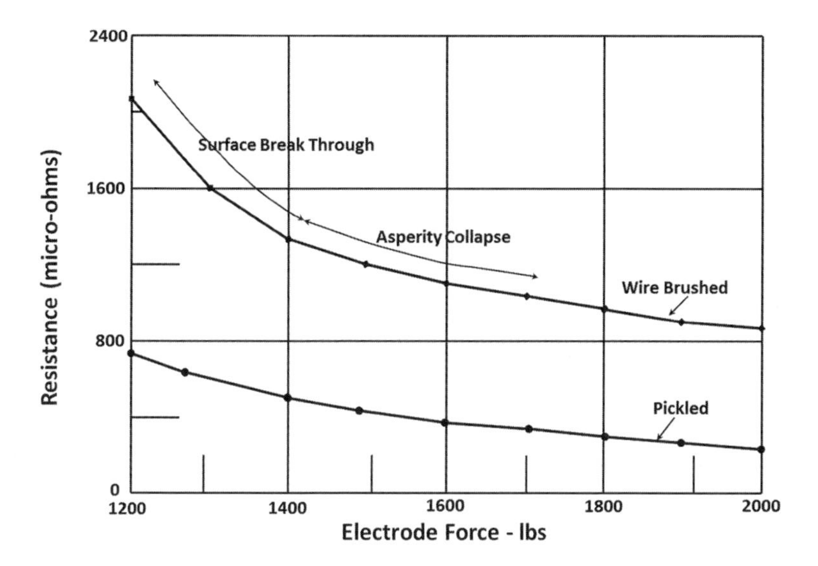

Figure 3.20 The interfacial resistance and its relation to the surface conditions and mechanical pressure

One should be able to predict the needed welding lobe adjustment, i.e. moving the lobe upward or downward, or shifting the lobe left or right, based on an understanding of the electric resistance behavior from Figure 3.18. For example, increasing the roughness of the joined panels leads to an increase in the surface asperities where the panels will make the first contact as the mechanical pressure progresses, thus it affects the interfacial resistance value. Figure 3.20 shows the impact of the different roughness finishes on the interfacial resistance, which is typically evaluated in micro Ohms. Additionally, Figure 3.20 has the electrode mechanical force in the x-axis to show the effect of increasing the load on such asperities. One can conclude that

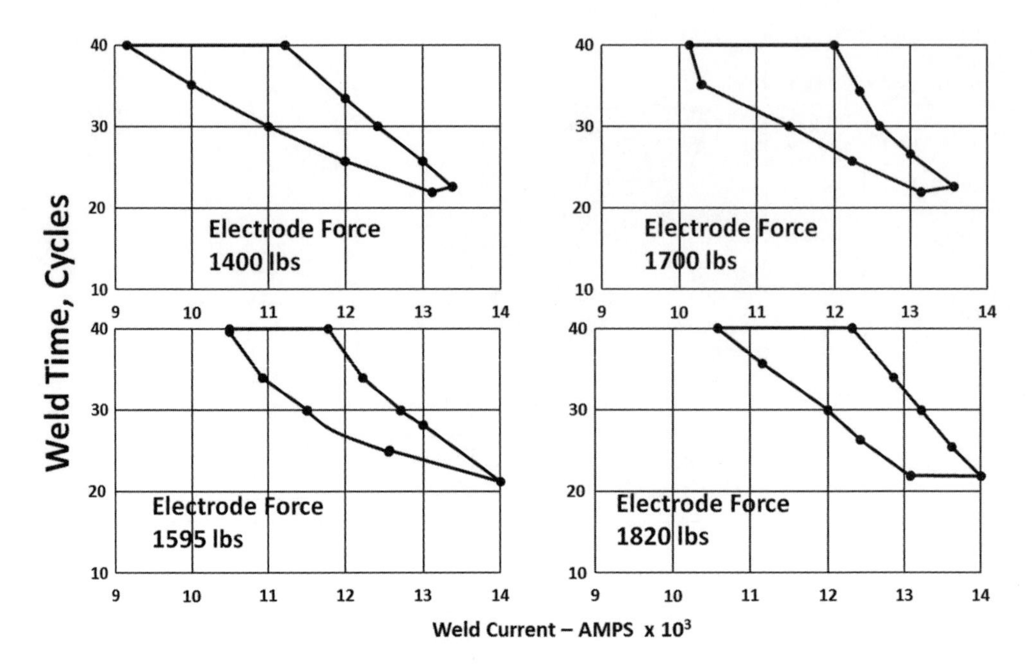

Figure 3.21 The welding lobe behavior for different mechanical pressure values

increasing the holding force leads to closer contact (more collapse of the asperities) and thus a smaller air gap and more contact, consequently, a decrease in the electric resistance. The consequent impact on the welding lobe warrants an increase in the welding current to deliver the same amount of heat needed to melt the material. One should always remember that the heating value depends on Ohm's law, i.e. any decrease in resistance should be compensated for with an increase in the current to ensure that the heat is the same.

Figure 3.21 (a)–(d) displays the adjustment done to the welding lobe, corresponding to the changes in the electrode force. As the applied force increases, the interface resistance decreases, so the welding lobes are shifted to the left to increase the welding current requirements. Additional illustrations of the weld nugget conditions at different welding conditions are shown in Figure 3.22.

Other adjustments of the welding lobe might be required when shifting between the different steel grades, for example, the adjustments needed for welding HSS and AHSS when compared with mild steel are shown in Figure 3.23. This figure shows that in the case of HSS, the lobe is shifted toward the left, i.e. to the low welding current values, however, the AHSS lobe is further shifted toward lower values of the welding current due to the internal higher values of electric resistance. Additionally, the lobe might be shifted up for more welding cycles to allow for longer heating and cooling periods, which reduce the tendency of such grades to form any precipitation.

Weld conditions	Number of Cycles	Weld Current	Resulted Weld
	1	15,000 amps	No nugget formed
	2	15,000 amps	No nugget formed
	5	17,000 amps	Partial Nugget
	7	19,000 amps	Nugget Formed
	9	19,500 amps	Expulsion
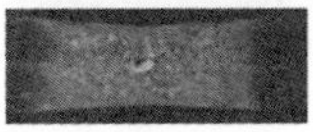	13	19,500 amps	Burn Through

Figure 3.22 The weld nugget conditions for different weld settings

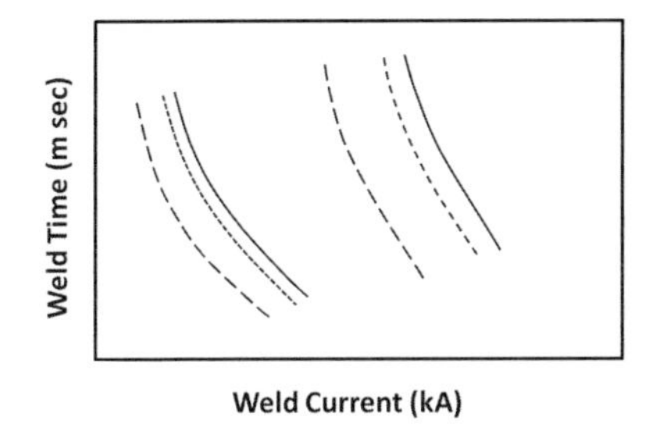

Figure 3.23 The weld lobe for HSS and AHSS when compared with mild steel

Other new spot welding schemes apply the weld cycles in different pulses to better control the heating of the AHSS panels. Additional techniques to control the welding of AHSS are through applying an initial value of mechanical squeezing pressure (e.g. 4 kN) before the weld phase, followed by a higher value of pressure (~5 kN) once the current is turned on and till the end of the welding process. Other new advancements in spot welding control are based on supplying two different welding current values, at the same mechanical pressure. The first welding current is responsible for forming the nugget while the second current flow comes after the weld is complete as a form of post-weld heating treatment. Some of these different variations of spot welding are displayed in Figure 3.24.

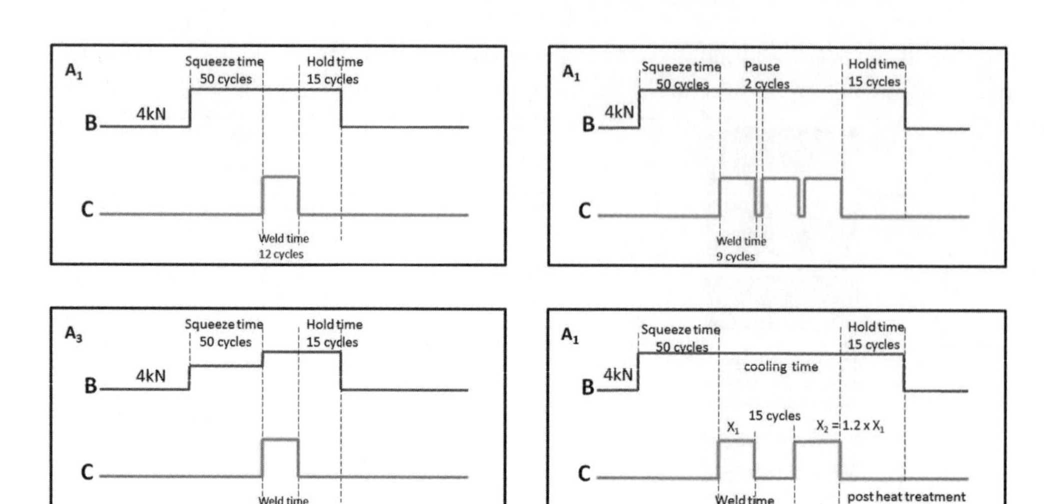

Figure 3.24 Different welding schemes designed to better control the heating and cooling rates

Other new developments in spot-welding practices are based on the use of shaped electrode tips that help in delivering higher current density with minimal misalignment, such as dome-shaped electrode tips. However, an important consideration when selecting an electrode is their service life, which is typically defined as the frequency (in number of welds) before they require dressing or replacement. For spot-welding hybrid joining practices, such as the combination of a spot weld and an adhesive bond, careful considerations of the total resistance should be accounted for when deciding on the weld current and time.

3.3 Robotic Fusion-Welding Operations

The body-weld process features a sequence of joining operations, done through the use of jigs, fixtures, while the actual welding activities are achieved through a robotically spot-welding process. This section addresses these main elements in detail, namely: (a) the jigs and fixtures in terms of their different types with a focus on the flexible fixture systems that can accommodate different vehicle models with different shapes and geometries; (b) the robotic welding process in terms of its control (sensors and control schemes), development of the robotic line (number of robots), and its actual functionality.

The body-weld fixtures help in holding the vehicles' body parts during the welding process to constrain its dimensions and fit. To better illustrate the body-weld fixtures' roles, types, and number, one should understand the specific sequence of the body-build process. The process assembles around 300–400 stamped panels into sub-assemblies, which include: (a) main sub-assemblies such as the under-body sub-assembly, the side members (left and right) sub-assembly; (b)the fractional sub-assemblies, such as the roof, the cowl, the upper and lower backs, and the engine compartment sub-assembly; (c) the closures or shell sub-assemblies, such as doors, trunk, fenders, and the hood. This sequence is described in Figure 3.25, which explains the parts involved into each sub-assembly. For example, the body-side sub-assembly is composed of the A, B and C pillars, the quarter panels. However, other vehicle designs may consolidate these panels into one stamping to reduce the joining effort and the stamping steps. This reduces the processing required but at the same time, it increases the development cost because it adds a new major die to be developed and validated. Additionally, the consolidated body-side outer design reduces the scrap utilization from other processes. So design engineers should consider these issues when evaluating the potential improvement from parts' consolidation.

Each of the joining processes shown in Figure 3.25 requires a different fixture type and size; some are stationary and others are robotically mounted. The discussion in regard to the welding fixtures will focus on the major fixtures and their flexibility in terms of their ability to accommodate different vehicle body types. However, the general rules in regard to welding fixtures relate to their ability to hold the panels in a constrained relative position to each other to result in an accurate weld placement and alignment without distortions.

Additionally, the ease of loading, locating, and clamping is essential for good fixture performance, which is typically achieved through the proper use of "Fixture points," "Support points," "Register points," and finally the "Clamping points." Also the fixture should be designed and configured not to obstruct the welding gun's access to the body-shell. Copper and aluminum alloys are typically used for welding fixtures to avoid welding spatter or splash sticking the fixture surface, thus affecting its performance.

In terms of fixtures' flexibility, different approaches have been developed and used by different OEMs. The development of flexible body-weld fixtures can lead to great savings on the OEMs' part because it reduces the effort, time and cost required to add a new vehicle model or change the geometry of an existing one. Typically, each time a new model is introduced, OEMs adopt one of the following strategies to manufacture it:

1. They stop current production to apply the necessary changes so that the new model can be assembled using the existing robotics and welding systems.
2. They dedicate a new line for the production of the new model; either in an existing plant location (if the space and capacity allow) or in a new location. Both of these implementations are expensive in terms of capital equipment, and time.
3. They adapt a current production line to make the new model, without stopping the production of current models. Naturally, this strategy is the most flexible because it allows the OEM to use existing investment to manufacture new models without any losses from stopping the production.

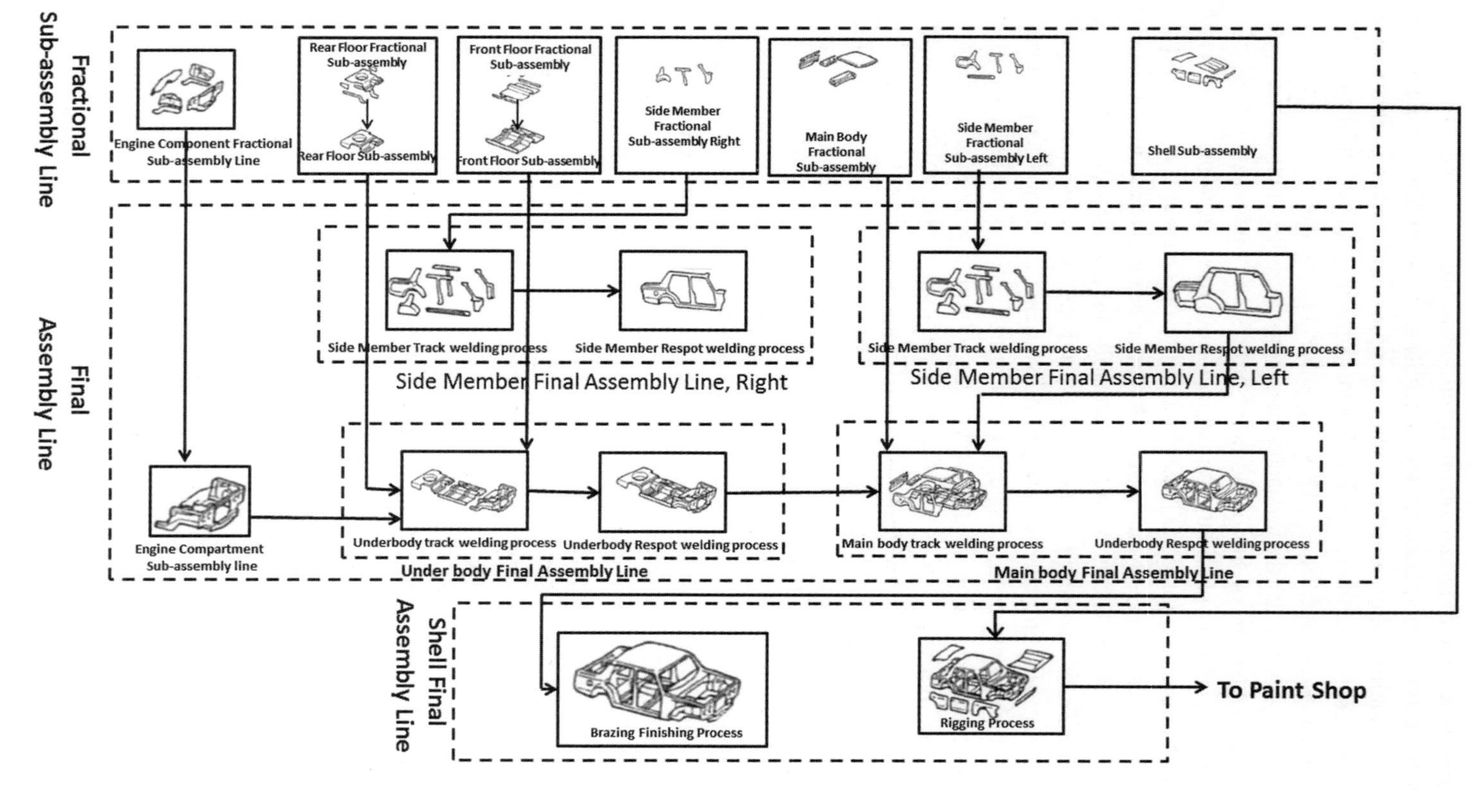

Figure 3.25 A typical sub-assembly flow within a body-weld line

One of the main challenges in adopting the third strategy comes from the body-weld area because the robots' trajectories and fixture systems are designed according to the current running vehicle models. Hence adopting a flexible welding line should provide the following functions:

- a geometry function, to ensure the precise positioning/locating of panels relative to each other, for each vehicle model;
- a handling function, to ensure the precise and sequenced transportation of panels, fixtures, and body-shells of the different vehicle models;
- a tooling function to ensure the suitable processing (number of spot welds or tack-welds, location of welds, etc.) for the different vehicle models.

The following text introduces some of the practices in fixture and framing systems that allow for flexible body welding lines.

One advance in the flexible fixture designs is based on the numerically controlled (NC) locator system, to help change the data when running different vehicle models. The data change through the control of the NC program. This system found applications in the tack welding phase, which is the phase where temporary joints are done to hold the body parts together without having the dimensional conformance between the different joined panels. The NC system features a large number of servo-motors (~300) that are controlled by the NC program according to the vehicle CAD model to change the data, hence the relative position of the panels, each time a new vehicle model is introduced. The system allows for good flexibility due to the available degrees of freedom for the servo-motors, however, the large number of motors translates into high capital investment and maintenance costs, in addition to the system complexity.

Another implementation of a flexible fixture system is based on the utilization of rotary devices, where each face of the device features a different fixture. Hence for each vehicle model, the device rotates so that the suitable fixture is selected. The rotary fixture selection is a simpler solution than the NC locator, however, it has limited number of fixtures that can be mounted and thus is limited to a few (typically three) vehicle models. The rotary fixture system is shown in Figure 3.26.

A third flexible fixture is possible through the use of an automatic fixture selection system, where the fixture is changed every time the vehicle model changes, allowing for greater flexibility in terms of the number of models at lower equipment cost. However, the use of such system requires a batch production mode because of the time needed to select, transport, and mount the suitable fixture, in addition to removing and storing the old one.

Another famous, fixture system solution was introduced by the Fiat Motor Company, to help automate the tack welding line, through the use of a *Robogate* system. This system features multiple tack welding routes with the fixtures' sliding mechanisms between these routes, thus different vehicle models can be tack welded

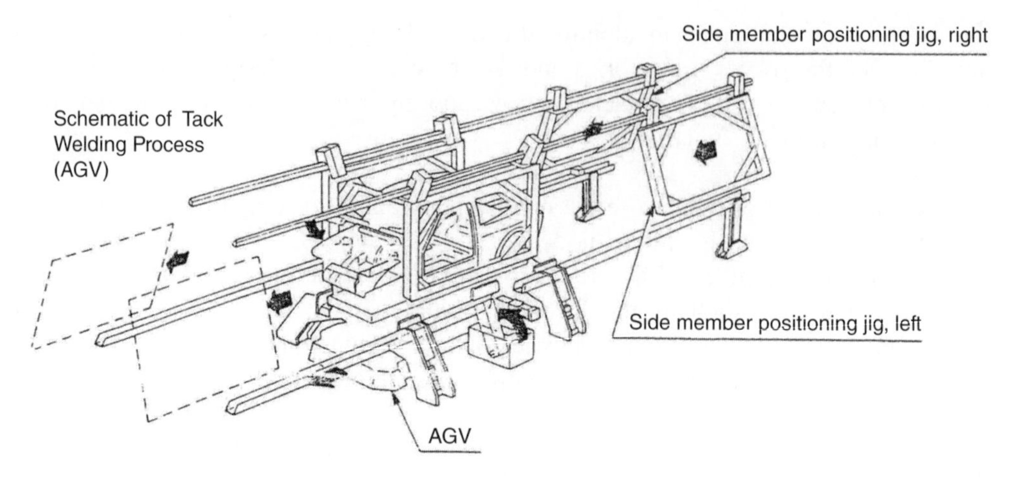

Figure 3.26 The rotary fixture system

in parallel. The *Robogate* system is an older system with the main disadvantage that it has a large footprint because of the parallel tack welding lines. The *Robogate* system implementation comes in two different flavors; the first uses automated guided vehicles to manage the transportation of the panels, whereas in the second implementation the parts flow along the tack welding stations in sequence. These two implementations are shown in Figure 3.27.

Additional flexible body-weld systems have been introduced by the Toyota Motor Corporation, called the Toyota Flexible Body-Line, and the Nissan Motor Corporation. The Nissan system is known as the Nissan Intelligent Body Assembly System IBAS.

In the Toyota flexible framing system, a jig-pallet circulation system is developed for fixing and framing the different body shapes, where each shape has its unique jig while a programmable logic controller functionality is loaded onto the pallet along with the jig to allow for the transfer of the control functions of shape recognition and transfer routes. A high speed transportation system carries the pallets of the different jigs, utilizing servo motors. Additionally, the Toyota system relies on off-line programming of the spot-welding robots to define each vehicle trajectory path. Figure 3.28 displays the Toyota jig-pallet fixture for the under-body sub-assembly. The jig-pallet system is used for the under-body, the roof, and the body-side sub-assemblies. Additionally, the Toyota system uses a serial layout of its spot welding to reduce the number of robots involved; however, the serial production line is only effective at low production quantities (typically at or below 25,000 units per month). A parallel layout will be more effective at high production volumes, in terms of the number of robots needed to complete the welding process.

Additional improvements, introduced by the Toyota flexible body-weld system, include the use of common spot-welding guns for all vehicle models, thus reducing the effort required in changing guns per model. For specialized welds (one-sided

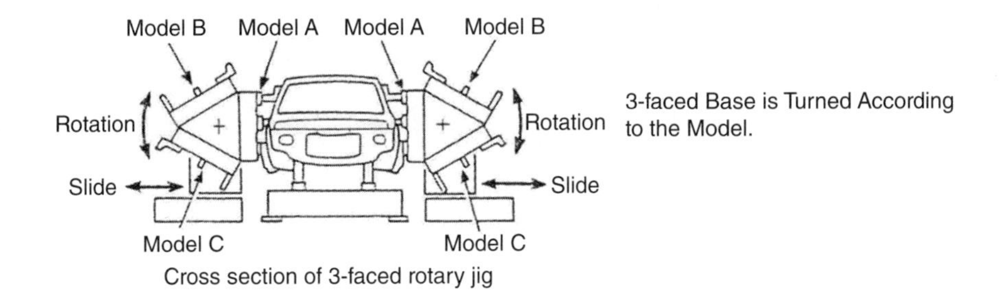

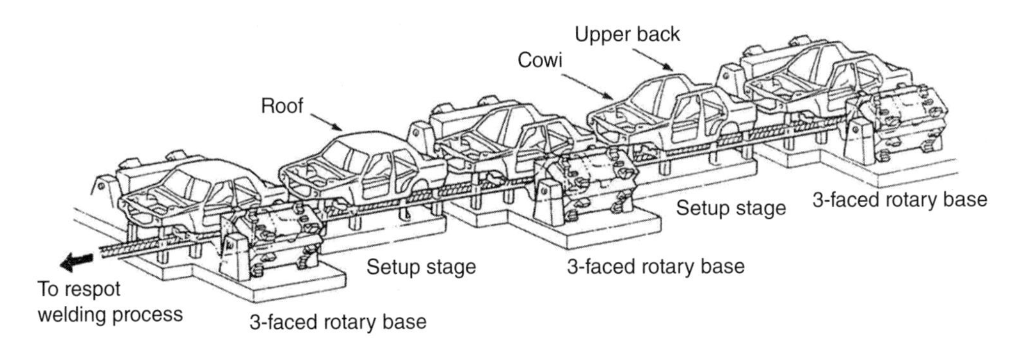

Figure 3.27 The *Robotgate* fixture systems

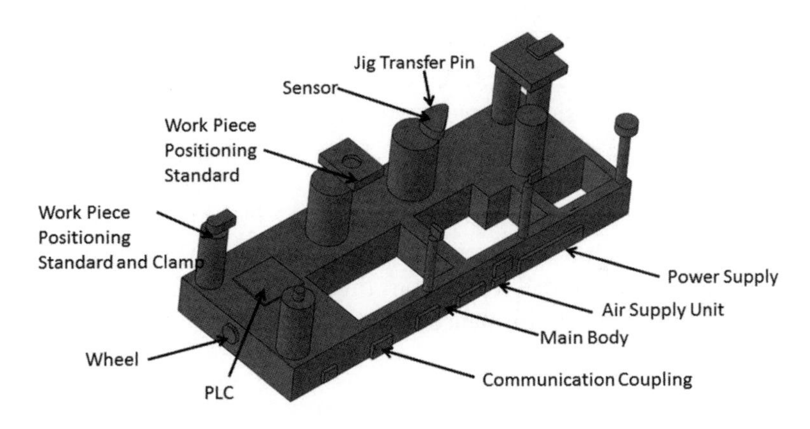

Figure 3.28 The Toyota flexible jig-pallet system

access) that require specialized guns, spot welding is replaced by an arc welding scheme.

The Nissan IBAS utilizes the NC locator framing system for some of the sub-assemblies (e.g. the body-side and the floor), through the use of 35 positioning robots, so tack welding can be done at 62 locations. The IBAS system features a different variety of sensors to feedback the dimensional accuracy to the NC system.

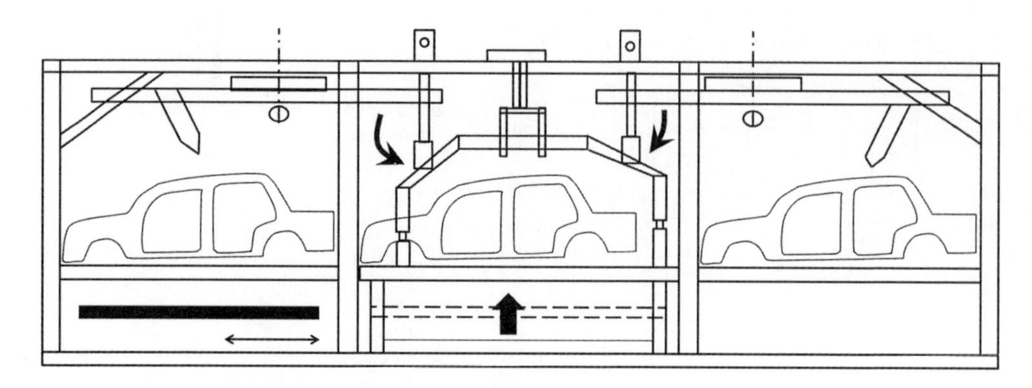

Figure 3.29 The preciflex system

Furthermore, the Nissan IBAS features a fault diagnosis functionality to check and evaluate the system status and provide feedback information to the NC locator. The IBAS uses a CAD teaching system that supports the NC off-line programming for new vehicle models. Also, the CAD information simulates the system's (robots') performance and computes its cycle time, and checks for any interference/collisions between the robots and/or body shell.

Another flexible framing innovation has been developed by Renault and is called the Preciflex. This framing system employs a triangular frame where the under-body, the shell front, and the shell rear geometrical-framing features are installed at the lower, upper front, and upper rear faces of the triangle, respectively. The Preciflex can also change the frame side geometrical references per the vehicle model being tack-welded; the change can be at all three sides or just one, allowing for further flexibility when welding different models but with same under-body platform, i.e. the lower side is not changed. Other advantages of the Preciflex system include their modular design allowing for quick prototyping of vehicle shells, this will also help in reducing the stamping die validation time if the functional-build practice is used for the dimensional conformance. The Preciflex system is displayed in Figure 3.29.

3.3.1 Robotic Spot Welders

The spot-welding process is implemented through industrial robotics for the following reasons:

- The large number of spot welds needed to assemble the body-shell, around 4000–5000 welds, thus manual operations result in slow production speeds. The typical available time for a spot-welding robot for each vehicle shell is around 20 seconds, in which the robot applies around 6 spot welds in 6 different locations.

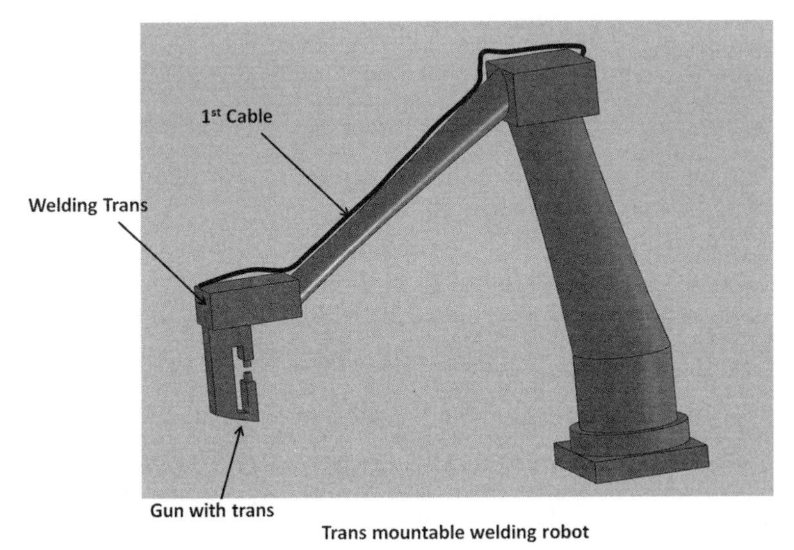

Figure 3.30 A typical spot welding robot

- The greater accuracy achieved with robotic welders over the human (manual) implementation, in terms of deviation in weld positions, gun alignment, and number of applied welds. Typical values for manual spot welding can reach up to +/– 20 mm in deviation from the specified weld location, while the robotics application range from +/– 0.5–1.0 mm. In regard to welding electrode alignment, a manual welder deviates in about +/– 15% from the correct (vertical to the surface) inclination, while the robotic welder has a +/– 1% deviation.
- Robotic welders can be integrated into the production main control unit, resulting in quicker and better adjustments of the welding conditions. Also, the integration between the robotics and the PLC units allows for automated flexible framing and fixing applications. Additionally the robotic controllers issue an alarm if the robot or the welding process fail, to the main production control unit or the ANDON controller to indicate the failure location, thus aiding in detecting and resolving any failures.
- Spot welding of the different panels requires a large number of degrees of freedom, and is not feasible with manual operations due to ergonomics (reach, weight, and frequency of repetition) and spatial accessibility.
- The current spot-welding robot integrates all the welding equipment, tooling and accessories in one unit. This includes the welding gun, the transformer (integrated into the robot), the air (pneumatic pressure) and water lines, the power supply, and the controller. Figure 3.30 displays a robotic welder featuring the location of the different welding equipment on it.
- The reliability of the welding robotics in terms of Mean Time Between Failures MTBF is around 50,000 hours.

- The economy of welding robotics is much better than that in the manual case; a welding robot's hourly rate is approximated at $6/hr, while that of a body-weld line worker is at $30–40/hr.

However, the robotic spot-welders should be prepared for welding a specific body-shell variation through a pre-production set-up process. The set-up typically includes two phases; the first starts around 1 year before the actual production which is the design stage that includes determining the weld conditions (i.e. welding lobe), the selection of the welding gun, the robot accessibility to the different weld locations. The second phase is called the teaching phase which typically starts around 4 months before the start of production or the launch of a new model. The teaching phase include the actual robot programming, installation (if any new robots are needed), and the try-outs. The arrangement of a robotic cell within the body-weld line is based on computing the number of spot welds needed and their location in order to compute the time needed for the robot end-effecter to move from one spot to the next. Furthermore, the weld configuration at the different positions within the shell in terms of number of panels, the total joined stack thickness, and the two-sided versus one-sided accessibility, all decide on the welding gun and time required. For example, six spot welding robots are responsible for welding the upper body of a vehicle shell while the under-body typically requires an additional two arc spot-welding robots to deal with areas with one-sided access. Also, the selection of the robot size (load-carrying capacity) and type (degrees of freedom) should be done according to the body shell CAD file. In addition, raised platforms can be used to elevate certain robots to enable the robot manipulator to reach certain locations. Other mounting configurations include floor mounted and inverted robots.

Also, each robot size will govern its work envelope (i.e. its motion footprint), which decides on the robot reach; compact robots might be desirable if the space around the vehicle shell is limited. The speed of a welding robot is typically described in terms of the number of degrees an axis can move around in one second. Predicting the motion of the robot manipulator is important not only from a welding accessibility point of view but also from safety considerations as in the case of robot collision with the body-shell, collision with human operators, collision with other robots in the cell. The robots' motion can be predicted through its manipulator's kinematics, which can be described according to Equations 3.5, 3.6 and 3.7, in terms of its Cartesian coordinates (x, y, z):

$$x = [L_1 + L_2 \sin\theta_2 + L_3 \sin(\theta_2 + \theta_3)] \cos\theta_1 \tag{3.5}$$

$$y = [L_1 + L_2 \sin\theta_2 + L_3 \sin(\theta_2 + \theta_3)] \sin\theta_1 \tag{3.6}$$

$$z = L_2 \cos\theta_2 + L_3 \cos(\theta_2 + \theta_3) \tag{3.7}$$

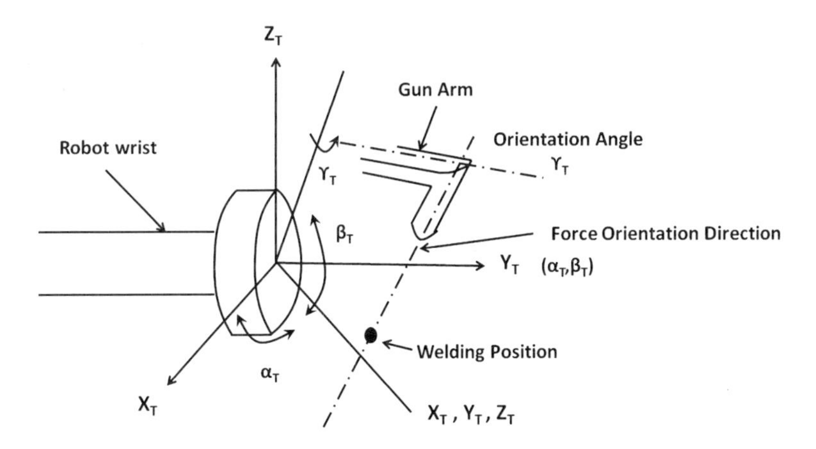

Figure 3.31 A robotic manipulator coordinate system

where, θ_1, θ_2, and θ_3 are the joint configuration and the L_1, L_2, and L_3 are the length of the links. Thus, to determine a spot weld location and conditions, the following information should be programmed: the spot weld position (x, y, z), the squeezing force (of the welding gun), the gun orientation which is typically defined in terms of two angles; α and β, and the orientation angle of the robot arm holding the gun Φ, this arrangement is displayed in Figure 3.31. Additional checks should be performed after the programming is done in terms of interference between the robot motion path and the body-shell; this is done typically by tracing the electrode motion and when it clamps and releases the panels.

To help determine the suitable spot welding gun for each weld location, the engineers should compute the weld conditions in terms of weld current, weld time (cycles), and the required electrode pressure based on the panels' thickness, material type, and the number of panels in the stack. Once a weld gun is selected based on the weld condition requirements, the interference check is done based on the gun inertia and weight characteristics. This sequence of events for selecting a weld gun is depicted in Figure 3.32.

Each robotic welder features a variety of on-board sensors to monitor its welding conditions; for arc welding the main characteristic of interest is the arc voltage and current. The current is typically measured using a Hall Effect-based sensor or a current-shunt sensor. Additionally, for arc-welding, following or tracking the applied weld seam (or bead) is important and is typically done in one of two ways: the through-the-arc sensing and using the optical sensing approach. For automotive arc welding, the optical sensors are widely used through off-the-shelf sensor packages; the optical sensor projects a laser beam ahead of the welding arc, while a charged couple device camera collects the beam reflection to decide on the bead location and orientation. Other sensors for MIG welding can monitor the wire feed rate to evaluate the deposited metal as the welding progresses.

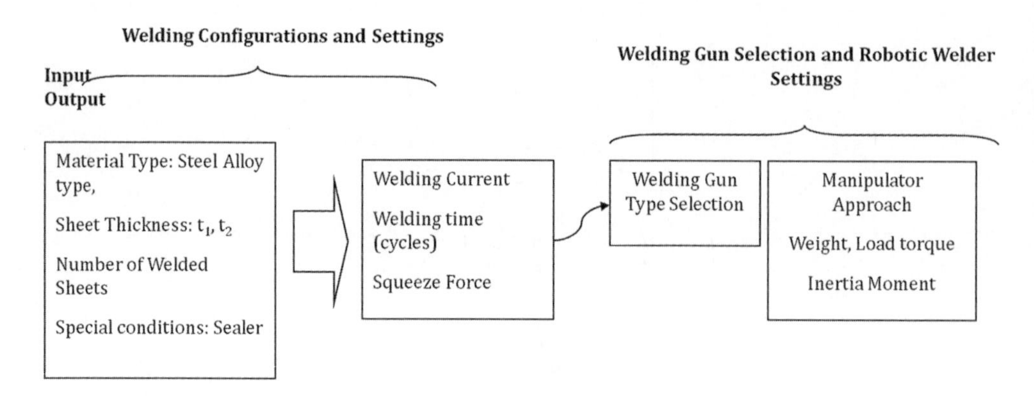

Figure 3.32 The flow of the spot welding process settings and calibration

One can trace the body-weld line activities by inspecting the overall robotic welding line which is shown in Figure 3.33 in addition to the accompanying information flow in Figure 3.34.

3.4 Adhesive Bonding

Joining through adhesive bonding is gaining more acceptance within the automotive industry due to its added advantages, such as its ability to join different materials such as joining high strength steel with aluminum or magnesium panels without worrying about thermal distortions, compatibility or galvanic corrosion issues; this ability has facilitated the introduction of lightweight materials (lower density than steel) into the automobile body structures. Other reported benefits from adhesive bonding can be summarized as follows;

- The application of adhesive bonding is a robust process that can be automated and integrated within the body-weld production line at low cost.
- Adhesively bonded structures are typically hemmed (more inertia) thus it improves the overall stiffness (by almost 20% improvement in torsional stiffness) and crash resistance of bonded joints.
- The adhesive bond formulations can be customized to facilitate its application step, in addition to improving its performance metrics.
- The adhesive bonded joint is not visible, thus it does not have any negative impacts on panels' appearance and design, and can be done for *Class A* surfaces.
- Adhesive bonding enables modular designs and wider use of platform strategies because of its lower requirement (than fusion welding) on specific production sequences.
- Adhesive bonding has less effect on the overall dimensional conformance of the BiW, than that of the fusion welding practices.

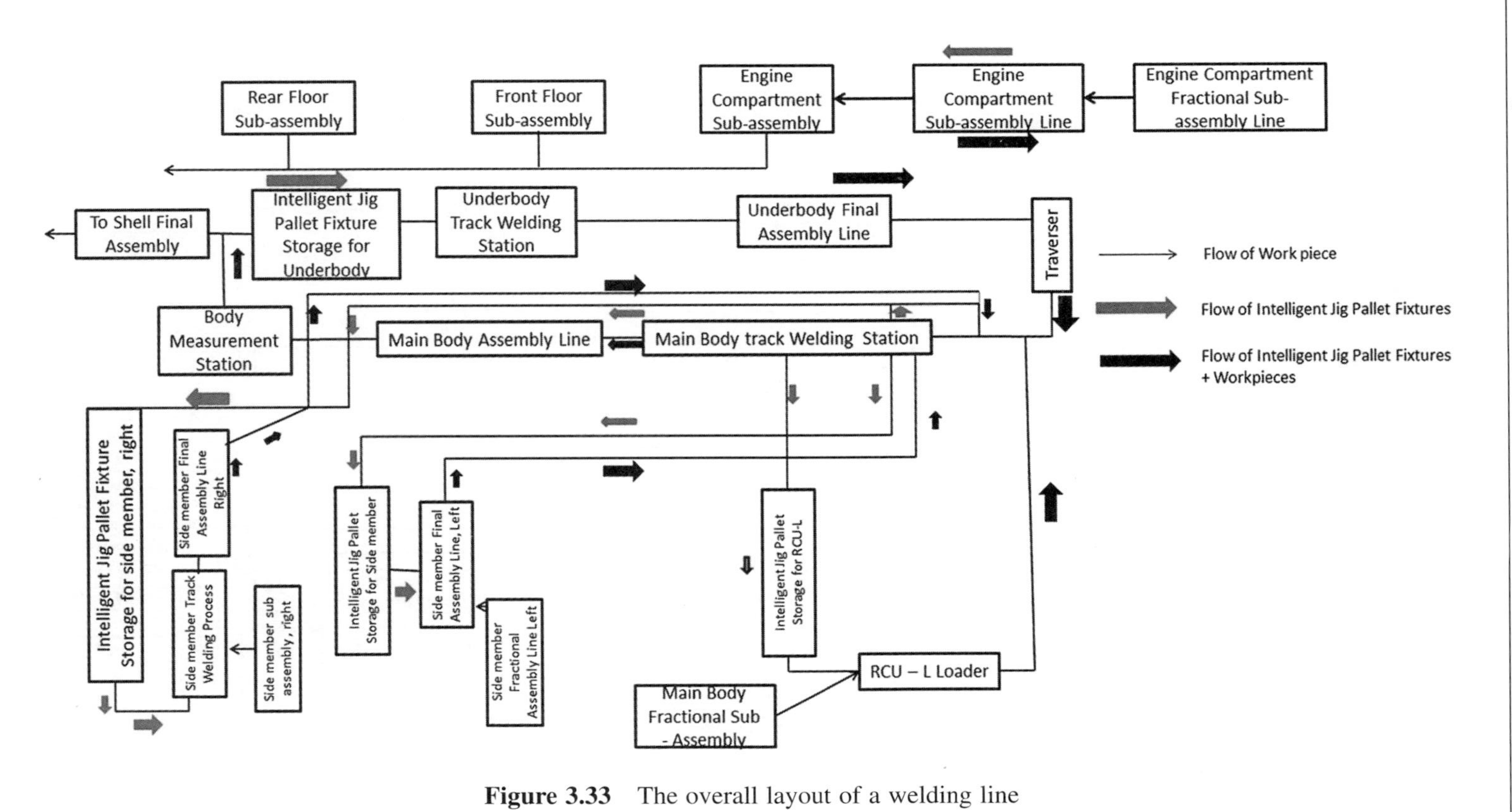

Figure 3.33 The overall layout of a welding line

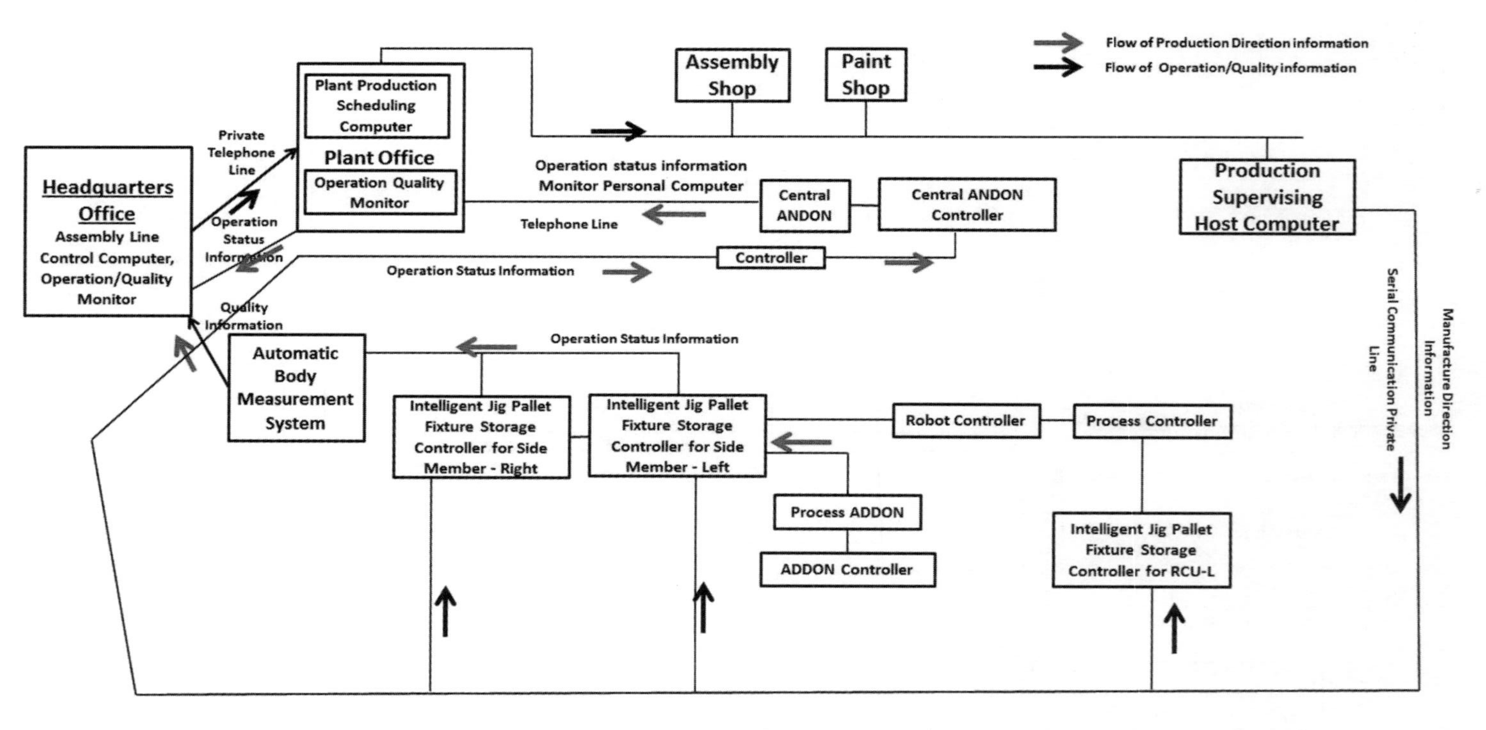

Figure 3.34 The information flow in a welding line

To provide a numerical example of the current adhesive bond status within the automotive industry [28], the new S-Class Coupé has more than 100 m of structural bonds in BiW applications; also the BMW 7 series has more than 10 kg of structural adhesives applied. Within the automobile body structure, adhesive bonding is used for: (a) anti-flutter bonding or (b) structural bonding. Anti-flutter adhesives are used to compensate for tolerances (as in large gaps as in windshield installation), and to absorb vibration. On the other hand, structural adhesives are used to join panels, hemming, and in stiffening the joined panels.

Usefully, the automotive adhesives might be classified based on their curing characteristics [29]: (a) rapid set at room temperature adhesives, which include anaerobic and cyanoacrylate adhesives, with applications for gaskets, bonding of electrical and electronic items, thread sealing, bonding of caps in cylinder head covers and gear boxes; (b) rapid set at high temperature; epoxides and phenolic adhesives fall into this category, finding applications in bonding of heat exchangers and hem flanges; and, finally, (c) moisture-evaporative synthetic resin sealants. Other adhesives' classification might be used depending on the adhesive's intended application and hence its attributes, such as structural and non-structural adhesives.

3.4.1 Basics of Adhesive Material Selection

The choice and utilization of the adhesive bonding technology should be done while considering the current automotive manufacturing practices and sequence. For example, steel mill and stamping lubricants have a direct impact on the sheet metal surface energy, hence this affects the wetting and adhesion performance of the adhesives. Thus the choice of the adhesive must take these lubricants into account, or if an additional surface preparation step is needed, its location and cost should be computed. Another important consideration for selecting the adhesive type and formulation is due to the fact that the adhesive bond should be able to withstand the conditioning/cleaning processes during the painting conditioning phase, without washing out, getting contaminated, or carrying/absorbing any chemicals or particles.

Additionally, the location where the adhesive bonding is introduced affects its performance characteristics and the production steps. For example, the adhesive bond in the trim assembly area (final assembly) is applied to painted surfaces; it was found that the adhesive performs differently (its effect on crash-worthiness tests) if it is applied to different paint colors or paint-film layers, hence the OEMs tend to mask the areas where the adhesive is to be applied so it doesn't get painted, to guarantee consistent surface conditions. The adhesives used in the body-weld area are typically applied to surfaces covered with lubricants and oils.

Additionally, the adhesive viscosity should also be carefully selected because it affects the robotic application (temperature, pump and nozzle configurations) and control (flow rates) because if the joints' fill rates are below a certain limit, the joint strength might be affected. On the other hand, if the fill rate is above a certain limit, then it will squeeze out from the joint, causing tooling and equipment contamination. Also the adhesive viscosity behavior inside the paint curing ovens must be considered along with its curing characteristics. The compatibility between the adhesive material and the cosmetic sealer is also important, otherwise the adhesive might interact with the sealer. An adhesive bond is typically defined as a substance capable of holding at least two surfaces together in a strong and permanent manner, while a sealer is described as a material capable of attaching to at least two surfaces, thereby, filling the space between them to provide a barrier or protective coating.

From a production safety point of view, the adhesive should not produce any chemicals or volatiles through its application, handling, and through its service life.

The above discussion of the different requirements of the adhesive material can be further quantified into specific material characteristics to help select the suitable adhesive for the intended application, while meeting the production demands. Such characteristics are: the adhesive material aging (or shelf life), temperature tolerance, load bearing ability, UV sensitivity, viscosity (and its temperature profile), adhesion characteristics (surface energy), thermal expansion coefficient, shrinkage, moisture absorption, curing characteristics, thermal and electrical properties, ability to withstand crash conditions (stiffness), ability to be removed (for recycling and repair), and finally its cost. So, before selecting an adhesive material, engineers should evaluate the adhesive material from three perspectives: (a) the service performance metrics; (b) production processes and sequence compatibility; and (c) the materials to be joined and the joint design.

Automotive adhesives are typically classified into a 1K or a 2K adhesive, where 2C or 2K refers to two component adhesives that include a binder (base, resin) and a hardener (catalyst). The resin is the principal component in an adhesive, and controls its main properties such as the curing window, or load-bearing capability. On the other hand, the hardener is added to the resin to promote certain curing reactions such as cross-linking. Other added components might include:

- accelerators: to accelerate or de-accelerate (control) the curing rate, the storage life (pot life, shelf life) of the adhesive;
- solvents: used to help disperse the adhesive to a certain consistency or control its viscosity;
- catalysts: used to promote the resin cross-linking and solidification;
- plasticizers: used to provide flexibility and/or elongation;
- extenders: used to reduce the adhesive concentration to reduce the formulation;
- fillers: might be added to control viscosity.

Typically adhesives have a limited shelf-life and hence it ages, so different agents are further added to delay this aging; these include antioxidants, anti-hydrolysis

Table 3.7 The different types of adhesives

Adhesive material type	Family	Examples
Elastomers	Natural rubber Synthetic rubber	Rubber Polyolefins
Thermoplastic	Vinyl polymers Polyesters Polyethers	Polyvinyl acetate Polystyrene Polyphenolic
Thermoset	Epoxy Phenolic resin Polyaromatics	Epoxy polyamide Phenol, resorcinol Polyimide

agents, and adhesive stabilizers. The different types of the adhesive bonding materials are tabulated in Table 3.7 [30].

3.4.2 Basics of the Adhesion Theory and Adhesives Testing

Several theories exist to explain the adhesive bonding mechanisms; these include the adsorption theory, simple mechanical interlocking, diffusion, electrostatic interaction, and the weak-boundary layers theory.

The adsorption theory states that adhesion is a result of the molecular contact between two substances and the surface forces that develop between them and the adhesion results from the adsorption of adhesive molecules onto the substrate and the resulting attractive forces, usually designated as secondary or van der Waals forces. For these forces to develop, the respective surfaces must not be separated more than five angstroms in distance. Therefore, the adhesive must make intimate, molecular contact with the substrate surface. The mechanical theory of adhesion states that for the adhesion to occur, the adhesive must penetrate the cavities on the surface, displace the trapped air at the interface, and lock on mechanically to the substrate. Thus, the mechanical interfacing theory might require roughening of substrates to provide "features" for the substrate, and by virtue of roughening increases the total effective area over which the forces of adhesion can develop. In both of these theories, the adhesive must make intimate contact with the substrate which is only possible if the substrate surface is able to spread the adhesive molecules, which occurs if the substrate surface has a higher surface energy than that of the adhesive material. The substrate ability to spread the adhesive material is typically called "wetting." Wetting is the first step in the adhesion process. To illustrate this, Figure 3.35 shows an epoxy adhesive applied to different substrates, each with different surface energy; if the adhesive has a higher surface energy relative to the substrate, then it will bubble over the surface.

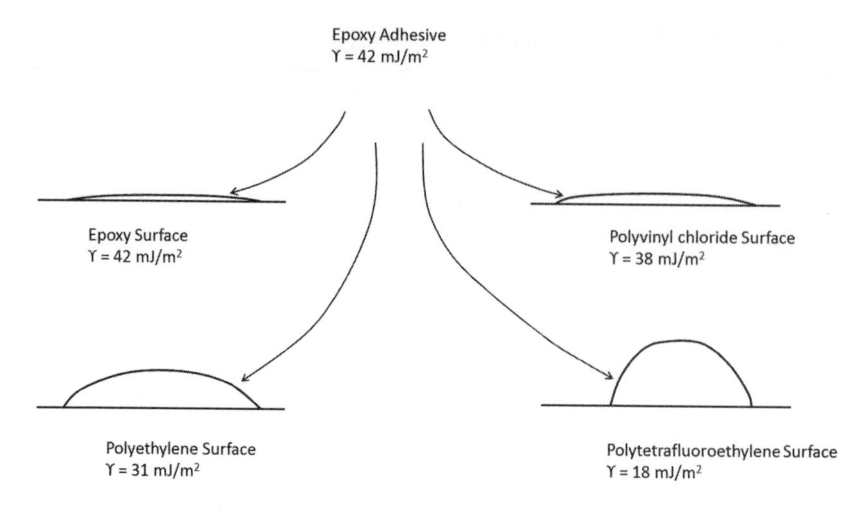

Figure 3.35 An epoxy adhesive applied to different substrates

The adhesive bond strength can also be evaluated using several destructive and non-destructive test procedures, however, the best method to evaluate the contact between the adhesive and the substrate is through conducting a contact angle measurement. The forces holding an adhesive to a substrate or maintaining the cohesive integrity of a solid can be measured as the work necessary to separate the two surfaces beyond the range of the forces holding them together. In one case, the surfaces are the adhesive and substrate; in the other, they are like molecules in the bulk of the material. "This force is dependent on the intermolecular forces that exist in the material and upon the intermolecular spacing. It is sometimes referred to as the surface energy" [30].

In contact angle measurement, a drop of adhesive is applied to the substrate surface. The drop is allowed to flow and equilibrate with the surface. The contact angle is then measured using a device called a goniometer, which is a protractor mounted inside a telescope. The angle that the drop makes with the surface is the contact angle. The contact angle can then be used to evaluate the different interaction between the substrate and the adhesive surface energies/forces. The equilibrium of these forces is shown in Figure 3.36 and is described mathematically in Equations 3.8 and 3.9.

$$\gamma_{LV}\cos\theta = \gamma_{SV} - \gamma_{SV}\gamma_{LV} \tag{3.8}$$

$$\gamma_{SV} = \gamma - \pi_e \tag{3.9}$$

Furthermore, the contact angles can be used to estimate a critical threshold for surface energies or surface tension needed for wetting to occur. This formulation was proposed by Zisman [31], where he proposed that a critical surface tension, γ_c, can be

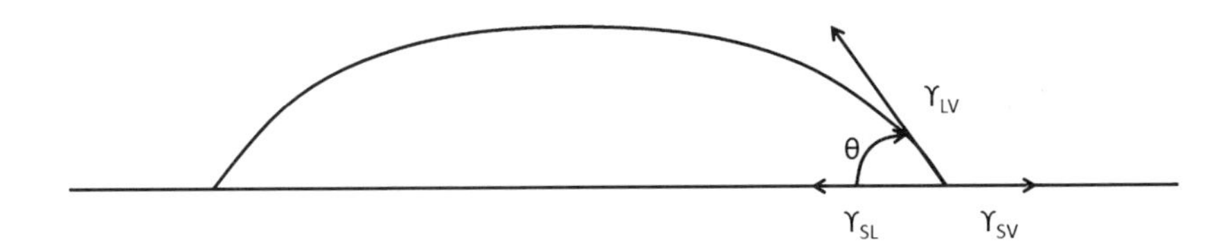

Figure 3.36 The balance of surface tension forces

estimated by measuring the contact angle of a series of liquids with known surface tensions on the surface of interest. These contact angles are plotted as a function of the γ_{LV} of the test liquid. The critical surface tension is defined as the intercept of the horizontal line $cos\,\theta$ with the extrapolated straight line plot of $cos\,\theta$ against γ_{LV}.

For automotive applications, one should ensure that the adhesive material is not sensitive to the manufacturing process chain variations. So in the case of the adhesive bond applied in the hemmed flange, which joins most of the automobile body closures, e.g. door inner and outer panels, hood inner and outer panels, one should check for the hemmed ability to do the following: resist water ingress, fill the gap (flow in the joint), provide uniform appearance (bubbling), and be compatible with the after-hem sealants, and the electro-deposited coating baths. Some standardized testing methods exist to quantify some of these functionalities such as the dye penetrant test, which is used to check for the joint tightness (water ingress). However, other adhesive functionalities are controlled and affected by the adhesive flow and curing characteristics, in other words, its composition, glass transition T_g temperature, cure rate while in storage, its sensitivity to humidity, and lastly its volatiles evaporation rates. To quantify and investigate these attributes under production variations, Fourier transform infrared spectroscopy (FTIR), differential scanning calorimetry (DSC), and thermo-gravimetric analysis (TGA) procedures can be done to accurately describe the adhesive's behavior. Even though automotive OEMs might not apply all these tests but rely on known adhesive formulation, the adhesive material's interaction with the manufacturing and handling environment will only be quantified through these tests. The following presents a complete study of the effect of the different process variations on the automotive, adhesively bonded hemmed joints [32].

The process variations addressed in this study are:

- Variations in the mixing ratio between the resin and the hardener, by the robotic dozer. A robotic dozer deviates from the nominal mixing ratio around +/– 6%.
- Variations in the holding time and conditions (temperature and relative humidity RH), which affect the 2K reaction-curing rate.
- Variations in the E-coat oven ramp-up conditions and its effect on the volatiles evaporation rates.

To analyze the first variation (mixing ratio), an FTIR test is done to analyze the FTIR peaks that might be indicative of the resin or hardener materials, then such a peak can quantitatively compare the epoxy content (resin) to a calibrated (nominal) sample. The FTIR indicates that even though the dozer fluctuates within +/– 6%, this does not have a major impact on the adhesive performance from the hemmed joint point of view, i.e. the adhesive had the same degree of cure.

For the second variation, the holding (storage) time is defined as the time between the start of the 2K mixing (reaction) and its final cure in the E-coat oven; such a storage time fluctuates depending on several factors. Such factors include: the location of the press-shop (hemming station) relative to the paint-shop, the die-tooling changes, and the production (demand) rate. A typical fluctuation is found to be around 1–2 days, however, if the supplier is going through a tooling change, then this fluctuation will be much higher. To study this variation, DSC curves are prepared for samples collected from different holding times to monitor the impact of the storage time on the adhesive glass transition temperature T_g, and its melting temperature T_m, mainly its curing characteristics. Additionally, a TGA procedure is conducted to investigate the effect of the storage conditions, mainly the temperature and the relative humidity (RH). To do this, the samples are placed inside an environmental chamber, where summer and winter conditions are imitated for the duration of both holding times. The weather conditions are based on data from the national weather service center for local conditions; summer ($T = 38°C$, RH 85%) and winter ($T = 25°C$, RH 50%). The samples are then kept in the environmental chamber for 1 and 2 holding days, and then transferred directly to the TGA chamber. The TGA evaluates the weight loss as the samples cure, to monitor its solid and volatiles contents. From the TGA analysis, it is found that the volatiles' release rate is higher for winter conditions than that of the summer sample. This is due to the higher reaction cure-rate for the summer sample, caused by the higher temperature. Further analysis indicates that the current adhesive is insensitive to the humidity variations, because an FTIR procedure showed no difference in the adhesive composition.

For the third variation, the final cure of the 2K adhesive takes place in the electrocoat (E-coat) oven; however, such ovens' curing-profiles are calibrated upon vehicles' model change. A typical curing profile is composed of a ramp-up, steady and cool-down regions. Typical adjustments are done through changing the ramp-up region, to accelerate/decelerate the curing process. The TGA analysis is used to monitor the effect of three ramp-up rates (10, 20, and 30 °C/min) through utilizing these rates to cure the adhesive while recording its weight loss. The weight loss is recorded via volatiles' release in an air purge, to simulate actual production conditions. This means that by accelerating the curing process inside the E-coat oven, less time is available for the adhesive to release its volatile content compared with the slower cure profile. The non-released volatiles will ultimately evaporate when the joint is completely sealed (with after-hem sealant), which might affect the sealant adhesion and its cure. Even though the volatile release difference is small, the current 2K adhesive is a high solid (>90% solid), so a more drastic difference is expected for adhesives with smaller solid percentage. The implications of such volatile percentages might affect the final cure of the combined adhesive and after-hem sealant, hence the final hemmed joint performance.

3.5 Welding and Dimensional Conformance

The fusion welding process utilizes high temperatures to perform the weld, which lead to thermal expansions and contractions as the metal heats and cools leading to dimensional distortions. Weld distortions are typically classified as:

- *Lengthwise shrinkages,* which happens if the weld is applied along the length of the panel, while the panel is not secured properly. Then the panel will bow upward from both ends as the weld cools down. This distortion can be avoided through the use of proper fixing and clamping tools, and by making the welds along the neutral axis of the joint.
- *Cross-wise or transverse shrinkage*, which is related to the panels welded through a butt joint without proper clamping. The panels will contract, getting closer to each other at their end.
- *Warping*, which is also called the weld deposit contraction, and is most common in V-butt or U-groove joints. Warping is mitigated through setting up the panels so they contract in an opposite direction to that of the deposit warping. If additional clamping is added to avoid warping, then internal stresses might lead to a weaker weld.
- *Angular distortion*, especially in fillet welds because such welds develop residual stresses in both the longitudinal and transverse directions. Thus, when a fillet weld is applied in a T-shaped joint, the vertical panel is pulled toward the weld location.
- *Spot welding* tends to causes heat distortions in the area next to the weld, however, the size of this distortion is comparable to the size of the weld nugget.

Distortion and warping deviation can best be avoided through the proper design of the weld joint and the correct selection of the weld conditions, in addition to utilizing suitable clamping forces and fixtures' designs. For example, strong-backs are temporary stiffeners that might be used to avoid distortion.

The automotive body-weld lines feature projection-based sensors to monitor for the relative position of the joined panels to avoid any missed welds or any dimensional distortions in the final BiW structure and shell. Additionally, each body-weld line will have a coordinate measuring machine (CMM) room to measure the BiW fit and dimensions. The CMM measures the precise locations of the different features within the BiW by recording its (x, y, z) coordinates relative to a reference location/feature. The CMM measures the points by using a probe that is either manually or automatically controlled. Figure 3.37 shows a full-size CMM machine dedicated to automotive dimensional validation.

Figure 3.37 A full-size CMM machine

3.6 Advances in Automotive Welding

Recent advances in the automotive welding technologies have emerged to resolve some of the technical challenges in current practices and to introduce new potentials in joining dissimilar materials.

3.6.1 Friction Stir Welding (FSW)

One of these advances is the friction stir welding (FSW) process, where the joint is established, at temperatures below the substrate melting temperature, through an intensive plastic deformation at the interface. The plastic deformation is generated by the frictional force of a rotating pin that transverses a butt joint, shown in Figure 3.38. The main advantages of the FSW include:

- The FSW results in a very low distortion, due to the lower joining temperature and the welding configuration in terms of the sample clamping.
- Fully mechanical process, without any major thermal effects.
- No fume, porosity, or spatter.
- No melting of base material.
- Cost effective for suitable applications.
- Mechanical properties equal to or better than traditional welds.
- Can be used to weld dissimilar materials.

On the other hand, the FSW joints have not yet been tested for long-term service conditions, and can result in a reduced strength in heat treatable alloys. Also, the FSW still suffer from a thermo-mechanically affected zone equivalent to HAZ in

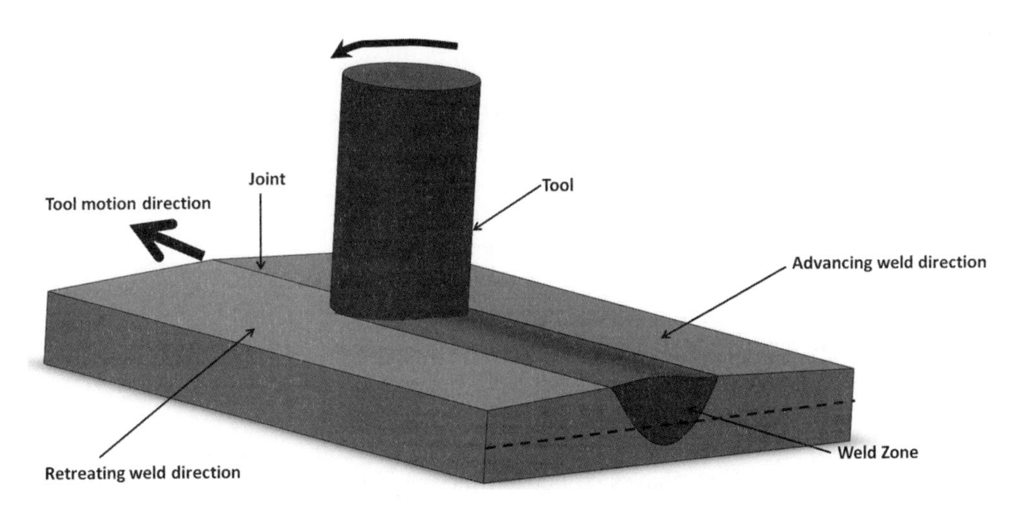

Figure 3.38 A friction stir welding set-up

fusion welding, and the FSW is limited in terms of the weld geometries and requires large clamping forces. The FSW has been used to join aluminum (cast and wrought), dissimilar materials (Al 2219; AZ91), magnesium, zinc and copper. For the automotive applications, it can be used to weld wheel rims, suspension arm struts and body components.

3.6.2 Laser Welding

A typical BiW shell features approximately around 40 meters of pinch weld flanges, requiring 800 or more spot welds. The pinch weld flange widths can be significantly reduced if a continuous weld seam is applied. MIG welding bead size limits its applicability to such flanges, so laser welding is a better candidate to replace some of these spot welds. Eliminating these flanges can yield a weight savings of up to 30–40 kgs from the total vehicle net weight. Potential locations where laser welding can be used are displayed in Figure 3.39. Other potential advantages of the laser welding technology include:

- Laser welding does not require regular maintenance like spot welding, where the electrode tips are dressed and replaced on regular basis, thus increasing the welding line up-time.
- Laser welding does not require a two-sided access as is the case in spot welding, which provides more design flexibility in creating the joints' geometries and material thicknesses.
- Laser-welded joints are found to perform better than spot-welded joints, in terms of shear and tensile tests

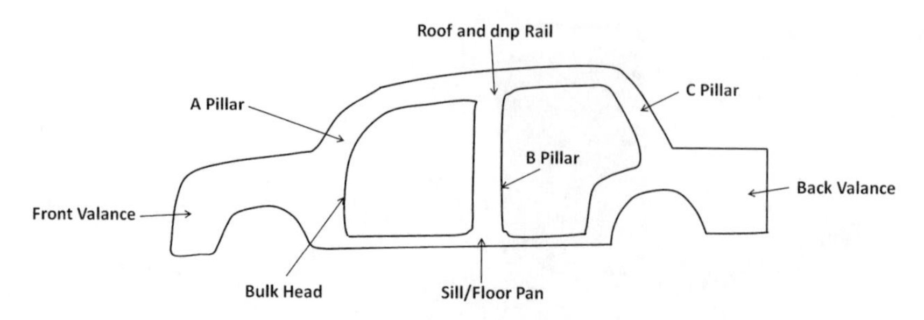

Figure 3.39 The different locations where laser welding can be used

- Laser welding has faster welding speeds around 2 m/minute, while the spot welding cycle includes different phases (squeeze time, weld time, cooling time) leading to slower welding speeds.
- Laser welding provides minimal distortion in the welded panels, when compared with spot or MIG welding.

The limited applicability of laser welding within the automotive body-weld lines is due to the intensive requirements on joint edges' preparation, in addition to the required expertise in manipulating the weld conditions for the different thicknesses and steel grades. Additionally, one of the main limitations of laser welding is the need to apply it using robotics. One of the current applications of laser welding in the automotive industry is in welding the upper part of the rear quarter panel to the vehicle roof panel.

3.6.3 Weld Bonding

The weld bonding process is a joining technique that combines spot welding and adhesive bonding. This hybrid joining technology tries to produce joints that can withstand diverse loading schemes, while having structural damage tolerance characteristics. Weld bonding is one variation of several other hybrid joining practices that include rivet bonding, and weld brazing [33, 34]. The weld bonding process aims to combine the favorable attributes from both spot welding and adhesive bonding in one joint. The adhesive bonding is based on joining the panels over a large surface area so it spreads the stress and hence reduces any stress concentrations, in addition, the adhesive bonding does not generate any mechanical distortions or micro-structural changes to the joined panels. For spot welding, its favorable performance characteristics can be summarized through its ability to withstand high stress loads (higher than adhesive bonded joints), carry tensile, shear or compressive loads applied in or out of plane. Also, the spot-welded joints are more stable than adhesive bonded ones at high temperatures. However, spot welding can result in a heat-deformed zone

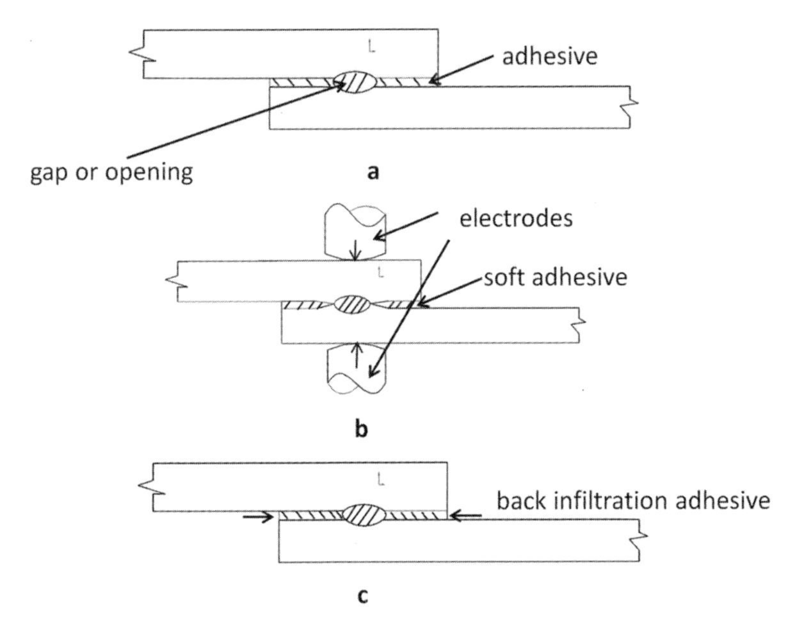

Figure 3.40 A typical example of the weld bonding process

adjacent to the weld nugget, and the spot-welded joint will typically contain residual stresses. So the final sought properties from a weld bonded joint are as follows:

1. increased in-plane tensile shear and/or compressive buckling load-carrying capability under certain joint geometrical configuration;
2. improved out-of-plane load-carrying ability when compared to adhesive bonding only;
3. increased tolerance of transient elevated temperature increase for in- and out-of-plane loading compared to adhesive bonding alone, including increased resistance to static fatigue by creep (or stress-rupture);
4. enhanced load and stress distribution when compared to spot welding alone;
5. increased fatigue life when compared to spot welding only;
6. increased energy absorption capacity;
7. increased environmental compatibility and durability under different chemical or thermal forcings.

Figure 3.40 shows one possible implementation of the weld-bonding process.

Other advances in the welding technologies include different variations of the solid-state bonding techniques such as the roll bonding or the accumulative roll bonding (ARB). Roll bonding or accumulative roll bonding joins the materials by applying intensive plastic deformation at their interfaces through a rolling or extrusion process. Additionally the ARB process results in refined grain micro-structure of the final joint.

The mechanical joining technologies will be discussed in Chapter 5 on the automotive final assembly.

3.7 The Automotive Joining Costing

This section utilizes two specific examples from the BiW panels to illustrate the body-weld sub-operations that relate to building sub-assemblies that will ultimately be fed into the main welding line. The first example features joining a door sub-assembly using adhesive bonding (at the hemmed joints) and spot welding, while the second example discusses the MIG welding of an automotive frame. For each example the different process steps are described and a costing scheme is also discussed.

3.7.1 Joining an Automotive Frame

The automotive frame sub-assembly to be joined is shown in Figure 3.41, illustrating the different weld locations and types. Additionally, Table 3.8 displays the different components involved in this frame assembly, in addition to the different weld joint types and material thicknesses. After studying Table 3.8, one can use the MIG welding schedule to decide on the suitable welding conditions in terms of weld wire feed rates, shielding gas flow rate, and welding current; the suitable welding conditions for each joint location are given in Table 3.9. From Table 3.9 one can compute the total consumables in terms shielding gas in cubic ft, the electrode wire in Lbs, the electric power in kWhr. For example, the electrode consumption is equal to the weld time multiplied by the feed rate, as in Equation 3.10 and the weld time is given in Equation 3.11, while the electrode volume is shown in Equation 3.12, and the power consumption is described by Equation 3.13.

$$t_{weld} = \frac{l_{weld}}{weld_{speed}} \tag{3.10}$$

$$Elect_{cons} = t_{weld} \times feed_{rate} \tag{3.11}$$

$$Elect_{vol} = t_{weld} \times Elect_{cons} \times A_{weld} \tag{3.12}$$

$$P_{consu} = I_{weld} \times V_{arc} \times t_{weld} \tag{3.13}$$

Considerations to be taken include:

1. *The variable cost*, in terms of \$/meter of MIG weld applied, which can be evaluated from the total cost divided by the number of MIG weld meters in this case

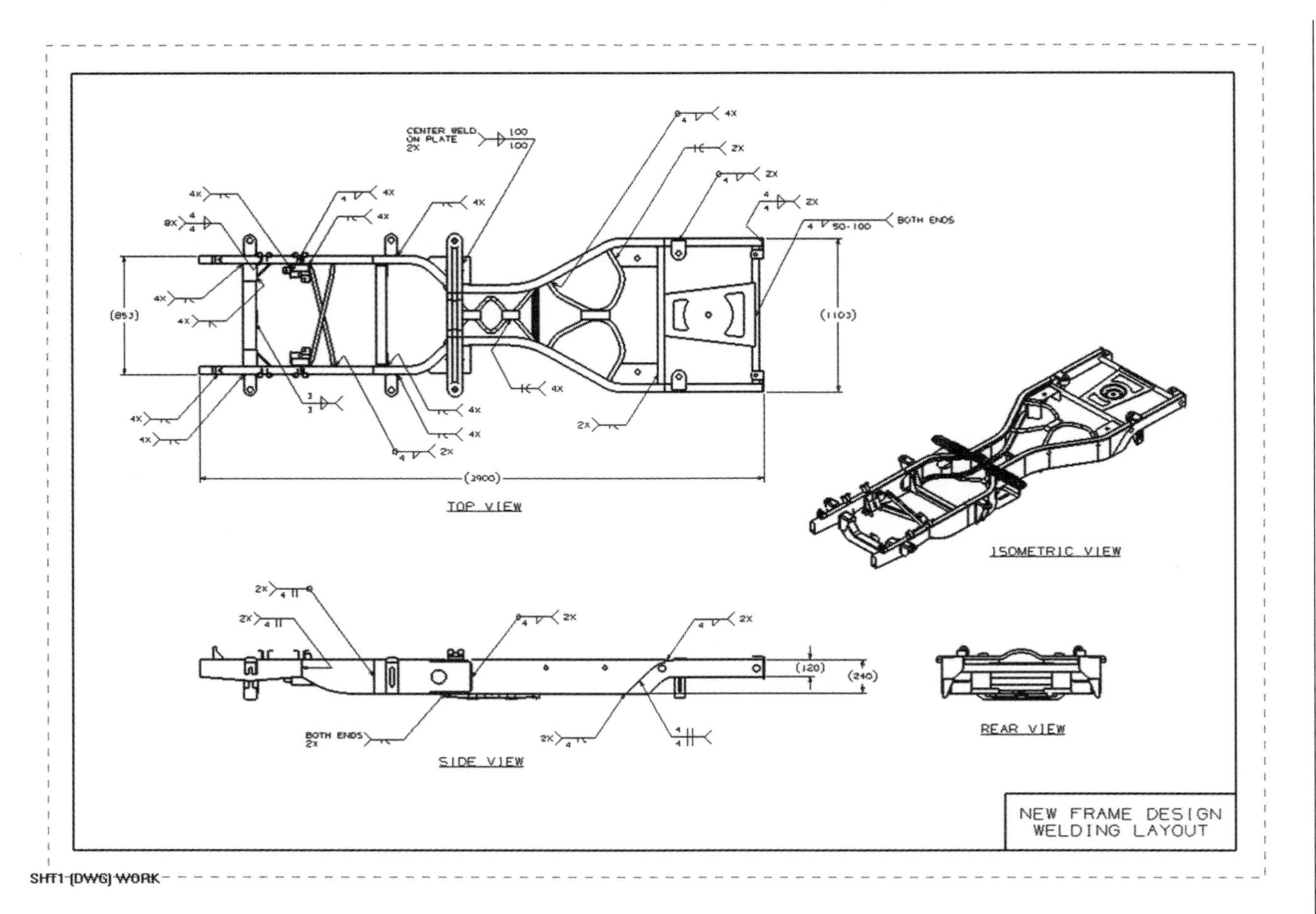

Figure 3.41 The frame design

Table 3.8 The different components involved in this frame assembly

Part ID	Component Name	Thickness (mm)	Joint No.	Joint Components (From/To)	Weld Type	Weld Length (cm)	Weld Length (in)
1	S/M Frame Frt RH	2.25	1	1 to 7	Butt Joint	25	9.84
2	S/M Frame Frt LH	2.25	2	2 to 4	Butt Joint	25	9.84
3	C/M #1	2.00	3	3 to 1 and 3 to 2	Fillet	60	23.62
4	S/M Frame CTR EXT LH	2.25	4	4 to 8	Butt Joint	48	18.90
7	S/M Frame CTR EXT RH	2.25	5	7 to 9	Butt Joint	48	18.90
8	S/M Frame CTR LH	2.90	6	8 to 15	Butt Joint	84	33.07
9	S/M Frame CTR RH	2.90	7	9 to 16	Butt Joint	84	33.07
10	Reinf C/M #3 LH	1.40	8	10 to 8	Fillet	66	25.98
11	Reinf C/M #3 RH	1.40	9	11 to 9	Fillet	66	25.98
12	C/M #3	1.40	10	12 to 8 and 12 to 9	Fillet	25	9.84
13	Reinf Asy-Frame CTR LWR	1.20	11	13 to 13 and 13 to 13; 13 to 8 and 13 to 9	Fillet	70	27.56
14	Reinf Asy-Frame CTR RR	1.50	12	14 to 14; 14 to 8 and 14 to 9	Fillet	60	23.62
15	S/M Frame RR LH	1.20	—	—	—	0	0.00
16	S/M Frame RR RH	1.20	—	—	—	0	0.00
17	Bkrt-Spring MTG LH	3.50	13	17 to 15	Fillet	42	16.54
18	Bkrt-Spring MTG LH	3.50	14	18 to 16	Fillet	42	16.54
19	C/M #6	1.20	15	19 to 15 and 19 to 16	Fillet	25	9.84
20	C/M #2	1.25	16	20 to 27 and 20 to 27	Fillet	36	14.17
22	C/M #5	1.30	17	22 to 15 and 22 to 16	Fillet	25	9.84
23	C/M #4	2.25	18	23 to 32 and 23 to 7	Fillet	36	14.17
25	Tors Bar MNT LH	2.00	19	25 to 10 and 25 to 11	Fillet	50	19.69
26	Tire Carr	1.20	20	26 to 19 and 26 to 22	Fillet	42	16.54
27	C/M #2 Brkt	2.00	21	27 to 8 and 27 to 9	Fillet	72	28.35
28	Eng MNT Brkt LH	3.50	22	28 to 2 and 28 to 3	Fillet	50	19.69
29	Eng MNT Brkt RH	3.50	23	29 to 1 and 29 to 3	Fillet	50	19.69
30	UCA MNT	3.50	24	30 to 1 and 30 to 2	Fillet	25	9.84
31	Shock MNT Brkt	3.50	25	31 to 7	Fillet	18	7.09

Table 3.8 *(continued)*

Part ID	Component Name	Thickness (mm)	Joint No.	Joint Components (From/To)	Weld Type	Weld Length (cm)	Weld Length (in)
32	LWR Cntrl Arm MT	4.00	26	32 to 2	Fillet	12	4.72
33	JNC Bumper RH	3.50	27	33 to 7	Fillet	12	4.72
34	Body MNT #2 Brkt	3.00	28	34 to 4 and 34 to 7	Fillet	60	23.62
35	Bumper Brkt FRT	4.00	29	35 to 1 and 35 to 2	Fillet	40	15.75
36	IDLER Arm MNT	3.00	30	36 to 1 and 36 to 3	Fillet	25	9.84
37	Body MNT #1	3.00	31	37 to 1 and 37 to 2	Fillet	40	15.75
38	C/M #1 Reinf	2.00	32	38 to 3	Fillet	15	5.91
39	JNC Bumper LH	3.50	33	39 to 4	Fillet	25	9.84
40	JNC Bumper Reinf RH	3.50	34	40 to 7	Fillet	40	15.75
41	JNC Bumper Reinf LH	3.50	35	41 to 4	Fillet	30	11.81
42	Rear Bumper Brkt LH	3.00	36	42 to 15	Fillet	40	15.75
43	Rear Bumper Brkty RH	3.00	37	43 to 16	Fillet	40	15.75
44	Axle MNT LH	3.50	38	44 to 3	Fillet	25	9.84
45	Axle MNT RH	3.50	39	45 to 29	Fillet	15	5.91
46	Reinf Brkt Eng LH	3.50	40	46 to 47 and 46 to 4	Fillet	36	14.17
47	Reinf Brkt Eng RH	3.50	41	47 to 29	Fillet	15	5.91
48	Reinf Tapping Plate	3.50	42	48 to 15	Butt Joint	60	23.62
49	Reinf Frame S/B Cap	2.50	43	49 to 16	Butt Joint	60	23.62
50	Shock MNT Brkt	3.50	44	50 to 17 and 50 to 18	Fillet	35	13.78
51	Body MNT-4	3.00	45	51 to 15 and 51 to 16	Fillet	50	19.69
52	CM3 Reinf LH-2	1.20	46	52 to 8 and 52 to 12	Fillet	45	17.72
53	CM3 Reinf RH-2	1.20	47	53 to 9 and 53 to 12	Fillet	45	17.72
54	Track Bar Brkt	3.00	48	54 to 15	Fillet	42	16.54
55	Lwr R-Cntrl Arm Brkt	2.50	49	55 to 16	Fillet	42	16.54
56	Upp R-Cntrl Arm Brkt	2.50	50	56 to 15 and 56 to 16	Fillet	25	9.84
57	C/M #7	1.80	51	57 to 13 and 57 to 13; 57 to 8 and 57 to 9	Fillet	36	14.17

Table 3.9 The suitable welding conditions for each joint location

Part ID	Component Name	Electrode Diameter (mm)	Welding Current (A,dc)	Wire Feed (in/min)	Wire Feed (cm/min)	Travel Speed (in/min)	Travel Speed (cm/min)	CO_2 Gas (cfh)	CO_2 Gas Required (cf)
1	S/M Frame Frt RH	1.1	330	350	889	105	267	32.5	0.05
2	S/M Frame Frt LH	1.1	330	350	889	105	267	32.5	0.05
3	C/M #1	1.1	330	350	889	105	267	32.5	0.12
4	S/M Frame CTR EXT LH	1.1	330	350	889	105	267	35.5	0.11
7	S/M Frame CTR EXT RH	1.1	330	350	889	105	267	36.5	0.11
8	S/M Frame CTR LH	1.1	270	340	863.6	180	457	22.5	0.07
9	S/M Frame CTR RH	1.1	270	340	863.6	180	457	22.5	0.07
10	Reinf C/M #3 LH	1.1	270	340	863.6	180	457	22.5	0.05
11	Reinf C/M #3 RH	1.1	280	350	889	190	483	22.5	0.05
12	C/M #3	1.1	280	350	889	190	483	22.5	0.02
13	Reinf Asy-Frame CTR LWR	1.1	280	350	889	190	483	22.5	0.05
14	Reinf Asy-Frame CTR RR	1.1	280	350	889	190	483	22.5	0.05
15	S/M Frame RR LH	1.1	280	350	889	190	483	22.5	0.00
16	S/M Frame RR RH	1.1	280	350	889	190	483	22.5	0.00
17	Bkrt-Spring MTG LH	1.1	280	350	889	190	483	22.5	0.03
18	Bkrt-Spring MTG LH	1.1	280	350	889	190	483	22.5	0.03
19	C/M #6	1.1	280	350	889	190	483	22.5	0.02
20	C/M #2	1.1	280	350	889	190	483	22.5	0.03
22	C/M #5	1.1	280	350	889	190	483	22.5	0.02
23	C/M #4	1.1	325	360	914.4	130	330	32.5	0.06
25	Tors Bar MNT LH	1.1	280	350	889	190	483	22.5	0.04

Table 3.9 (*continued*)

Part ID	Component Name	Electrode Diameter (mm)	Welding Current (A,dc)	Wire Feed		Travel Speed		CO_2 Gas (cfh)	CO_2 Gas Required (cf)
				(in/min)	(cm/min)	(in/min)	(cm/min)		
26	Tire Carr	1.1	280	350	889	190	483	22.5	0.03
27	C/M #2 Brkt	1.1	325	360	914.4	130	330	32.5	0.12
28	Eng MNT Brkt LH	1.1	325	360	914.4	130	330	32.5	0.08
29	Eng MNT Brkt RH	1.1	325	360	914.4	130	330	32.5	0.08
30	UCA MNT	1.1	325	360	914.4	130	330	32.5	0.04
31	Shock MNT Brkt	1.1	325	360	914.4	130	330	32.5	0.03
32	LWR Cntrl Arm MT	1.1	325	360	914.4	130	330	32.5	0.02
33	JNC Bumper RH	1.1	325	360	914.4	130	330	32.5	0.02
34	Body MNT #2 Brkt	1.1	325	360	914.4	130	330	32.5	0.10
35	Bumper Brkt FRT	1.1	325	360	914.4	130	330	32.5	0.07
36	IDLER Arm MNT	1.1	325	360	914.4	130	330	32.5	0.04
37	Body MNT #1	1.1	325	360	914.4	130	330	32.5	0.07
38	C/M #1 Reinf	1.1	325	360	914.4	130	330	32.5	0.02
39	JNC Bumper LH	1.1	325	360	914.4	130	330	32.5	0.04
40	JNC Bumper Reinf RH	1.1	325	360	914.4	130	330	32.5	0.07
41	JNC Bumper Reinf LH	1.1	325	360	914.4	130	330	32.5	0.05
42	Rear Bumper Brkt LH	1.1	280	350	889	190	483	22.5	0.03
43	Rear Bumper Brkty RH	1.1	280	350	889	190	483	22.5	0.03

(*continued overleaf*)

Table 3.9 (*continued*)

Part ID	Component Name	Electrode Diameter (mm)	Welding Current (A,dc)	Wire Feed (in/min)	Wire Feed (cm/min)	Travel Speed (in/min)	Travel Speed (cm/min)	CO_2 Gas (cfh)	CO_2 Gas Required (cf)
44	Axle MNT LH	1.1	325	360	914.4	130	330	22.5	0.03
45	Axle MNT RH	1.6	380	210	533.4	85	216	32.5	0.04
46	Reinf Brkt Eng LH	1.1	325	360	914.4	130	330	32.5	0.06
47	Reinf Brkt Eng RH	1.6	380	210	533.4	85	216	32.5	0.04
48	Reinf Tapping Plate	1.1	280	350	889	190	483	22.5	0.05
49	Reinf Frame S/B Cap	1.1	280	350	889	190	483	22.5	0.05
50	Shock MNT Brkt	1.6	380	210	533.4	85	216	32.5	0.09
51	Body MNT-4	1.1	280	350	889	190	483	22.5	0.04
52	CM3 Reinf LH-2	1.1	280	350	889	190	483	22.5	0.03
53	CM3 Reinf RH-2	1.1	280	350	889	190	483	22.5	0.03
54	Track Bar Brkt	1.1	280	350	889	190	483	22.5	0.03
55	Lwr R-Cntrl Arm Brkt	1.1	280	350	889	190	483	22.5	0.03
56	Upp R-Cntrl Arm Brkt	1.1	280	350	889	190	483	22.5	0.02
57	C/M #7	1.1	280	350	889	190	483	22.5	0.03

Table 3.10 The basic assumptions for welding costs

Hour of Production / year	5000
Vehicles per year	220000
Typical Robot Efficiency (%)	70.00
Labor Wage (Wages and Benefits) $ per hr	30.00
Total Number of Material Handeling Robots	10
General Robot Rate $ per hr	6.00
Lifetime Program Investment (years)	5.00
Labor Wage/hour (in $)	30
No of Labor	25
No of Robots used	28
Shielding Gas Cost for one cyclinder ($)	500
Volume of one cylinder (ft^3)	300
Power Cost ($ per kWh)	0.20
Cost of Shielding Gas per bottle ($)	40.00
Electrode Cost ($ per lb)	2.55
Tooling investment ($)	2000000
Building, Maintenance and other cost ($)	800000

around 20 meters of MIG. Also, one can use the $/vehicle to help estimate the added cost if more vehicles are to be made.

2. *The fixed cost*, in terms of building maintenance and insurance, and tooling investment, etc. The fixed cost is computed here based on typical information from Table 3.10.
3. *The number of robots* needed to complete the frame welding, given the number of production hours per year, the production volume (number of frames per year), and the up-time calculation. The up-time is decided based on computing the available time i.e. number of productive hours—the time needed for maintenance (e.g. changing gas shielding cylinders, electrode wire, etc.)—the timed needed for robots to move from one point to the other without welding (robot inefficiency).
4. For example, the number of frames to be made in one minute, which is the same as the number of vehicles produced in one minute, is based on Equation 3.14.

$$No.Vehicles_{min} = \frac{hrs_{year} \times 60}{No.Vehicles_{year}} \tag{3.14}$$

Conducting the calculation results in a total MIG weld cost of approximately $40/ frame based on a frame sub-assembly area that has around 28 robots (between

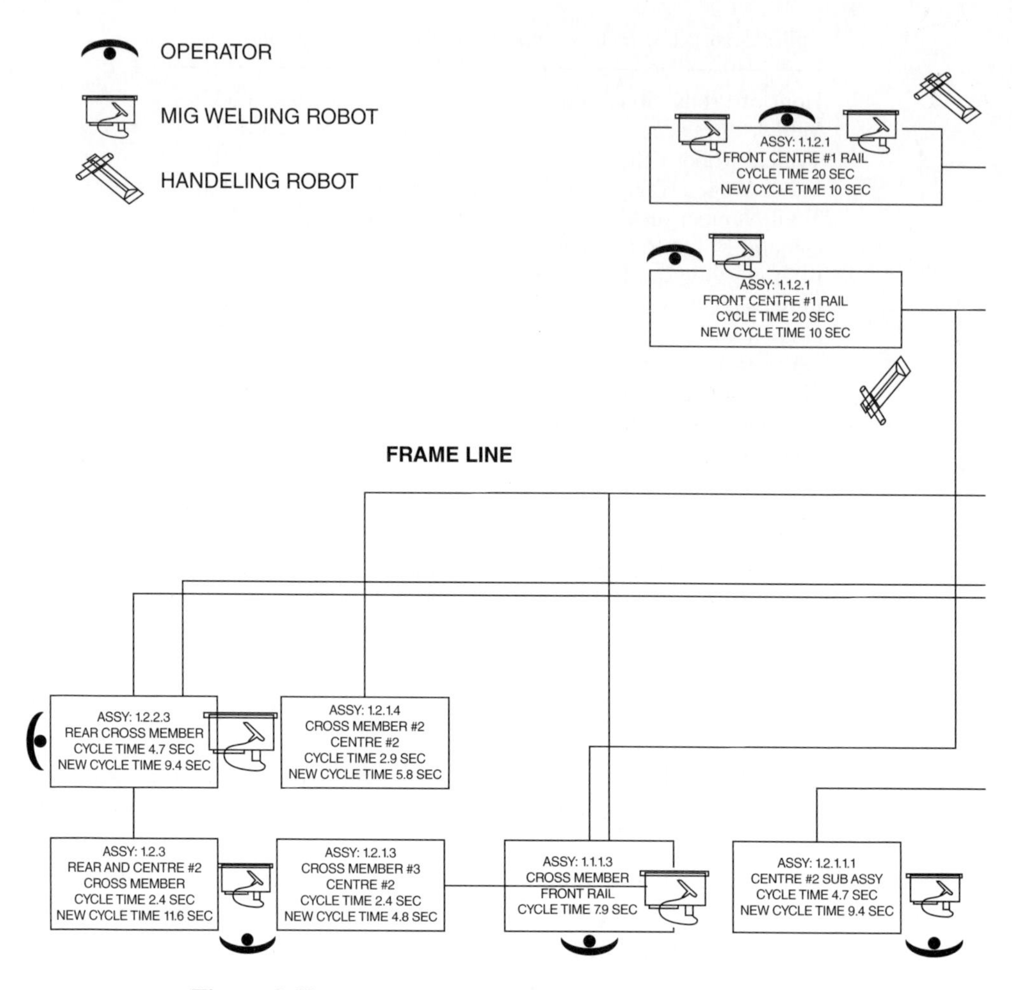

Figure 3.42 A proposed layout for the frame welding line

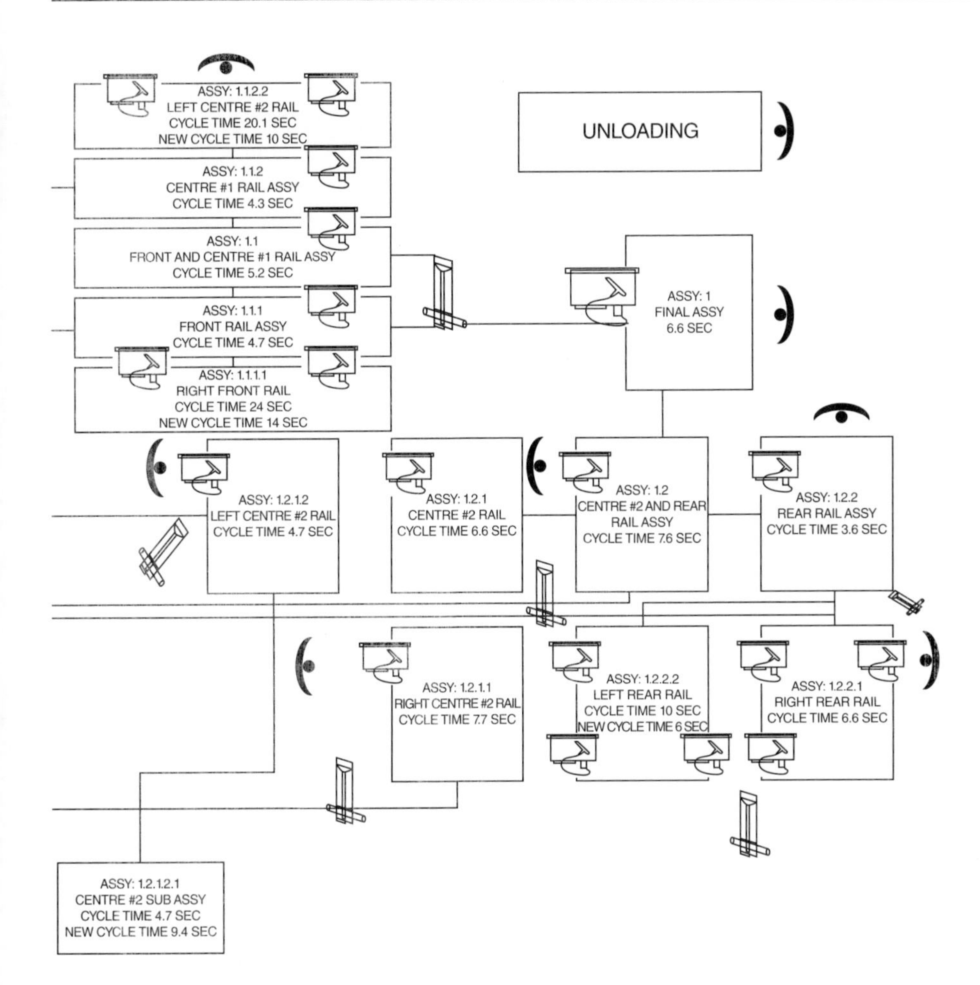
ASSY: 1.1.2.2
LEFT CENTRE #2 RAIL
CYCLE TIME 20.1 SEC
NEW CYCLE TIME 10 SEC
UNLOADING
ASSY: 1.1.2
CENTRE #1 RAIL ASSY
CYCLE TIME 4.3 SEC
ASSY: 1.1
FRONT AND CENTRE #1 RAIL ASSY
CYCLE TIME 5.2 SEC
ASSY: 1
FINAL ASSY
6.6 SEC
ASSY: 1.1.1
FRONT RAIL ASSY
CYCLE TIME 4.7 SEC
ASSY: 1.1.1.1
RIGHT FRONT RAIL
CYCLE TIME 24 SEC
NEW CYCLE TIME 14 SEC
ASSY: 1.2.1.2
LEFT CENTRE #2 RAIL
CYCLE TIME 4.7 SEC
ASSY: 1.2.1
CENTRE #2 RAIL
CYCLE TIME 6.6 SEC
ASSY: 1.2
CENTRE #2 AND REAR RAIL ASSY
CYCLE TIME 7.6 SEC
ASSY: 1.2.2
REAR RAIL ASSY
CYCLE TIME 3.6 SEC
ASSY: 1.2.1.1
RIGHT CENTRE #2 RAIL
CYCLE TIME 7.7 SEC
ASSY: 1.2.2.2
LEFT REAR RAIL
CYCLE TIME 10 SEC
NEW CYCLE TIME 6 SEC
ASSY: 1.2.2.1
RIGHT REAR RAIL
CYCLE TIME 6.6 SEC
ASSY: 1.2.1.2.1
CENTRE #2 SUB ASSY
CYCLE TIME 4.7 SEC
NEW CYCLE TIME 9.4 SEC

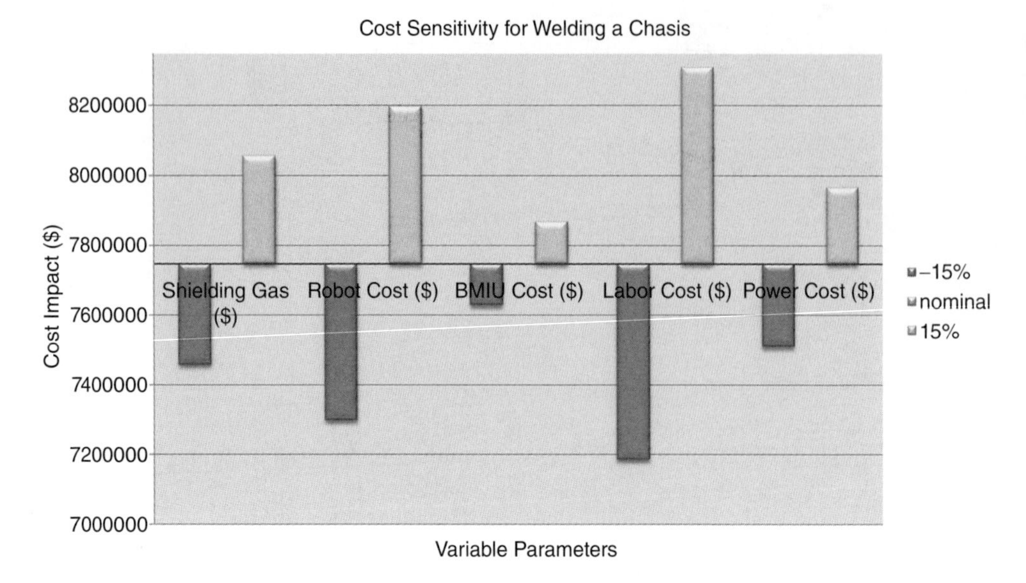

Figure 3.43 A welding line cost sensitivity plot

welding and material handling robots). The number of robots is decided on based on the production rate (task time) and the available welding time. Also, one can propose a robotic layout for this sub-assembly area as depicted in Figure 3.42. Further cost analysis includes a sensitivity analysis based on whether the cost of the contributing factors fluctuated between +/– 15%, which is plotted in Figure 3.43.

3.7.2 Sub-assembling Automotive Doors

In this case study, the calculations are done for a standard door composed of a door inner, door outer panels, a window frame panel, in addition to some reinforcements such as the crash bar and a hinge reinforcement panel. The joining process is composed of spot welding and adhesive bonding. The adhesive bonding is applied at the hemmed flanges that join the inner and outer panels. A typical door sub-assembly is shown in Figure 3.44. Following the door design for each weld location can establish the weld conditions from the spot-welding schedule (Table 3.11) and the spot welding design guidelines in regard to electrode diameter, spacing and flange dimensions as depicted in Table 3.12. Additionally the cost calculations can be done using a similar approach that of the previous case study (frame), in addition to accounting for the adhesive material cost, final calculations are shown in Table 3.13.

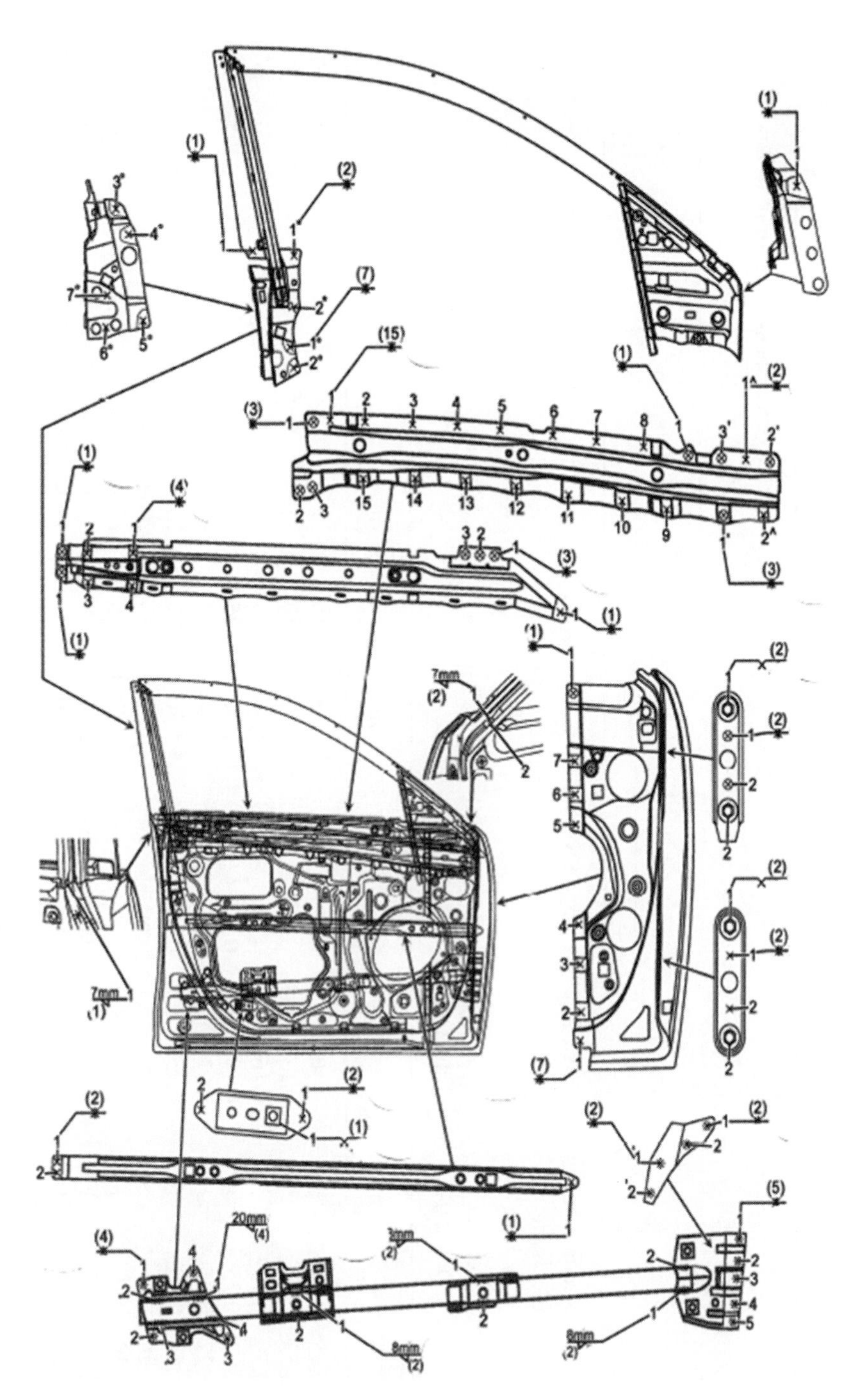

Figure 3.44 A typical door weld

Table 3.11 Spot-welding schedule

Thinnest Material Thickness		Recommended Electrode Contact Diameter		Fused Zone Diameter Expected		Recommended Flange Overlap (Minimum)		Recommended Single Spot-Weld Spacing		Expected shear strength in lbs/spot (kg/spot)					
										Aluminum Alloy 5052-H34		Carbon Steel 70 ksi		Stainless Steel 90 ksi	
in	mm	in	mm	in	mm	in	mm	in	mm	lbs	kg	lbs	kg	lbs	kg
0.010	0.25	0.13	3.0	0.10	2.5	0.38	10.0	0.75	19	154	70	286	130	374	170
0.021	0.53	0.19	5.0	0.13	3.0	0.44	11.0	0.75	19	264	120	704	320	1034	470
0.031	0.79	0.19	5.0	0.16	4.0	0.44	11.0	1.00	25	528	240	1254	570	1760	800
0.040	1.02	0.25	6.5	0.19	5.0	0.50	13.0	1.00	25	748	340	2024	920	2794	1270
0.050	1.27	0.25	6.5	0.22	5.5	0.56	14.0	1.00	25	1122	510	2970	1350	3740	1700
0.062	1.57	0.25	6.5	0.25	6.5	0.63	16.0	1.20	31	1584	720	4070	1850	5280	2400
0.078	1.98	0.31	8.0	0.29	7.5	0.69	17.5	1.50	38	2068	940	5940	2700	7480	3400
0.094	2.39	0.30	8.0	0.31	8.0	0.75	19.0	1.80	46	2596	1180	7590	3450	9240	4200
0.109	2.77	0.31	8.0	0.32	8.0	0.81	20.5	2.20	56	3036	1380	9130	4150	11000	5000
0.125	3.18	0.38	10.0	0.33	8.5	0.88	22.5	2.50	64	3564	1620	11000	5000	13200	6000
0.180	4.57	0.38	10.0	0.35	9.0	1.00	25.0	2.50	64	3784	1720	11660	5300	14520	6600

Table 3.12 The spot-welding design guidelines in regard to electrode diameter, spacing and flange dimensions

Panel #	Material Thickness (mm)	Recommended Electrode Contact Diameter (mm)	Fused Zone Diameter Expected (mm)	Recommended Flange Overlap (Minimum) (mm)	Recommended Single Spot-weld Spacing (mm)
1	2.0	8.0	7.5	17.5	38
2	2.0	8.0	7.5	17.5	38
3	1.2	3.0	2.5	10.0	19
4	1.2	3.0	2.5	10.0	19
5	1.7	6.5	6.5	16.0	31
6	3.5	10.0	8.5	22.5	64
7	3.5	10.0	8.5	22.5	64
8	1.2	3.0	2.5	10.0	19
9	2.9	10.0	8.5	22.5	64
10	1.7	6.5	6.5	16.0	31
11	1.7	6.5	6.5	16.0	31

Table 3.13 The result of the final calculations for spot welding

Vehicles per year	220000
No. of Doors	4
NO. of Labor	5
Doors per years	880000
Lifetime programe investment (years)	5
Perimeter of the door (mm)	3750
Area to be applied with adhesive for a perimeter of length (mm)	6.8
Volume of space to be applied with adhesive (ml)	25500
Lord 320/322 Epoxy ($) for 50 ml	9.41
Time taken for adhesive applying (sec)	31
Time taken for hemming (sec)	18
Anti Flutter (ml)	6
No of Robots required/vehicle	46
No of spot welding robots Robots required/vehicle	6
Total Cost of Robots/door	0.001832669971
Total Power consumed (Spot + MIG) kW	33.91
Power Consumed/door (kWh)	0.621366754
Power cost/door ($)	0.12

Exercises

Problem 1

In your own words, explain the following statements about sheet metal joining:

- Continuous dressing (maintenance) of the spot electrode tips is necessary to ensure a good weld.
- Arc welding weakens the welded assemblies, even though the weld itself is as strong as the based metal.
- The pulse material transfer in arc welding is considered more favorable to the globular mode.
- Shielding gas CO_2 is not used for welding aluminum or magnesium assemblies.
- The selection of automotive adhesives is governed by wide range of requirements.
- The spot weld pitch should be at least 10 times the panel thickness.
- The automotive substrates' surface energy should be evaluated before joining adhesives are selected.
- MIG welding application requires a shielding gas.

Problem 2

Using the space provided, plot the resistance curve for a typical spot welding, while indicating the different stages of the nugget growth.

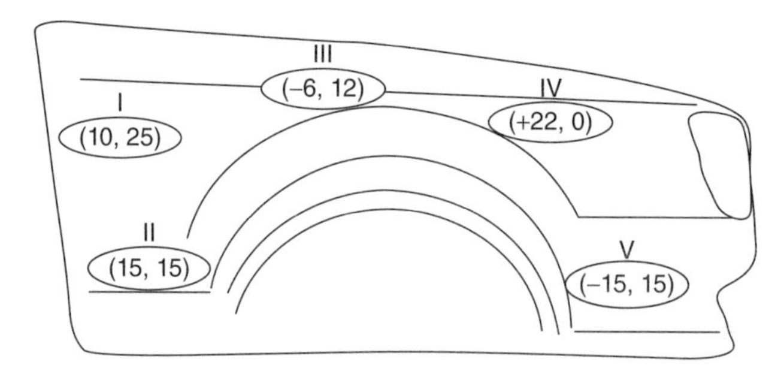

Problem 3

After inspecting the product design shown below, answer the following questions:

1. Decide on the spot-welding configurations, which include: electrode-tip diameters, welding pitch S, welding spacing from the edge W.
2. What would you change if the upper panel is made of copper while the lower panel is made of steel (provide *two* options)?
3. Decide on the suitable welding lobe from A, B, C, or D for the following materials: high carbon steel, aluminum, if you know that lobe B is used to weld a low-carbon steel.
4. For a material weldable with lobe A, decide on the minimum and the maximum welding power in watt, if the interface resistance is evaluated at 30 micro-Ohm.
5. If the spot welds were replaced with a MIG weld: (a) re-design the part shown, and then compute (b) the consumed electrode volume, and (c) the weld tolerance.

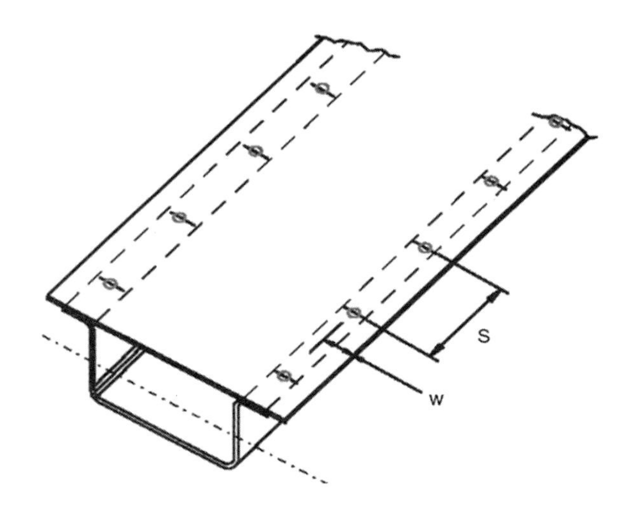

Lower panel thickness = 0.05 in

Upper panel thickness = 0.05 in

Problem 4

Based on the welding lobe shown below, and given following information: the interface resistance is evaluated at 40 micro-Ohm, the electrode force = 150 Kg_f, and the panels being welded are made of mild steel.

1. Evaluate the minimum amount of energy in Watt needed to form a nugget.
2. If the electrode force was increased from 150 Kg_f, to 200 Kg_f leading to a decrease in interfacial resistance to about 25 micro-Ohm, draw the new welding lobe on the sketch below.
3. If the material being welded is changed from mild steel to Aluminum.

First, describe the weld (nugget) condition if the same welding lobe was used. Second, suggest the change to the spot-welding scheme (lobe and procedure) to guarantee a good weld.

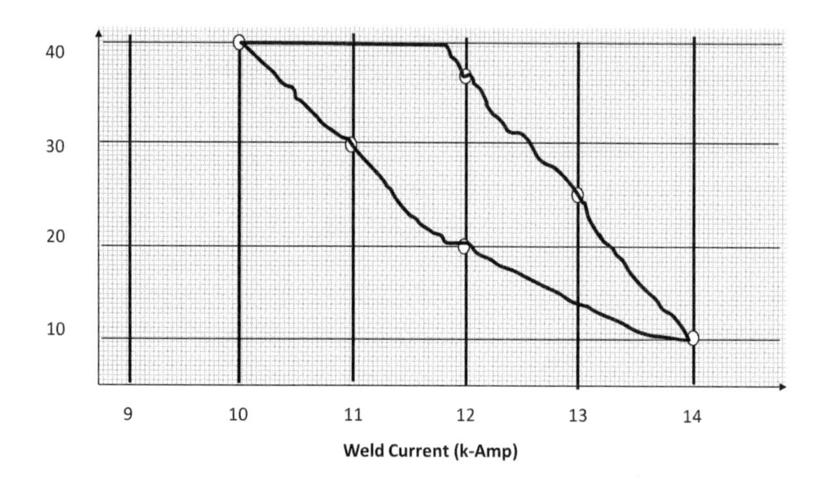

Problem 5

Based on the welding lobes (A and B), answer the following questions:

- Which welding lobe has a *higher* electrode pressure force?
- Which welding lobe is *more suitable* for welding stainless steel, if the minimum required current is 11 k-Amp?
- Which welding lobe is *more suitable* for welding low carbon alloy steel?
- If lobe A was used to weld a material typically welded using lobe B, comment on the weld conditions.
- Evaluate the minimum amount of energy in *Joules* needed: (a) to form a nugget for each lobe; and (b) to weld expulsion; given that the interfacial resistance is evaluated at 40 micro-Ohm.

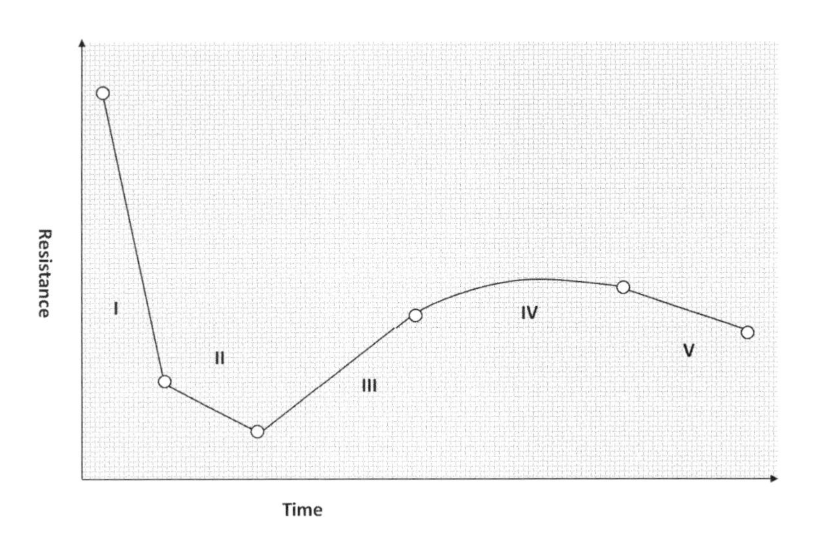

Problem 6

After inspecting the resistance curve for a typical spot welder (substrate is mild steel), answer the following questions:

1. Indicate (on the graph) the *weld expulsion* point.
2. Indicate any possible change(s) on the graph below, for the following scenarios:
   i. the electrode force increased from 150 Kg_f to 200 Kg_f;
   ii. the panel roughness decreased;
   iii. the material changed from mild steel to aluminum.
3. If regions III, IV were extended over longer time, what would be the effect on the spot-welding lobe?
4. If region I extended over a longer time, what would be the effect on the spot-welding lobe?

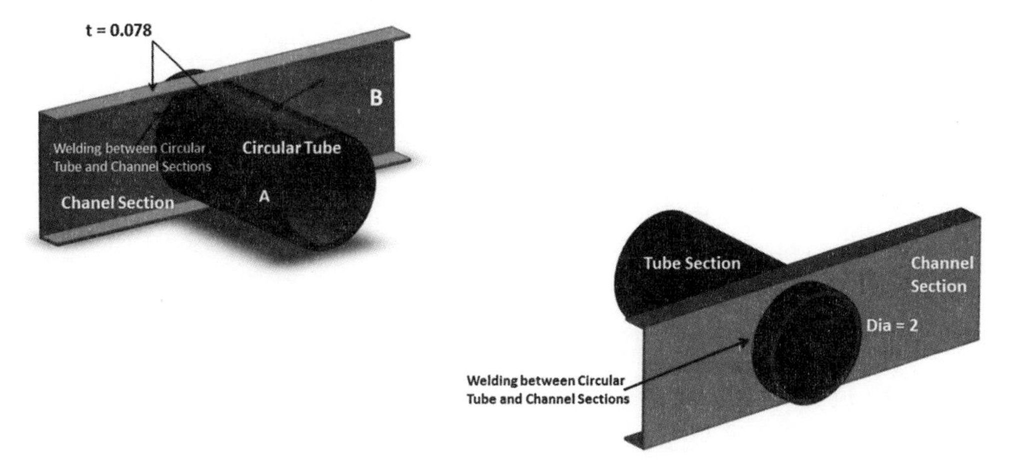

Problem 7

Knowing that tube A is joined to channel B through arc welding, answer the following questions:

1. What is the amount of electrode (filler) in pounds, needed to complete this weld?
2. Draw the proper welding symbol.
3. Indicate the weld tolerance.

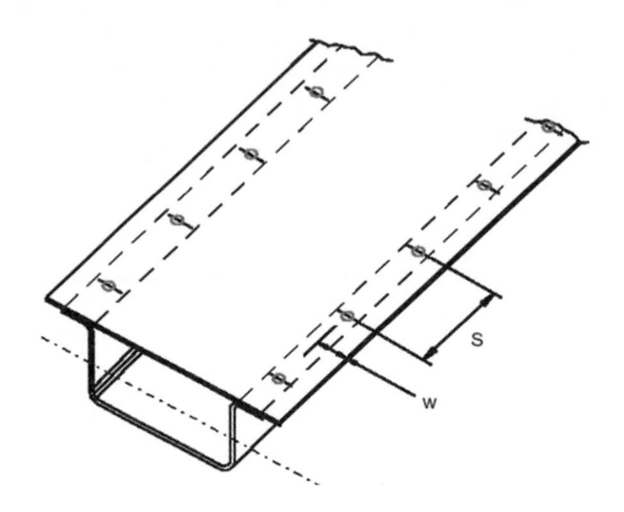

Problem 8

Explain the corrective action for following weld problems:

- lack of penetration;
- porosity in weld bead.

4

Automotive Painting

4.1 Introduction

The painting process applies the different protective layers needed to ensure that the BiW is corrosion-resistant. This includes not only the protective paint films on the vehicle shell but also the necessary under-body sealer, PVC and wax applications. Also the painting process controls the final look of the vehicle shell through its color and gloss characteristics. The main layout of the paint line is shown in Figure 4.1, which displays the different activities within the paint-lines.

4.2 Immersion Coating Processes

The first process in the painting sequence is the conditioning and cleaning of the BiW arriving from the body-weld line. The cleaning process is intended to clean and remove all the steel mill oils, stamping lubricants, and other welding sludge and contaminants. These contaminants not only affect the appearance of the applied paint films but also the paint layers adhesion and integrity, because any surface contaminants affects the panels' surface energy.

The second step in the pretreatment process (after cleaning) is called the conditioning process which helps to promote the paint film adhesion, and reduce the metal to paint reaction, particularly with galvanized surfaces, and finally improving the corrosion resistance, more specifically the blister resistance. Metal surfaces constitute different contaminants with different chemistries such as the ones shown in Figure 4.2.

The conditioning process is typically composed of the following functional steps:

- cleaning
- rinsing

The Automotive Body Manufacturing Systems and Processes, First Edition. Mohammed A. Omar.

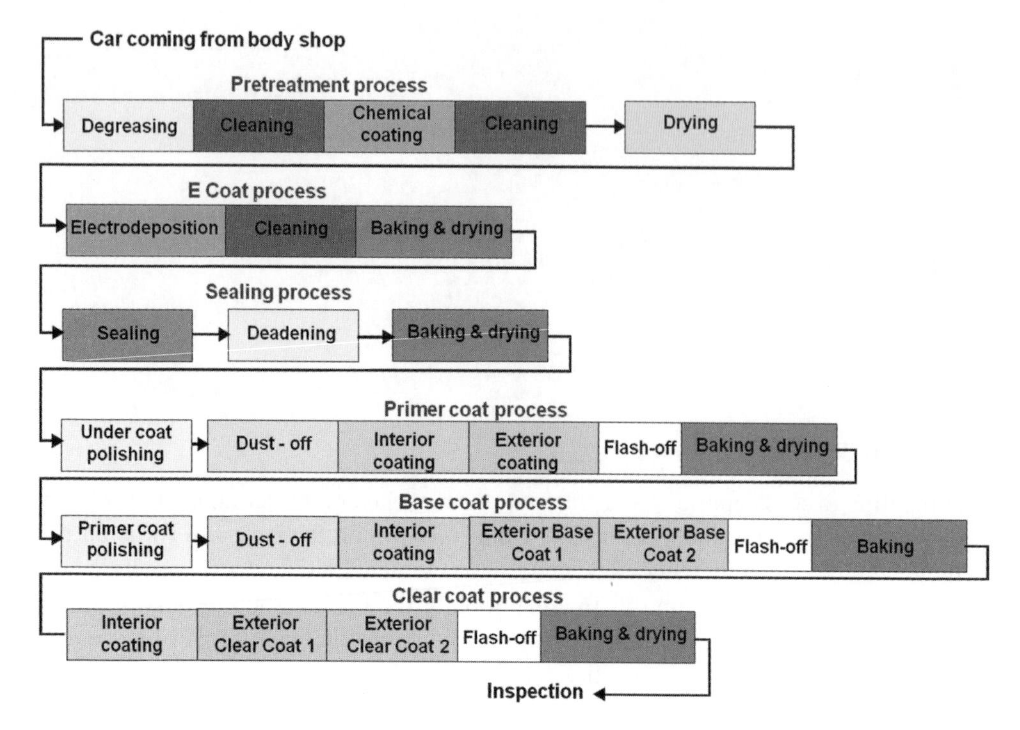

Figure 4.1 Typical activities in a painting booth

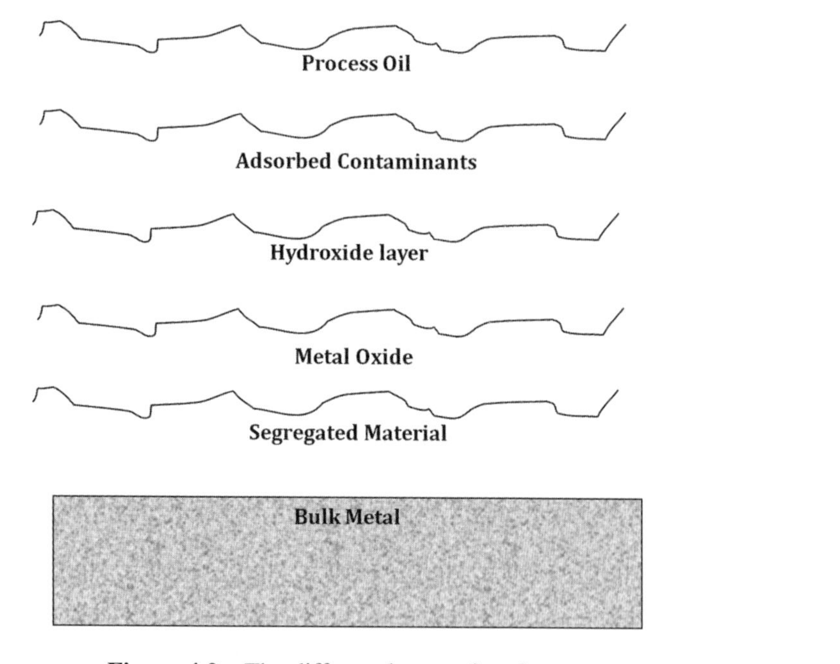

Figure 4.2 The different layers of surfaces

- conditioning
- conversion coating
- rinsing
- post-treatment, and de-ionized water rinsing.

4.2.1 Cleaning

The cleaning process can be done using different cleaning agent formulations but the most common cleaning solution is based on an aqueous cleaning method, which is typically composed of water in addition to cleaning detergents with amounts of acid or alkali. This method has no solvent emissions and can be readily applied to different panel materials (steel, aluminum), in addition to its effectiveness for different types of oils and contaminants. The aqueous cleaning can be done in two implementations: through spray cleaning and/or dip (or immersion) cleaning. In spray cleaning, the panels are pressure-spayed to impact a mechanical displacement action, thus removing any oils or contaminates. Spray cleaning is fast due to its mechanical scrubbing action; however, this process is not suitable for complicated geometries because of its limited reach and coverage of holes, channel/tubes cross-sections.

In immersion cleaning, the BiW is lowered into the cleaning agent bath, where the aqueous solution covers all the panels' surfaces and penetrates through hidden corners or holes. However, the dip cleaning approach is slower and continuous monitoring of the bath chemistry is required. Other considerations when selecting either of these two implementations is the equipment capital cost and investment, for example, the spray system requires a variety of pumps, valves, and pipe installations in addition to its higher energy consumption when compared with the dip baths.

Most of the automotive OEMs utilize a dip-based cleaning with spray intermediate steps to combine the advantages of the two approaches. In general, the aqueous cleaning process is composed of the following phases [35]:

1. The *wetting* process, which helps in reducing the panel surface tension using surfactants so that the oil and contaminants disappear from the panel surface due to their higher surface tension.
2. *Emulsification*, which is the next step after the wetting that results in suspending the oils and contaminants that just lift off from the surface, as a stable layer to facilitate its removal.
3. *The saponification* process constitutes the reaction between the alkaline cleaning agents with the oils and contaminants layer to result in soaps.
4. The *solubilization* process which helps in the dissolution of soluble soils.
5. The *displacement* step is designed to remove the suspended oil layers, soaps, and the solidified soils.

6. The *agitation* process includes the pressure-spray to further help in the displacement process and further remove any contaminants still on the panel surface.
7. The *sequestration* process is the deactivating step for the salts which might react with the cleaning agent or degrade the performance of the wetting surfactants
8. Finally, the *deflocculation* phase is where the soils are broken down and suspended to maintain a stable dispersion, thus preventing their agglomeration and any possible re-deposition over the clean panel surfaces.

One of the most common aqueous solution formulations is the alkaline detergent cleaning formulation, which is used to clean different substrates including different types of metal (aluminum, steel) in addition to plastic panels. These formulations are gaining more acceptance in the automotive industry because of their low phosphate and hazardous air pollutants (HAP), as these can affect the paint booth environmental audits negatively. The alkali concentration controls these formulations' ability to remove the oils, additionally, the higher concentration lead to better soil/oil dispersion which consequently leads to a longer service life and a cleaner surface performance.

In the case of cleaning aluminum substrates, OEMs might use specific formulations to ensure good cleaning performance, aluminum panels can be readily cleaned using aqueous chemical cleaners. Additionally, non-silicated alkaline cleaners can also be used.

The alkaline solution can be formulated using a variety of alkaline formulations such as sodium hydroxide, sodium carbonate or silicates. Additives are also added to the formulation to remove any hard water effects due to certain salts and to help in the emulsification phase and reduce the foaming through changing the solution cloud point. The cloud point of a cleaner is the temperature above which foaming is insignificant, the typical value for a cleaner is around 44 °C. Other aqueous cleaner formulations include emulsion-based cleaners, which have kerosene or similar organic solvents in their formulations. Such cleaners are suitable for removing a limited range of fatty oils and are considered flammable due to the solvent.

4.2.2 Rinsing

The rinsing process is done to reduce the amount of the remaining cleaner formulation on the BiW surfaces after the cleaning process is completed, in addition to eliminating the carry-over from one bath to the next, because the cleaning process suspends the oils and contaminates in their baths and any carry-over of these suspended layers will lead to re-polluting the surfaces and might also to degrading the next conditioning step performance. The rinsing is vital to ensure that each conditioning bath contains the designed chemistry. For example, any residuals from the cleaner formulation can lead to incomplete reactions in subsequent steps. To ensure good rinsing action, multiple rinsing stages might be used. The rinsing process sis typically achieved through a recirculating de-ionized water DI/RO rinse, with flow rate requirement of typically 5 cubic cm (cc) for each panel square foot. The de-ionized water is prepared through the use of rechargeable ion-exchangers, to remove all the positive charge carriers, i.e. the cations such as the sodium, calcium and iron. Also a second exchanger bed is designed to remove the negative charge carriers i.e. the anions, such as chlorides, sulfates, and phosphates.

4.2.3 Conversion and Phosphate Baths

The BiW structure will go through a phosphate bath to receive an insoluble salt-like conversion coating layer after the conditioning phase. The conversion coating is called "conversion" because the iron layer is converted through chemical reactions in the tank, into an iron phosphate or a zinc phosphate inorganic layer. The iron or zinc phosphate acts as a shield between the fresh metal substrate and the electrolyte solutions (air, water) to prevent the establishment of a corrosion cycle at the panels' surfaces. Additionally the phosphate coating provides a rougher surface profile which helps increase the surface area, thus promoting subsequent paint layers' adhesion to the surface. The conversion coating features a variety of surface pores, and undercuts that help in forming features into which the paint can flow and adhere to the surface. Furthermore, some conversion coating formulations leave a trace of acidic pH which helps in the paint adhesion; additionally the conversion can also neutralize any residuals of an alkaline pH. Alkaline solutions can degrade the paint resins and are typically utilized in paint removers and strippers' formulations.

Additional desirable effects achieved from the conversion coating films are due to their thermal expansion coefficient which serves as a buffer to ease the expansion and contraction differences between the paint and the substrate, this reduces the build-up of residual stresses at the paint–substrate interface. Also, the conversion coating reduces the lateral creep corrosion especially in the areas around the edges and scratches. Lateral creep is caused by the fact that when the steel oxidizes, it generates hydroxyl ions that are alkaline in nature, thus such ions degrade the paint film adhesion to the steel substrate leading to under-film corrosion. However, having a trace of acidity from the conversion coating film, it neutralizes any alkaline generation at those edges, preventing the corrosion from creeping under the paint layers.

The phosphating process is achieved by immersing the BiW in an acidic phosphate bath to convert the steel upper layer into one of two phosphate compounds, the first is the iron phosphate which requires thin layers of conversion coating, to deliver acceptable performance in terms of corrosion resistance. However, the iron phosphate is the more cost effective choice. The other phosphating choice is the zinc phosphate which has thinner films but at the same time has superior performance over that of the iron phosphate films. Most of the automotive OEMs utilize the zinc phosphating process due to its performance even though it is more costly than the iron phosphating scheme. The conversion coating layer is typically described in terms of the applied weight per unit surface area due to its rough profile which complicates any thickness measurements. Typical weights for iron phosphates are from 160–645 mg/m^2, for zinc coating typical weights are around 1,076– 3,764 mg/m^2.

The phosphating process starts with a chemical reaction/attack by the phosperic acid in the bath, on the clean steel surface. The phosphoric acid oxidizes the iron atoms and then makes them soluble. The newly soluble iron reduces the strength of the phosphoric acid by lowering its concentration at the area surrounding the panels' surfaces. Type 1 primary phosphate salt converts itself into a type 2 mono-hydrogen phosphate salt, which turns instantly into a type 3 tertiary phosphate. As described, the reactions are shown in Equation 4.1 [33]:

$$H_2PO_4^{-1} \xrightarrow{yields} HPO_4^{-2} \xrightarrow{yields} PO_4^{-3} \tag{4.1}$$

where, iron phosphate: $FePO_4$ and $Fe_3(PO_4)_2$, for zinc phosphate: $Zn_3(PO_4)$.

The subsequent reactions follow the sequence in Equations 4.2 through 4.6 [33];

$$2H^+ + Fe \xrightarrow{yields} Fe^{+2} + H_2 \tag{4.2}$$

$$2H^+ + 2Fe^{+2} + [O] \xrightarrow{yields} Fe^{+3} + H_2O \tag{4.3}$$

$$Fe^{+3} + PO_4^{-3} \xrightarrow{yields} FePO_4 \tag{4.4}$$

$$Fe^{+3} + 3(OH^{-1}) \xrightarrow{yields} Fe(OH)_3 \tag{4.5}$$

$$3Fe^{+3} + 3PO_4^{-3} \xrightarrow{yields} Fe_3(PO_4)_2 \tag{4.6}$$

The iron phospohating process can be applied in different variations: through a two-stage system, a three-stage system, or a five-stage system. Such variations combine two main phases, a cleaning phase and a phosphating phase. Also these two phases can be done in one tank as in the two-stage system. For the five-stage system a sealer application is added to the phosphating sequence and the two phases are separated in two different tanks. The different iron phosphate implementations are given in Figures 4.3, 4.4, and 4.5, for the two, three, and five stages respectively.

The difference in performance between the different implementations is dependent on the chemical agents' role in each of these tanks. For example, the two-stage system is the most economical due the low equipment investment; however, the agents used in it are acting as surfactants for cleaning and as phosphating agents at the same time, which reduces their performance characteristics. The best iron phosphate coating is achieved through the five-stage system.

On the other hand, the zinc phosphating process generates a coat of fine, dense crystal pattern. The most commonly used ways to apply the zinc phosphate is to use a six- or seven-stage system. The added stages are due to the need for a conditioning

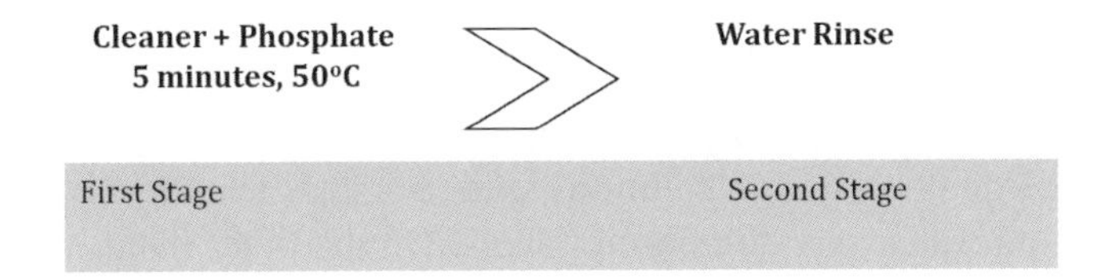

Figure 4.3 A two-stage phosphate system

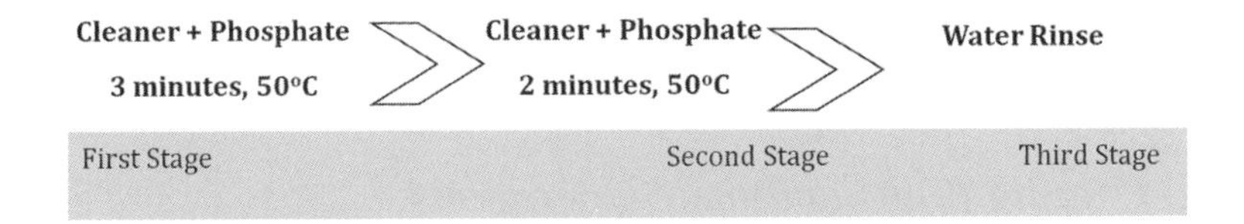

Figure 4.4 A three-stage phosphate system

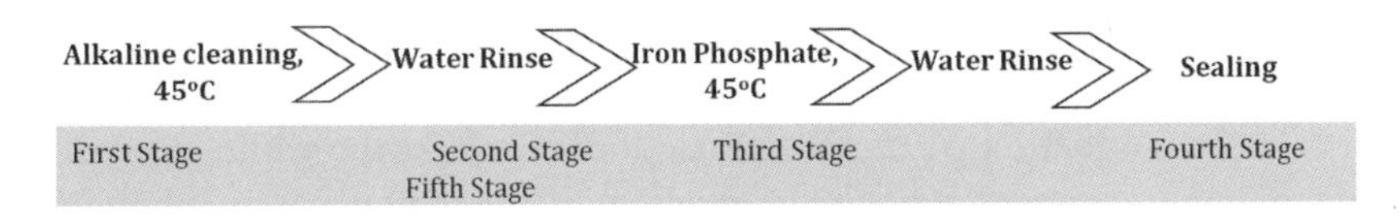

Figure 4.5 A four-stage phosphate system

process that aligns the zinc particles in the fine, dense pattern needed to achieve good corrosion resistance. The conditioners agents are typically made of titanium salt that is suspended in the conditioner bath and then collides with the zinc coat. The zinc coating process can be described in the equations sequence of 4.7 through 4.9:

$$Fe^0 \xrightarrow{yields} Fe^{+2} + 2e^- \tag{4.7}$$

$$2e^- + 2H_2O \xrightarrow{yields} 2OH^- + H_2 \tag{4.8}$$

$$4FE + 3Zn^{+2} + 6H_2PO_4^- + 6NO_2 \xrightarrow{yields} Zn_3(PO_4)_2 + 4FePO_4 + 6H_2O + 6NO \tag{4.9}$$

One of the main issues in both zinc and iron phosphate baths is the regular monitoring and maintenance of the baths at different stages. This is done to ensure that their chemical formulation in terms of the pH number and chemical agents' concentrations are within the operating limits. Additionally the bath temperature has a strong impact on the chemical reaction rates and the quality of the phosphate layer. Also, certain agents like those in the conditioning baths (titanium slat) age and degrade with time which means that they should be replaced by a fresh batch at regular intervals. The pH number describes the bath's alkalinity or acidity levels which consequently describe the bath effect on the subsequent paint films' adhesion.

4.2.3.1 Phosphating Aluminum

Iron, zinc, and/or chromium phosphate coatings can be used for phosphating aluminum substrates. Iron phosphating solutions tend to act as cleaning agents and deposit low weights of phosphate on aluminum panels. On the other hand, with zinc phosphating aluminum panels, some additives will first etch the aluminum surface to facilitate the coating process. The zinc phosphate layers on aluminum surfaces are clear-to-light-gray tone coatings. Also zinc phosphating of aluminum and steel panels can be done in the same bath, thus making zinc phosphate a more flexible choice, especially for hybridized BiW materials.

4.2.4 E-Coating Baths and their Operations

The electro-coating or E-Coating process is the process whereby an electrical current converts the soluble polar ionic resin into a neutral non-polar, i.e. insoluble, on the panels' surfaces, thus depositing the paint on them. Note that the direct electrical current passing into the E-Coat water tank will also decompose the water into hydrogen ions and oxygen gas (at the anode), and hydroxide ions and hydrogen gas (at cathode). The combination of these cations and anions at the cathode produces an insoluble polymer, causing paint to deposit until the surface is insulative, which happens around 13–45 um. This thickness limit can be further increased through increasing the passing electrical current. The E-Coating process is further followed by a spray phase to clean up any residues of the soluble resin still on the BiW panels, as shown in Figure 4.6. The sprayed liquid is typically supplied from the E-Coat tank

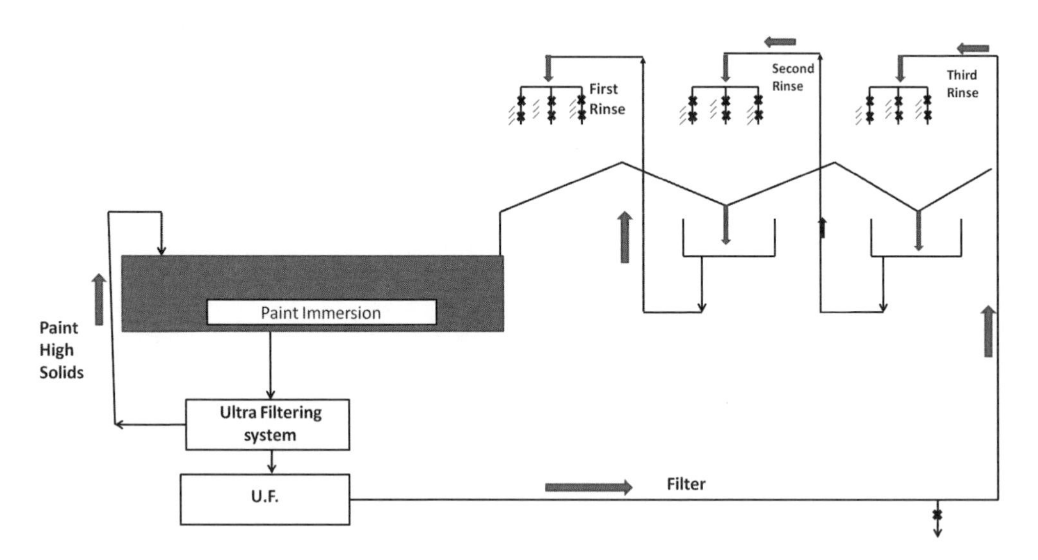

Figure 4.6 A schematic of an E-Coat bath system

and all the removed soluble resin will also be sent back to the E-Coat tank after it is filtered through a reverse osmosis or ultra-filtration process.

The returned liquid (from the spray) contains some solubilizer so the ultra-filtration process is designed not to remove such agents, otherwise the tank concentration will be changed. So the ultra-filtrate system focuses on the removal of any ionic contaminants.

The E-Coating process is considered an efficient way to deliver and deposit the paint on BiW different panels, both exposed and hidden. Also, the E-Coating can result in a very uniform paint film build-up. However, due to the reliance on the electrical current as way to help paint deposition, confined spaces and channels' cross-sections might restrict the electrical current effectiveness due to the Faraday cage effect. The Faraday cage effect prevents the electrical current from traveling inside a pipe more than half the diameter of that pipe. Again due to the E-Coating reliance on electric power, the power supply should be controlled such that it is in direct proportion to the panels' surface area covered with the solution at any given point in time. Typically 1 to 4 amperes (amps or A) are applied per square foot of panel surface area, resulting in an E-Coat film build-up of around *25 u*m thickness, deposited in approximately 80 to 90 seconds. The conveyor system carrying BiW shells into the E-Coat bath should be electrically grounded.

Typical E-Coat paint formulations used to come in two flavors: an anodic E-Coat or a cathodic E-Coat. A cathodic E-Coat resin is formulated so its polymer molecules are soluble acidic solutions. On the other hand, an anodic E-Coat resin is formulated to be soluble in alkaline formulations. In both cases the resins should be treated with specific appropriate solubilizers or agents to form positively or negatively charged molecular species called polymer ions. Acid solubilizers are needed for cathodic E-Coats, and alkaline solubilizers for anodic E-Coats. In the anodic system, the car body is made of positive electrodes. In the cathodic system, the car body is made of negative electrodes. Figure 4.7 shows the difference between the cathodic and anodic E-Coating processes.

Again, as the case with phosphate baths, the bath content should be monitored and maintained regularly. For E-Coating tanks, the main parameters that should be evaluated are: the solid total percentage, the pH number, the solubilizer concentration, the solution electrical conductivity, the tank temperature, the pigment and binder concentrations, and finally the tank voltage and amperage.

The E-Coat tank configuration should allow for large quantities of the E-Coat paint to keep up with the vehicle production rate, and of course be large and long enough to accommodate the largest BiW body style at the production rate. To provide a numerical example, for an automotive E-Coat bath, requiring an average of 2 minutes to complete the E-Coating process, with a conveyor speed of 4 m/min, the total tank length should be around 15 m long. A number of anodes are usually located vertically along both sides of the E-Coat tank, also other configurations exist where the anodes are aligned horizontally near the tank bottom too. A typical E-Coat tank is displayed in Figure 4.8.

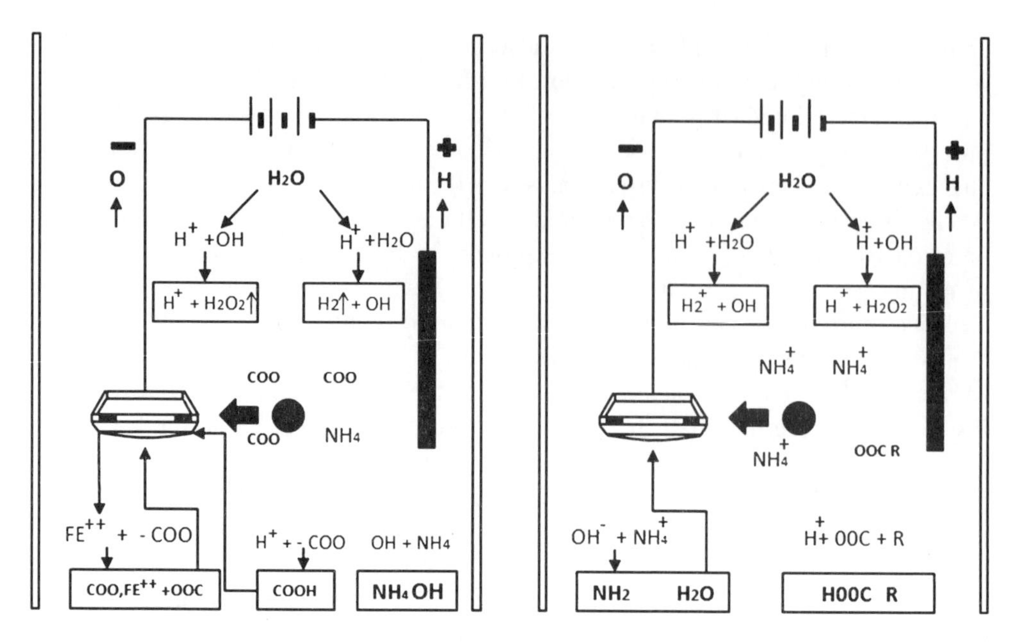

Figure 4.7 The difference between the cathodic and anodic E-Coating processes

Figure 4.8 A typical E-Coating bath. Reproduced by permission of Duerr Systems, Inc © 2010

The E-Coat process features several advantages that motivate its use in the automotive industry, with the following main ones:

- The E-Coat immersion is a fast process, leading to a high production rate, and an efficient painting process for the BiW panels.
- The E-Coating process can consolidate other paint subsequent films such as the primer, because the E-Coat can act as a primer paint, which leads to greater productivity at lower cost.
- The E-Coat has no HAPs or volatile organic compounds (VOC) emissions.
- The E-Coat film build-up can be controlled in thickness over the total surface area by manipulating the electric current in the tank.

However, its main limitations are related to the high capital cost and the need for continuous maintenance and monitoring activities. The immersion-based coating (E-Coat and phosphating systems) can be implemented using one of two BiW handling variations: the first is based on lowering the body shell in the tank, while the other is based on rotating the shell as it get submersed into the tank. The second scheme provides better gap filling and overall performance, its commercial name is RoDip a product of Duerr. More details about the material handling systems in the paint area will be discussed later in the chapter.

4.3 Paint Curing Processes, and Balancing

The paint curing process is necessary to solidify the liquid and cure the applied paint films. Liquid paint tends to collect dirt, sludge and other contaminants (such as hair, cloth fibers) from the paint area environment. Additionally liquid paint is more sensitive to temperature and humidity changes. So the painting area features a variety of ovens dedicated to different paint layers and the sealants. A typical paint area has an oven for the E-Coat film, an oven for the primer film, and an oven for the combined top-coat and clear-coat films. Additionally another oven is dedicated to the underbody sealant, and one for the repair area. The oven size, length and configuration depend on its intended application. Additionally the paint area ovens are typically composed of different zones; some zones are described as radiation zones and others as convection zones; this definition is based on the dominant mode of heat transfer in that zone. A typical oven layout is shown in Figure 4.9.

To illustrate the oven operation, the primer oven will be discussed in following text. The oven starts with a radiation zone; this zone is intended to solidify the upper layer of the paint film so it collects any oven sludge or contaminants, also the subsequent zones in the oven will feature air flow across the body so the upper layer should be solid to avoid any distortions in the paint. A group of radiation panels are aligned and oriented so they can provide uniform heating to the different BiW panels. The following two zones are based on a convection mode of heat transfer where

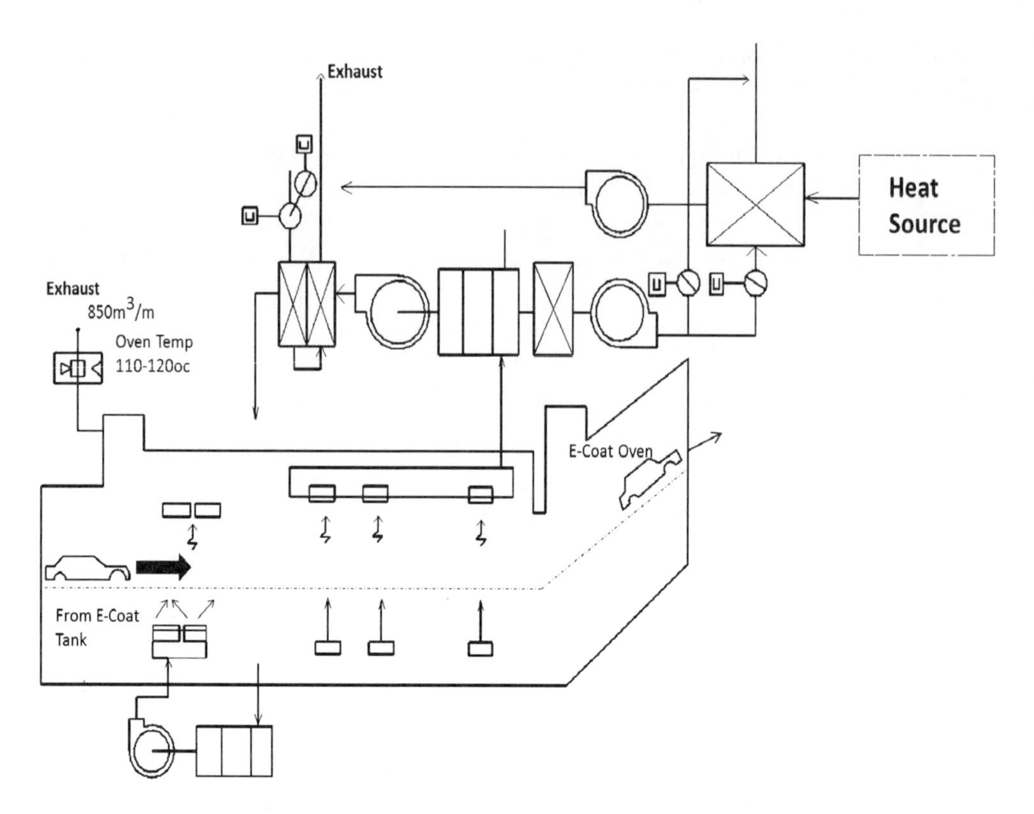

Figure 4.9 A typical automotive oven schematic

heated air is blown into the oven chamber to raise the temperature of the BiW. One of the main considerations in heat delivery in the oven is understanding the different heat inertias within the BiW panels; the thicker the panel, the more heat it will require to get heated. Additionally, these thicker panels will store the heat longer and might lead to over-baking of the paint applied over it. The BiW shell is carried through a carriage system that is made of thick steel members, so the heat delivery should account for these members and their locations. The convection zones aim to raise the BiW temperature and then hold it at that temperature as required by the paint curing window. The paint curing window is the plot of the holding temperature versus holding time that motivates a complete cross-linking of the paint chemicals so that the paint is fully cured. A typical paint cure window is plotted in Figure 4.10. The final zone within the curing oven is the cooling stage, where the BiW is cooled via circulating air; for low production the air seal can be used for the cooling stage. The complete oven heating and cooling history are shown in Figure 4.11.

Understanding the paint film interaction with the oven conditions in terms of applied temperature value and distribution is vital in ensuring good curing practices, without any under-baking, over-baking, or bubble formations. Lou and Huang in [36]

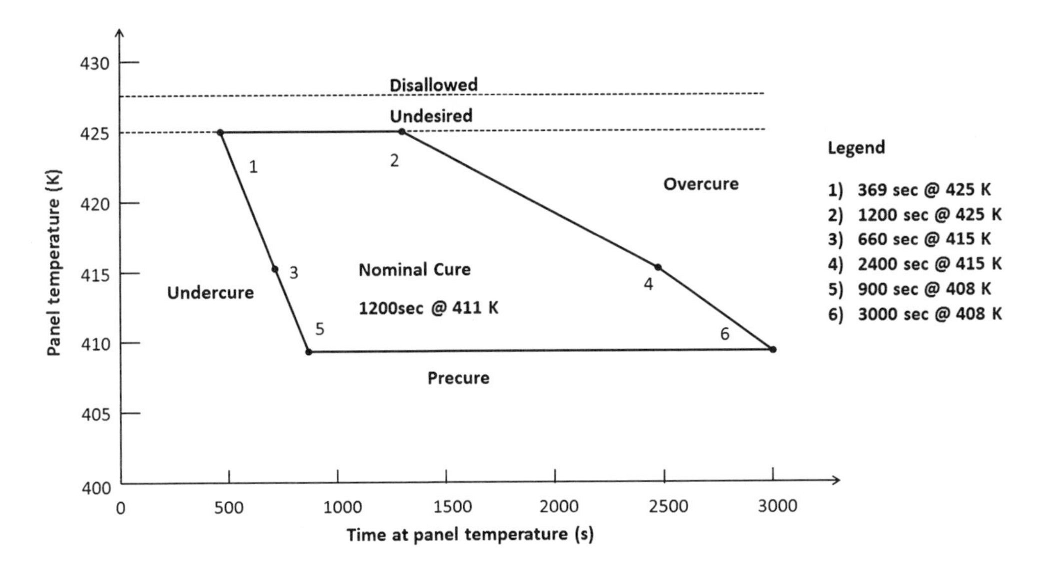

Figure 4.10 An example of a paint cure window

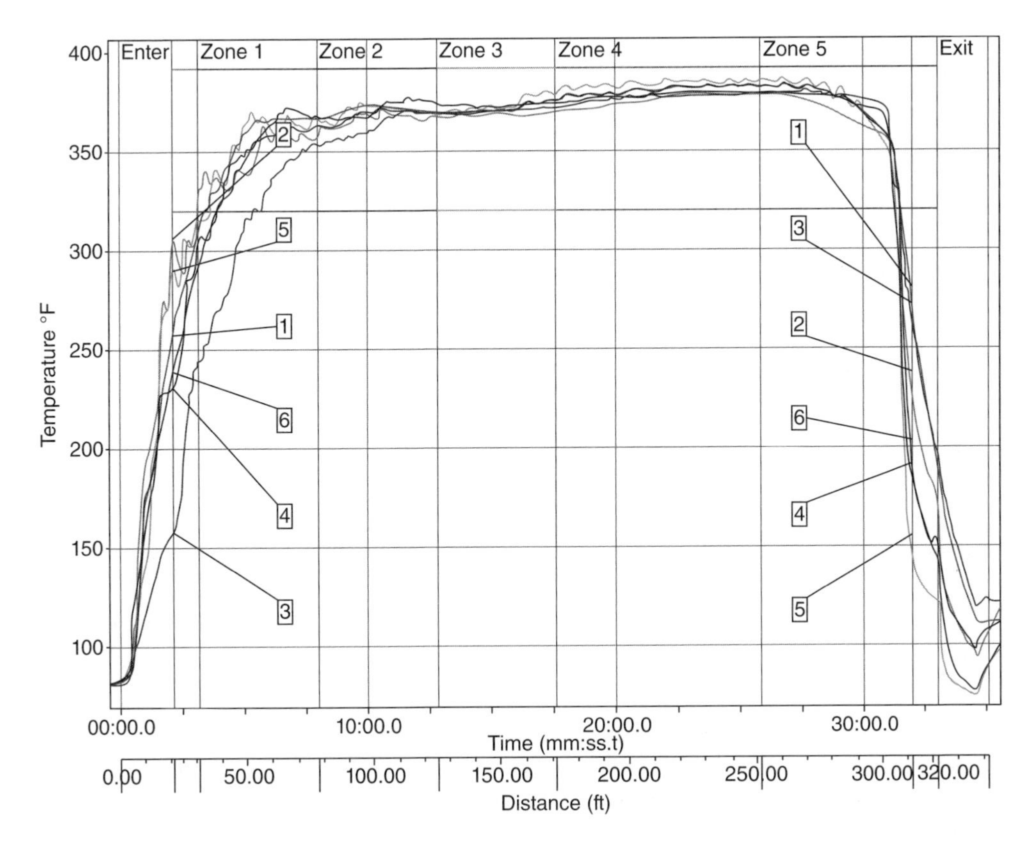

Figure 4.11 A typical oven time profile

have proposed a curing model for paint films that can be utilized in predicting the BiW panels' temperature, and their solvent removal, and cross-linking rate as a function of time. This model describes the temperature-time change as in Equation 4.10, correlating it with the oven conditions:

$$\rho C_p w \frac{dT}{dt} = \begin{cases} \vartheta \sigma \epsilon \times \left(T_{wall-i}^4 - T^4\right) + \alpha V_i^{0.7} \left(T_{air-i} - T\right) \cdots i = 1,..,N_{rad+conv} \\ \alpha V_i^{0.7} \left(T_{air-i} - T\right) \cdots i = (N_{rad+conv} + 1),..,N \end{cases} \tag{4.10}$$

where ρ is the panel density, C_p is its specific heat capacity, and w is the body panel thickness. The temperatures T_{air-i} and T_{wall-i} are the air and the oven wall temperatures for each zone i. The radiation view factor (or shape) factor is ϑ, while the Stefan-Boltzmann constant is σ and the emissivity is ϵ, and α is a constant, and V is the air velocity. Finally, the $N_{rad+conv}$ is the number of zones with radiation or convection modes of heat transfer, and N being the total number of zones.

Furthermore, more analyses can be done to correlate the temperature-time behavior with the cure rate or degree of cure, which can be evaluated using a differential scanning calorimetric through analyzing the exo-therms and endo-therms of the paint samples. For automotive OEMs, the paint film cure is typically evaluated based on the functional performance of the paint film, i.e. its scratch resistance, appearance, etc. New models for optimizing the cure rate characteristics have been introduced by [37]. At the same time, 3D simulation packages and models are available on the market to help predict the temperature-time profile inside an oven; for example, CADFEM provides full 3D simulation for the BiW curing, an example is given in Figure 4.12.

The current automotive manufacturing plants are producing different models on the same paint lines, which means different body styles with different geometries and panel reinforcements are being cured in the same ovens. This leads to a balancing step to prepare temperature-time profile for the different styles, and paint layer configurations. This balancing is done with the aid of a device called DataPaq, which is a simply a collection of thermocouples and data acquisition, in which the thermocouples are attached to a sacrificial BiW, while it's going through the oven to collect the panel temperature-time history. This practice is helpful because the thermocouples are mounted on different panels, to collect information about each one. However, new changes in automobile designs have motivated a search for alternative ways to evaluate the oven performance. Such new designs that have a strong impact on the oven temperature are more diversity in the BiW materials (with different thermal properties) with wider use of new deadeners' formulations such as the liquid applied sound deadener which requires specific bake characteristics, in addition to more use of the structural foam inserts which lead to an increase in the thermal inertia of certain zones within the BiW. Some companies have proposed the use of an embedded enclosure within the oven, equipped with infrared sensors to have a real-time, temperature measurement process.

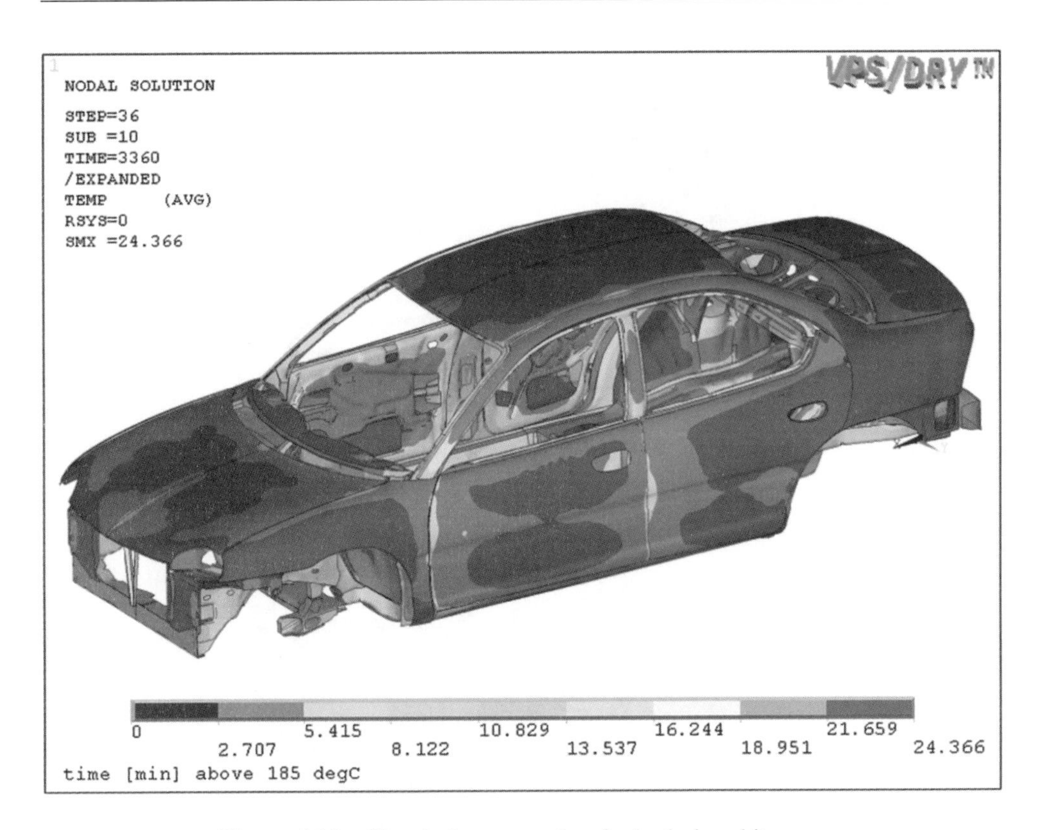

Figure 4.12 Simulation example of a body in white cure

Other alternatives to the traditional automotive ovens, have been introduced in the form of radiant-based systems, which include infrared curing lamp systems, and the ultraviolet curing. The infrared curing is based on delivering a high temperature to the surface using lamps, this direct heat delivery is very effective way of curing the paint however, the control of the lamp intensity, which follows a Gaussian profile, is difficult; additional challenges are due to concerns in regard to the system service life and consistency. On the other hand, the ultraviolet systems deliver a UV beam of light onto a paint that is impregnated with UV-absorbing pigments, so that an internal (within the paint) heating is initiated. The UV curing is effective and can be controlled through the UV source intensity, or by changing the concentration (typically termed as loading) of the UV pigments. The UV curing faces one main challenge, which is due to its radiation-based nature, in other words, it requires a line of sight into the panel to have it cured, so for hidden, complicated geometries the UV source should be manipulated to reach such corners. The UV curing is currently available in two implementations [38], a fixed (enclosure-based) and a robotic one as shown in Figures 4.13 and 4.14 respectively.

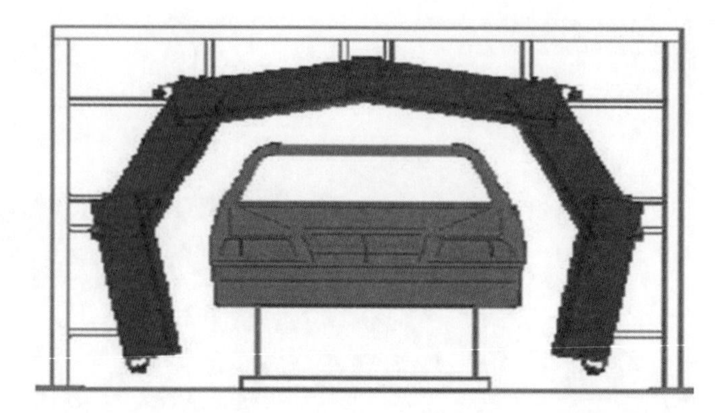

Figure 4.13 The UV curing fixed implementation

Figure 4.14 The UV curing robotic implementation

However, new advances have led to a new UV curing system which features a closed-loop robotic system capable of evaluating the paint DoC then moving the robotic manipulator with the UV source (a light emitting diode (LED)) mounted on it according to the curing requirements. This system schematic is displayed in Figure 4.15, and its two robotic implementations (sweep, and feedback) are shown in Figure 4.16.

4.4 Under-Body Sealant, PVC and Wax Applications

The paint area is also responsible for applying an under-body coating and a sealer application to the welded joints and any crevices within the BiW that might have a

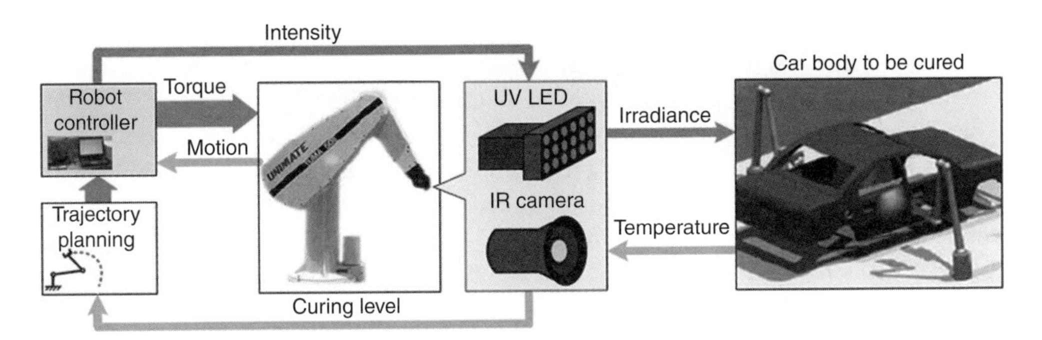

Figure 4.15 The proposed UV curing system

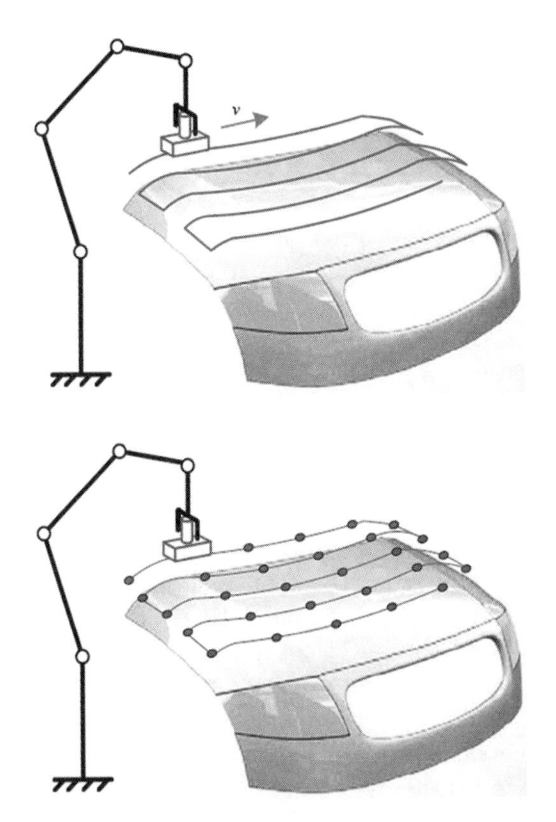

Figure 4.16 The two robotic implementations of sweep and feedback

high corrosion potential; additional advantages from such sealer applications include that this coating acts as a shield against the service environments' varying elements (weather, etc.), and it provides a resistance layer against mild impacts.

Sealants are pumped into gaps and seams. The placement of the seam seals should be designed and applied precisely, typically within 1 mm and the seal temperature should be also precisely controlled. The application of such sealants can be done manually or through a robotic applicator. For the robotic applicator, the flow-rate is programmed to achieve a specific fill rate within the gaps.

Additional coverage for the boxed section and cavities against corrosion is provided in the form of under-body wax. The solvent containing the preservative wax is sprayed into the BiW sections through dedicated nozzles. The end of the sealer application features an oven, to cure the applied sealant, additionally, the adhesive material applied in the hemmed door panels can be cured in such oven.

4.5 Painting Spray Booths Operations

The painting booth is a closely controlled environment, where the different sprayed paint films are applied manually or through robotic painters. In both, the air should be configured in terms of temperature, RH, cleanliness, and its flow rate. All this is done to achieve the suitable conditions for spray paint to be applied at the right thickness, achieving the right color, and flow characteristics over the painted BiW panels.

A paint booth is composed of different areas to deliver different functionalities; these areas are: the upper portion which is equipped with sophisticated air handling and filtering systems to provide a controlled air flow into the booth. The circulating air within the booth is conditioned to a very tight range, which increases the air handling requirements and consequently its power consumption, a typical paint booth costs around $5,000/ meter /yr in energy cost. The paint power consumption will be discussed in detail in Chapter 6 on ecology. The typical painting conditions are: air velocity around 30–60 cm/sec, air temperature from around 22 to 24 °C +/– 1deg, and an air humidity RH around 40–60% within a +/– 5% range.

The booth main space features the area where robotic painters operate, the width of such area should be as narrow as possible to reduce the requirement on air handling and reduce the space availability for paint particles to spread, hence reducing the cleaning requirements too. The lower portion of the booth contains a mixing area, where the excess paint is mixed with water for collection. This water has a chemical formulation that acts to affect the paint so it floats on the surface. Additionally the booth bottom area contains a scrubber (typically centrifugal) to separate the trapped paint from the water so that the water can be re-used and the paint is collected for disposal. A continuous air draft is also designed to pass from the upper through the booth area into the bottom portion to remove any paint over-spray into the water at the bottom. A 3D rendering of a painting booth is displayed in Figure 4.17, an actual picture in Figure 4.18.

Additional variations of the paint booth exist depending on the application, for example, certain designs will include flexible wall modules to allow for different connections into the booth. Booths utilizing the solvent-borne paints should also be

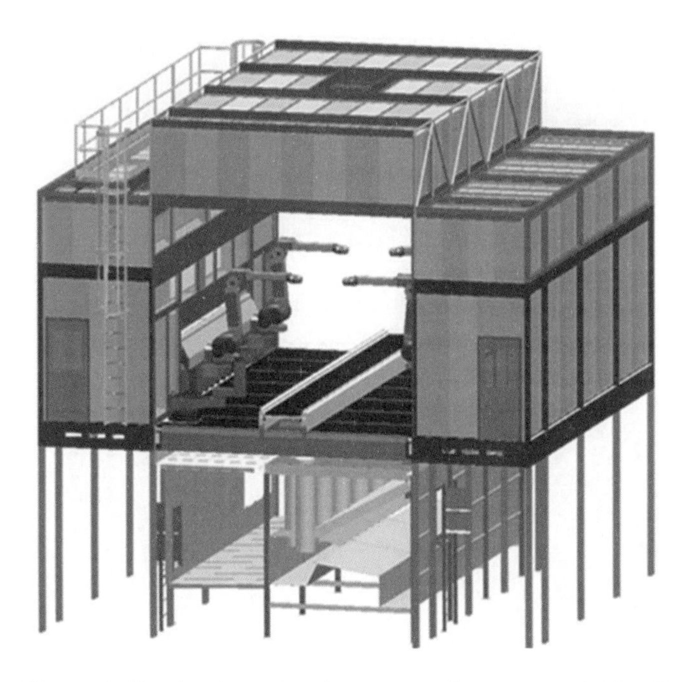

Figure 4.17 A schematic of an automotive spray paint booth

Figure 4.18 An automotive paint spray booth

Figure 4.19 A powder-coating booth

equipped with fire extinguishing systems, while those operating with waterborne paints should be designed so that the paint lines are isolated and grounded. The paint circulation system is typically composed of pumps, regulators, and a piping network. However, painting booths that are based on powder coating are different from the ones in Figures 4.18 and 4.19. The powder coating features different supply and paint collection systems, a powder coating booth is shown in Figure 4.19. However, one should note that most of the OEMs are using either solvent-borne or waterborne systems due to the knowledge-base and the accumulated experience established about these paints' formulations; in addition to some challenges in controlling the performance of powder coats and their color matching.

4.5.1 Spray Paint Applicators

Each body shell receives multiple paint layers as it progresses inside the different painting booths. The first film is the primer paint, which can be done either manually or through robotics. The primer paint is designed to deliver a controlled roughness profile for the panels to enhance the adhesion of the top and clear coats. The primer is applied at about 30–40 um, with 2–3 color variations. The primer coat is then cured and the vehicle shell passes through the top-coat booth, which can have a different paint film build-up based on the paint configuration; possible configurations are:

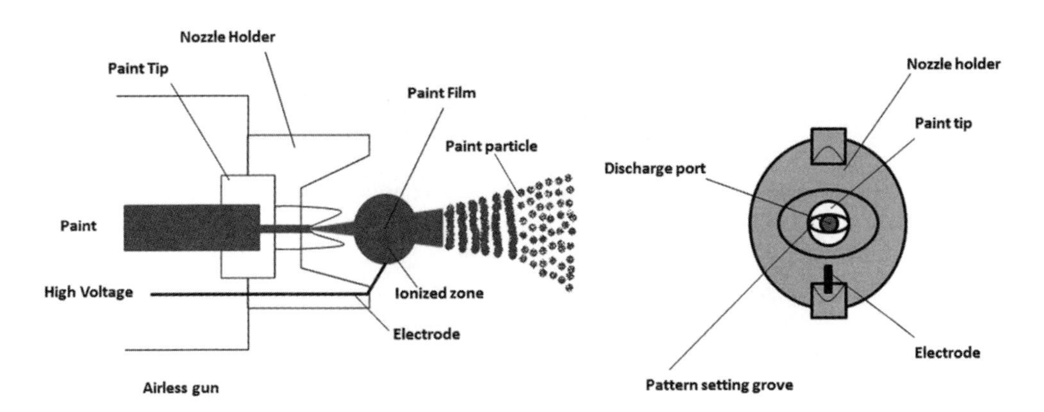

Figure 4.20 An airless spray paint applicator

- Solid color coat, which is a 1 top coat followed by 1 baking process, (1 C 1B).
- Metallic coat-film, which features a 2 coat (top-coat + clear coat) with 1 bake process i.e. (2 C 1 B).
- Pearl color coat-film, which has a 3 coat film (a base top-coat, a pearl-coat, and a clear-coat) followed by either a 1 bake or a 2 bake process (3 C 2 or 1B).

Typical paint film thicknesses are on average thus: the primer ~20–40 μm, the solid color (top-coat) around 30–40 μm, metallic mica around 15–20 μm, and the clear coat around 30–40 μm. These thicknesses translate into around 5–7 kg of paint weight applied on the vehicle. The detailed distribution for each coat weight is as follows; the E-Coat around 3 kg, the primer ~1.5 kg, the top-coat around 1.0 kg, and the clear coat ~1.5–1.9 kg.

Each of the above paint films requires a different spray paint applicator; these applicators are classified into the three methods that are used to atomize the paint:

1. *Airless*, which creates paint particles by directly pressurizing the liquid paint, using an airless paint guns. The air-less spray principle is depicted in Figure 4.20. This method of spray has low over-spray, and it is best used to apply thick films in few passes; however, the airless spray guns cannot deliver a Class A finish to the painted surface.
2. *Air spray*, which atomizes the paint by the use of a pressurized or compressed air. This atomization process delivers fine droplets (droplet diameter size around 15 μm) leading to a Class A finish surface and better control over the film thickness. However, due to the high pressure air, this spray method generates high level of VOCs from solvent-borne formulations. Some automotive OEMs utilize such an approach in painting metallic coats, because such coats require high paint speeds to align the metallic flakes parallel to the surface. Different variations of

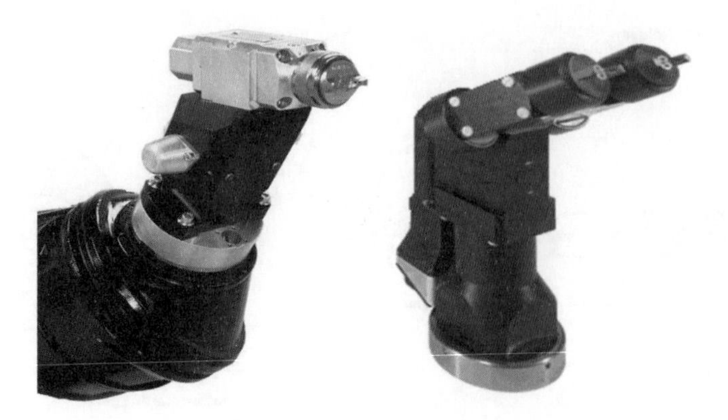

Figure 4.21 Dual and single-head spray applicators

such paint applicators exist, such as the ones in Figure 4.21, which shows a single-head spray gun and another dual-head one.

Air spray atomization can be done using a high volume of air supplied at low pressure (HVLP), this helps in shaping the atomized paint into specific patterns composed of low speed fine particles. This HVLP principle is considered one of the most efficient spray methods.

3. *Rotary guns*, which atomizes the paint using the centrifugal force generated by a rotating bell or disk applicator. The paint is introduced from groves (on the circumference) into a fast rotating bell, typical rotational speeds around (30,000–40,000 rpm), the high speed generates high centrifugal forces that breaks up the paint liquid into ligaments and finally into fine particles (approximate droplet diameter is around 20–25 um). Figure 4.22 shows a rotating bell type gun, and a plot of the effect of the bell rotating speed on the paint particles' size. The rotary bell applicator can also be equipped with shaping air stream to shape the paint flow pattern into controlled fan widths and hence control the droplet size distribution. Several advancements in the rotary guns have been introduced to improve its performance and so the rotary gun system is the most efficient from the transfer efficiency point of view.

   The transfer efficiency of a paint gun translates into the amount of paint spray reaching the BiW surface from that sprayed by the gun. The difference is considered an over-spray and is caused by the air movement in the booth (due to the downward draft) and the spray pattern. The transfer efficiency of a paint applicator is the most important performance metric along with the size of the atomized particles. To increase this efficiency, the paint applicators are equipped with an electro-static system to charge the paint particles as they leave the gun, so they attract to the BiW surface, which is grounded. The charging can be done through using a corona or a plasma-charging method. The electro-static charging is achieved by introducing an electrode into the atomization region, so that the paint

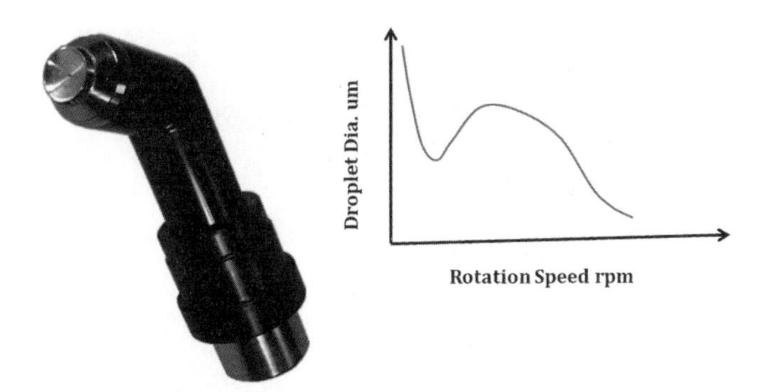

Figure 4.22 A rotary bell paint applicator with the effect of the rotating speed on the particle size

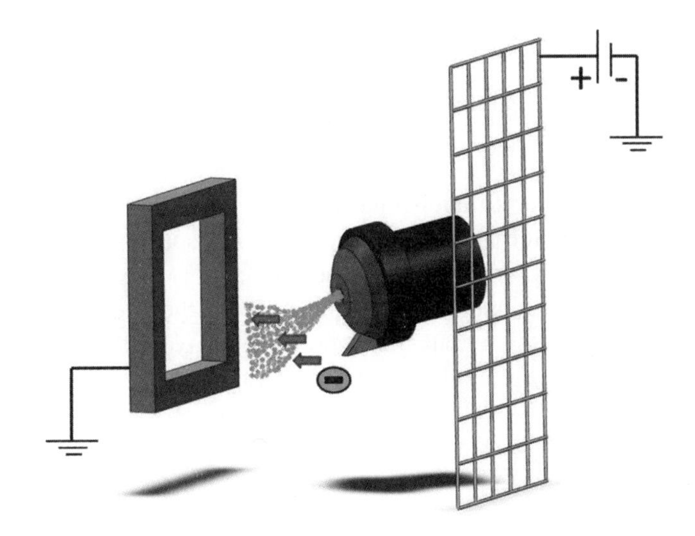

Figure 4.23 The electrostatic charging principle

particles pick up an electric charge, which motivates further break-up of the paint particles into smaller ones. This principle is illustrated in Figure 4.23, and an application of this principle is in Figure 4.24, showing an electrode installed in an air-spray gun. The electro-static charging has led to better paint application process at very high transfer efficiency. More about electro-static charging is given in Section 4.6.2.1.

For all the above paint applicators, controlling the atomization process to achieve fine particles with confined pattern, without high agitation to the paint solvent content leads to the best paint finish. For example, if the paint applicator is too close to the BiW, then the paint color shade will be brighter because less solvent has evaporated

Figure 4.24 Internal charging of the paint

from the paint particles as it traveled from the gun to the BiW surface, hence more solvent concentration is at the surface. If the applicator is too far away, then the color shade will be darker because the concentration of the color pigments is higher due to the more distance the paint travels, that is, more solvent evaporating from it. The same is true when the paint spray is increased, as the speed increases, the paint is agitated more, leading to more solvent evaporation.

The automotive painting operations rely heavily on the use of the air spray and the rotating paint applicator for all their spray paint layers and they are typically equipped with electrodes for an electro-static operation.

4.5.2 Painting Booth Conditioning, Waterborne, Solvent-borne and Powder-coating Systems

Due to the increased pressure on the automotive OEMs to utilize more environmentally friendly paint formulations, waterborne paints have become more popular within the industry over the solvent-borne paints. This section will start the discussion with an overview of the waterborne, solvent-borne and the powder coating general principles and chemical formulations then a more specific discussion of the automotive painting booth adjustments and operations will follow.

4.5.2.1 Waterborne Paint

Paint in general contains four main components: the pigment, the additives, the resin, and the fluidizing medium. The resins are considered the polymer within the paint formulation that forms the paint film. The resins are normally called the paint binders and most industrial paints are classified according to the resin type, which include

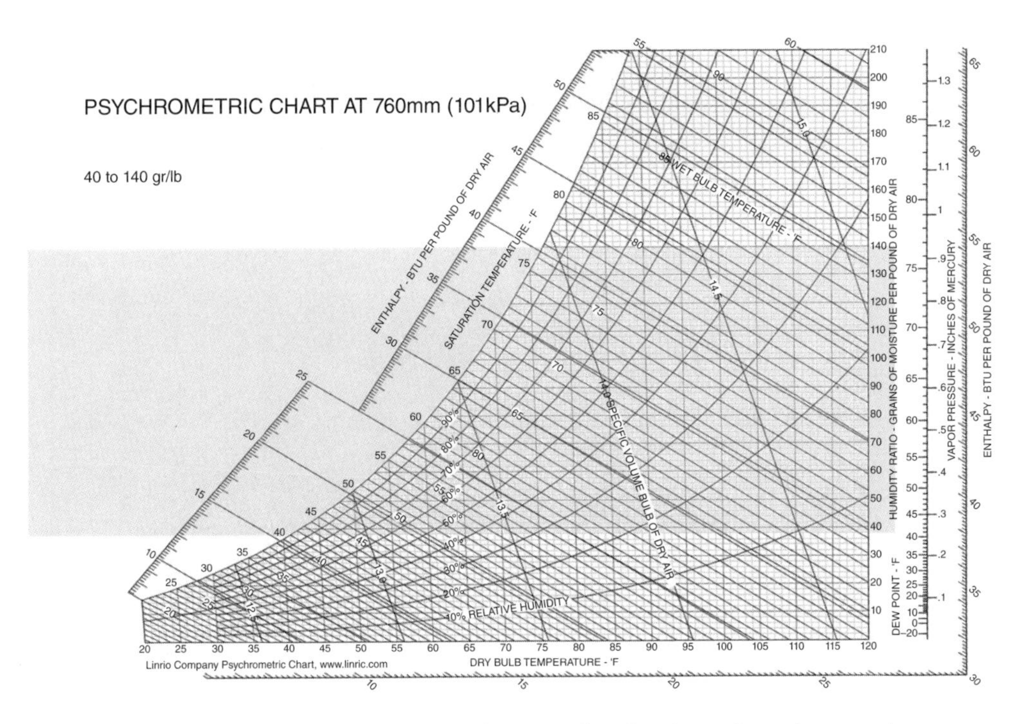

Figure 4.25 The operating range for spray booth using solvent-borne paint

the following families; the lacquers, the enamels, with the difference being in the way the paint is cured; the lacquers evaporate their solvent content while the enamels go through a cross-linking reactions. The pigments are designed to impart certain visual or functional characteristics, such as color pigments, metallic pigments, or UV pigments. The solvent is the fluidizing medium that allows the paint to be applied on surfaces, and is mainly classified into organic solvents and water-based solvents. The additives are added to modify or control certain paint properties such as aging, and flow.

So waterborne paints are those formulations that contain a water-based fluidizing medium in their chemistry; however, one should note that co-solvents are added to the formulation to allow the resin to be soluble in water. So, a waterborne paint is not 100% water-based. For example, the E-Coat formulations are waterborne paints due to their resin solubility in water.

The waterborne paints affect the operation of the painting booth due to the added water content within the paint, which means higher sensitivity to any changes in the RH within the booth environment. For example, the operating range for a solvent-based paint is shown on the psychometric chart in Figure 4.25 while that of a water-borne paint is given in Figure 4.26. Inspecting the two figures indicates the limited range for the booth environment in the case of waterborne paint, which translates into tighter air conditioning requirements. This limited range is set to ensure the

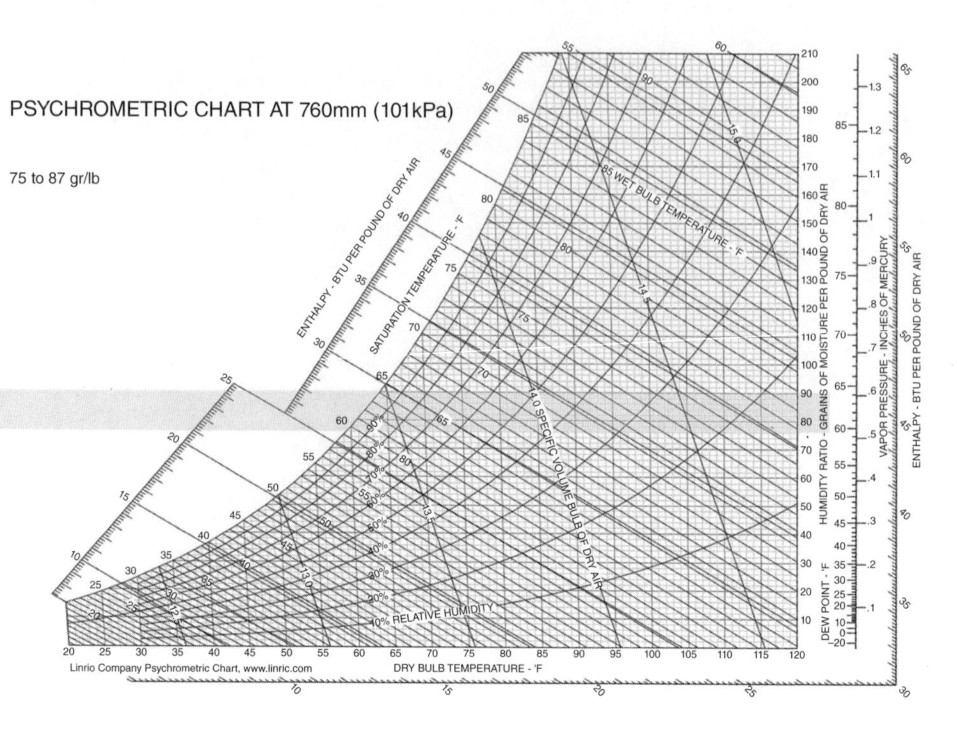

Figure 4.26 The operating range for spray booth using a waterborne paint

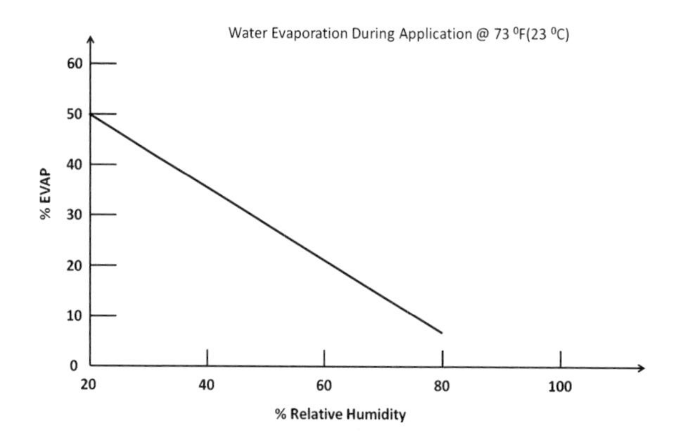

Figure 4.27 The relative humidity effect on the evaporation rate of water from a waterborne paint

amount of water evaporated from the paint as it travels from the applicator into the BiW surface; this is better described in the plot in Figure 4.27, which shows the water evaporation rate as a function of the booth's RH.

Additionally, having the paint with a high content of water affects the electro-static charging process because (a) the electric charge will tend to draw more and more

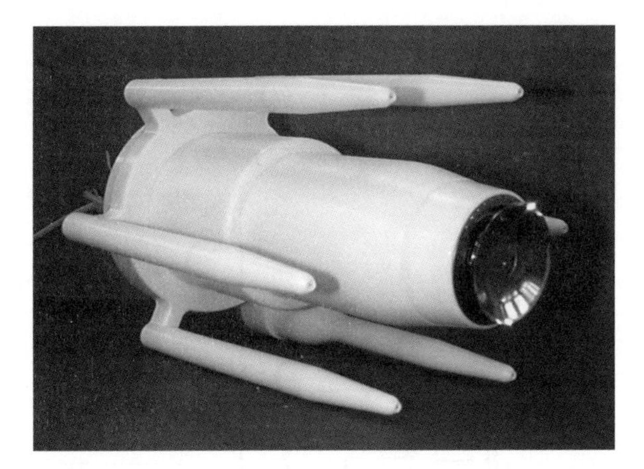

Figure 4.28 External charging electrodes for a bell rotary system

currents as the humidity increases; additionally (b) the electric charge might travel in the paint supply system, leading to safety concerns in terms of electrocution. So the usage of the waterborne paints not only adds new complications in terms of color matching and paint performance, but also it results in tighter control over the booth conditions, new production steps (flash-off), and further it requires modifications to the electro-static charging systems. However, most of the automotive OEMs have focused on such paint formulations to avoid any violations of the environmental regulations set by the Environmental Protection Agency, such as the VOC per applied paint Gallons limit.

From the above discussion, painting booths should be modified to accommodate waterborne formulations. The first modification is done by adjusting the conditioning requirements which ultimately leads to higher energy consumption. Additionally, the new booth should include a flash-off zone, to help evaporate part of the water content; this is done using an enclosure of infrared lamps where the BiW shell will pass through after it is top-coated.

In regard to the electro-static charging process, it can be modified in one of two ways; the first is based on insulating the complete paint supply system using voltage blocking devices or using an indirect charging method of the paint. The indirect charging system moves the charging electrode from the paint supply nozzle within the gun (Figure 4.24) and mounts it on the exterior of the paint applicator, as in Figure 4.28. This of course will require higher charging currents. The applicator in Figure 4.28 is developed by ABB under the name the Conductive Paint Electrostatic Spray system or COPES.

The other alternative is based on an innovative design of a new paint applicator (developed by ABB and Toyota Motor Corp.) under the name the Cartridge Bell System CBS. This system, as the name implies, utilizes a rotary bell paint applicator, with paint cartridges so the painting applicators does not have a direct connection

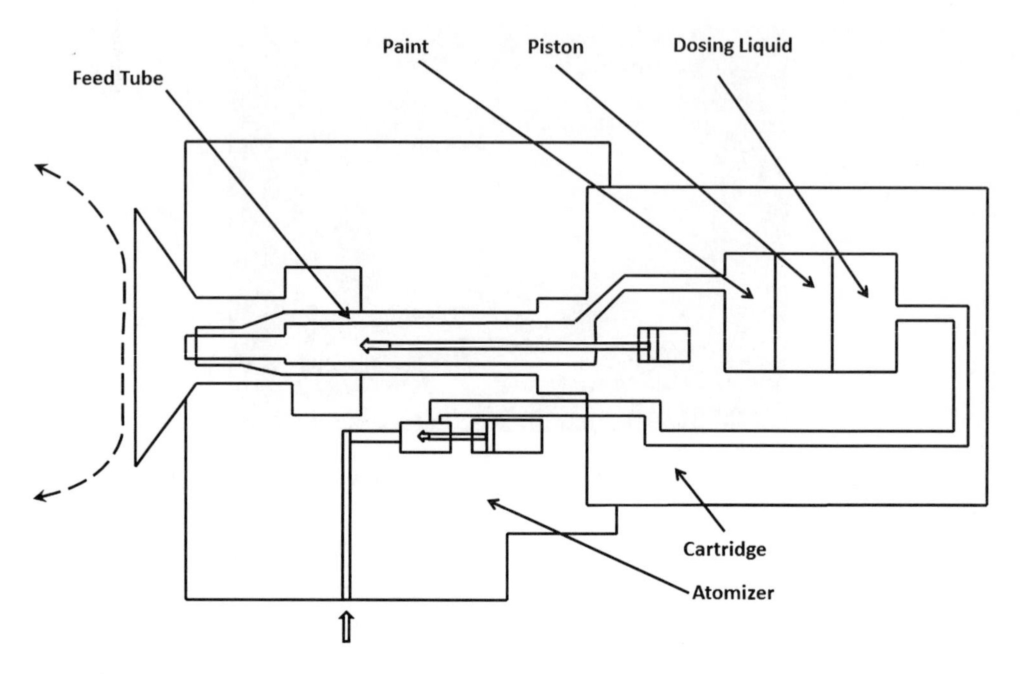

Figure 4.29 The cartridge bell applicator

with the paint supply line. The CBS system was initially developed to enable Toyota to paint different color BiW back to back without resorting to a color batching. Each cartridge contains a different color, which allows for faster color change operations, minimum cleaning requirements upon changing colors, and easier implementation of the electro-static charging of water-borne paints. The CBS applicator is displayed in Figure 4.29. Another color changer system is developed by Duerr named EcoLCC which is able to switch between colors in less than 10 seconds, the EcoLCC is shown in Figure 4.30.

4.5.2.2 Powder Coating

The other environmentally friendly alternative in terms of paint formulations is based on using powder-coating. In powder paint, the paint is in the form of finely ground powder that contains the same paint formulation (as solvent-borne and waterborne) but without a solvent, i.e. it is a dry paint. The main application of powder-coating paint in the automotive industry has been in the primer anti-chip coatings; however, OEMs are trying to expand the powder coating into the top-coat formulations. The challenges in using the powder coating for top-coat applications are related to the powder particle size which is around ~30 μm in diameter, hence it does not allow thin paint films or fine finishes needed for a Class A. Powder coating come in dif-

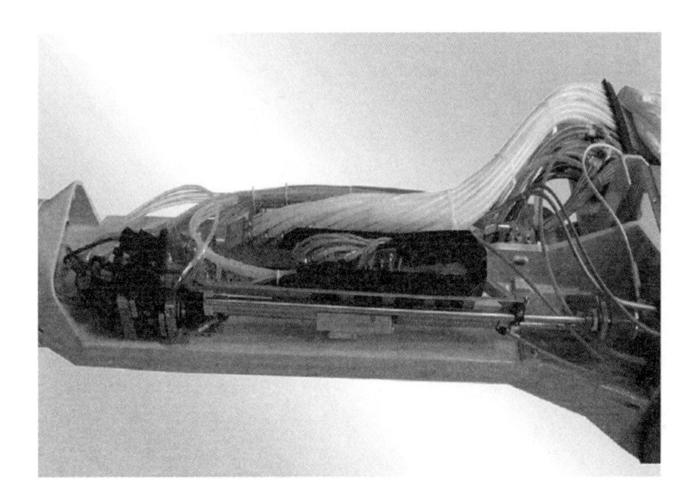

Figure 4.30 The EcoLCC system. Reproduced by permission of Duerr Systems, Inc © 2010

ferent configurations, each with specific advantages and disadvantages; Table 4.1 from [35] describes such formulations and their characteristics.

As the case with waterborne paint, powder-coating requires significant modification of painting booths initially designed for solvent-borne paints; in addition to the specialized paint applicators needed to apply the powder coat. For the automotive industry, the best application is done through electro-static powder-coat spray gun or a rotating bell, which are connected to a fluidized powder tank. A stream of dry air is pushed through the tank to supply the fine powder particles into the gun. Several OEMs including BMW and Volvo have used powder coating for their clear coats. The main advantages of the powder coating are: their low cost, low VOC emissions, and the ability to re-use the over-sprayed powder paint. The re-use of the powder paint reduces the plant water treatment activities. The main issues with powder coating are in their handling and application steps.

4.5.3 Paint Calculations

This section discusses the paint line basic calculations in terms of paint applied per vehicle style (i.e. surface area), the production rate (number of vehicles painted per unit time), and the number of atomizers needed to meet the production volume and rate. In addition to the basic computations, we provide the flow rate required from the paint applicator at a given transfer efficiency and a paint formulation solid content.

The flow from a paint applicator is governed by the film dry thickness t_{dry}, the paint solid content $S_{perecent}$, the applicator transfer efficiency η_{trans}, the gun velocity

Table 4.1 Main characteristics of powder coating

Characteristics Type	Weathering Resistance	Chalk Resistance	Corrosion Resistance	Chemical Resistance	Heat Resistance	Oven bake Resistance	Adhesion	Impact Resistance	Flexibility	Abrasion Resistance	Pencil Hardness Resistance
Acrylics	E	E	G	E	E	E	G	G	F	E	H-4H
Epoxy	E	P	E	E	VG	VG	E	E	G	E	HB-5H
Epoxy Polyester Hybrids	VG	F	VG	VG	G	E	E	VG	VG	E	HB-2H
Fluoro-Polymers	E	E	E	E	F	G	Require Primer	VG	E	G	HB-2H
Polyamides	VG	VG	E	VG	F	G	Require Primer	E	VG	VG	70-85 Shore D
Polyesters	VG	E	E	VG	G	VG	E	VG	VG	VG-E	HB-4H
Polyolephins	G	VG	E	Solvents P	F	F	Require Primer	VG	E	P-F	30-55 Shore D
Vinyls	VG	VG	VG-E	Variable	F	P-F	Require Primer	E	E	F-G	30-50 Shore D
E = Excellent VG = Very Good	G = Good F = Fair	P = Poor									

Notes: P = poor; E = excellent; G = good, VG = very good; F = fair.

Table 4.2 Typical percentage of solid content with the different automotive paint formulations

Type of paint	Paint N.V. (%)
Primer paint	55 ± 5
Metallic paint	20–40 ± 5
Clear paint	40–50 ± 5
Solid top coat paint (white)	45 ± 5
Solid top coat paint (Black)	40 ± 5
Waterborne primer paint	50 ± 5
Waterborne metallic paint	20 ± 5
Waterborne solid top coat paint	25 ± 5

V_{gun}, the applicator fan width (which can be controlled by the shaping air) W_{fan}, and the number of over-laps needed to complete painting a certain thickness L_{over}. The flow can be computed according to Equation 4.11:

$$CC = \frac{t_{dry}}{S_{perecent}} \times W_{fan} \times (V_{gun} \times 60) / \eta_{trans} \times L_{over} \tag{4.11}$$

Table 4.2 helps in establishing approximate values of the percentage of solid content in the different automotive paint coats.

Each of the paint applicators discussed previously has specific fan pattern widths, for example, the rotary applicators fan width ranges from 150–450 mm, while that of the airspray gun is around 100–500 mm.

The next step in paint calculation will be to evaluate the amount of paint needed to paint one vehicle. First, the vehicle should be transferred to the surface area to be painted and this depends on the size of the vehicle. Then the surface area is multiplied by the required dry paint-film thickness, which results in the dry paint volume that can be converted in wet (including solvent) paint volume by dividing it by the percentage of solid paint content. This is further illustrated in Equation 4.12;

$$Vol_{paint} = A_{surface} \times \frac{t_{dry}}{S_{perecent}} \tag{4.12}$$

Furthermore, one should take the amount of paint wasted as over-spray. This can be done by knowing the transfer efficiency of the paint applicator as in Equation 4.13.

$$Tot.Vol_{paint} = Vol_{paint} / \eta_{trans} \tag{4.13}$$

This equation provides the painting engineers with an estimate of the paint requirements based on the production volume for each body style. Furthermore, engineers need to evaluate the number of atomizers (or robotic painters) needed to meet production demands given the task time available. To do this, one need to know the production configuration, that is, the paint conveyor system, which can be a continuous moving conveyor, or a Stop & Go type conveyor. In either case, the painting time is the time duration that the BiW spends in the spray booth minus any wasted time (non-value-added time), such as the time needed to change colors or clean spray nozzles.

For the continuous conveyor case, the BiW spends the time t_{booth} given a part-to-part separation D_{p-p} and a conveyor speed V_{Conv}.

$$t_{booth} = \frac{D_{p-p}}{V_{Conv}} \tag{4.14}$$

In the Stop & Go case, the time t_{booth} is the same as the conveyor time, for example, the conveyor moves every 60 seconds, then the $t_{booth} = 60$ seconds, i.e. the booth will have 1 vehicle in it every 1 minute.

The available time for coating should also account for the robot non-value-added movement, i.e. movement without painting. This is typically known as trigger on and off efficiency. Then the actual available (value-added) time will be (Equation 4.15):

$$t_{available} = \frac{t_{booth} - t_{wasted}}{\eta_{robot}} \tag{4.15}$$

Finally, one can obtain the needed flow rate requirement by dividing the wet paint volume over the available painting time, which also helps in knowing the number of robots needed; this is obtained by dividing the flow rate required by that of the used paint applicator, a safety factor of 1 might be added.

Additional calculations can be done to evaluate painting lines' productivity, which might be useful in benchmarking the different painting lines at different facilities. For this calculation, one should focus on the paint line productivity which translates into surface area painted per unit time, i.e. ($m^2_{vehicle}$ / min). Also, the capital investment in the paint line should be considered in terms of number of robotic painters and the booth area; this can be quantified as: ($m^2_{vehicle}$ / min / m^2_{booth}) and ($m^2_{vehicle}$ / atomizer).

To illustrate the calculations with a quantitative example, we introduce the following example: Automotive paint engineers need to decide on the number of painting robots, and paint requirements for one year of production; to coat a new model with a total surface of $14\,m^2$ with a top-coat thickness of $40\,\mu m$. This OEM is using a 40% solid top-coat paint formulation, with a continuous conveyor type. The conveyor speed is rated at 3 m/min with part-to-part separation of 8 meters. The paint applicators used are rated at 60% transfer efficiency with robotic trigger on/off time of 80%,

Table 4.3 The paint calculations results

Information	Continuous Conv.	Stop & Go
Actual area to be painted (m^2)	14.00	14.00
Thickness after curing (μm)	40.00	40.00
Volume required-dry (CC)	560.00	560.00
Wet volume (CC)	1400.00	1400.00
Amount to be injected from the application (CC)	2333.33	2333.33
Cycle time (sec)	91.43	70.00
Useful time (sec)	81.43	60.00
Available time to paint the vehicle (sec)	65.14	48.00
Flow rate of the paint (CC/min)	2149.12	2916.67
Number of atomizer required	5.00	7.00

and a flow rate of 500 cc. Color change time can be considered to be around 10 seconds.

Will the number of robots change if the OEM conveyor is a stop & go at every 70 seconds, and why?

The sample calculations are given in Table 4.3 for both cases.

4.6 Material Handling Systems Inside the Painting Area

The painting area features a variety of conveyor and handling systems due to painting process requirements. For example, an overhead conveyor system is used at the beginning of the paint-line to allow the BiW to be immersed and sometimes rotated inside the conversion and E-Coat tanks. Additionally the conveyor systems within the paint area should be maintained and prepared to operate effectively under the different painting stations. For example, the overhead conveyor will pass through several chemical baths so it should be equipped with dirt trays and other measures to protect it from chemical attacks and from carrying the chemicals from one tank to the next, a typical E-Coat hanger is shown in Figure 4.31. Also the paint process features a variety of curing processes which add more preparation for the conveyor system lubrication so that it does not dry or evaporate while inside the ovens. Oven conveyor rails are designed to be heat-resistant with their control drives outside the booth. Different conveyor types exist for the immersion stages and include overhead or pendulum types. Advances in the immersion conveyor systems include the ability to rotate the BiW shell within the tank, Duerr introduced this technology under the name RoDip, depicted in Figure 4.32, while the typical dunk type system is given in Figure 4.33.

At the end of the immersion stages, the BiW shell is removed from the overhead hanger into an auxiliary conveyor that moves the BiW into a floor conveyor system,

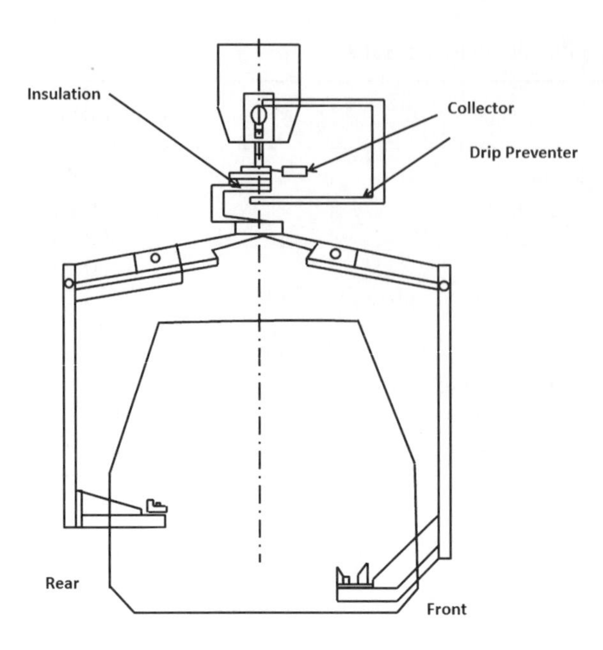

Figure 4.31 The E-coat hanger system

Figure 4.32 The RoDip system. Reproduced by permission of Duerr Systems, Inc © 2010

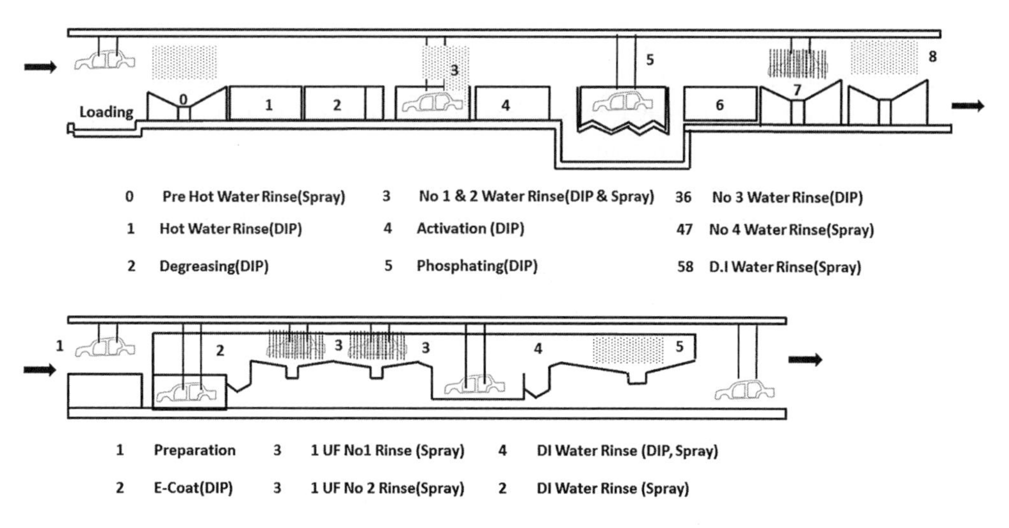

Figure 4.33 The dunk-based system

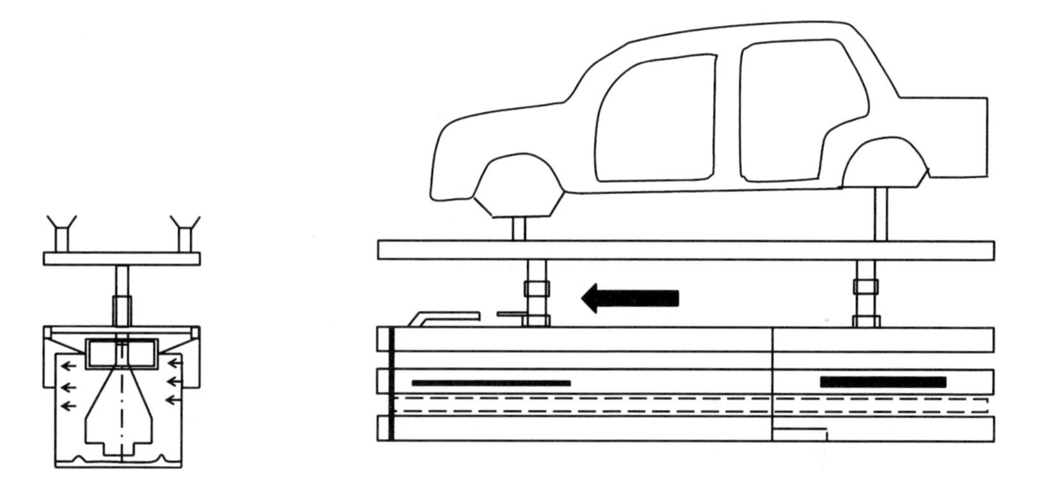

Figure 4.34 The floor conveyor system

shown in Figure 4.34. The floor conveyor provides more flexibility so that the painted BiW shells can be taken out of the production sequence to create color batches. The floor conveyor takes the shell through the sanding, inspection, and sealer stations.

The different paint conveyor systems can be actuated using different configurations that include power-pulled chain-based conveyor system, or powered chain and free conveyor system that utilizes pulleys actuated by chains. The power and free conveyor system allows one or more trolleys to be halted through production without stopping the conveyor chain itself. Finally, the inverted powered chain configuration for the floor-mounted conveyor is used; this type of conveyor is suitable for the final stage of the painting process, i.e. the top and clear-coating, because it tends to collect the dirt on the bottom of the conveyor which prevents it from getting into the BiW surface. However, this conveyor floor installation restricts access and establishes fixed points in terms of equipment allocations within the paint area. Other forms of transporting the shell within the paint area include skids and trolleys.

The floor conveyor system features several transport possibilities: cross-transfer, longitudinal transfer, and vertical transfer. Such transfers allow great flexibility in the paint booth handling, sorting, de-sequencing, and re-sequencing of the body shells.

4.7 Painting Robotics

The robotic painters introduce a repeatable, reliable means to deliver the spray paint on the BiW at high production rates, without any safety concerns due to the paint emissions. Additionally the robotic painters allow greater degree of control over the paint delivery in terms of the spatial motion and the paint flow rate. These robotics are fully integrated into the production line able to change painting programs upon changing body styles and colors.

The painting robotics, as in the case of the welding robots, should be programmed to design their end-effecter trajectories so they apply the adequate number of paint layers (over-lap) according to the desired film thickness and shell CAD geometry. Six axis industrial-style robots are typically used for the painting area offering six different degrees of motion: translation, reach, elevation, yaw, pitch, and roll. The last three axes are governed by the movement of the manipulator wrist. The yaw is the angular right and left motion, the pitch is the angular up and down, and the roll is the angular around motion.

Robot programming can be done in one of two styles: the point-to-point and the continuous path routines. In point-to-point, the robot arm is programmed to move in straight lines passing through a series of points spread in the painting path; this programming scheme is typically used for simple geometries that can be mapped into a series of spatial points. On the other hand, the continuous programming is still discrete, i.e. composed of spatial points to decide on the path, however, the distance between these points is much smaller. The continuous programming is intended for more complex geometries than those in point-to-point routine. Due to the combination of simple and complex geometries within a BiW, automotive OEMs program their robots in a mix of these routines to have the faster programming and lower computational requirements of the point-to-point method while still keeping the continuous path for the more intricate shapes within the shell.

The painting robotics feature different types of mounting and installation; some are over-hanging robots while other are floor-mounted. Reciprocating painters are also used for flat panels such the roof. The different robots' variations are displayed in Figure 4.35, showing a floor-mounted and over-hanging robot. When selecting and programming painting robotics, careful consideration should be given to each robot's working envelope to avoid any robot collision or interferences, because the booth space is limited.

A painting booth features a variety of robots to achieve the painting process according to the designed film build-up sequence, that is the number of laps to achieve the final thickness (over-lap), all this within the available paint time. A typical paint spray booth robotic configuration is given in Figure 4.36, which shows the variety in the number of spray applicators and the robotics. From this figure one can

Figure 4.35 The different types of painting robots used in spray booths

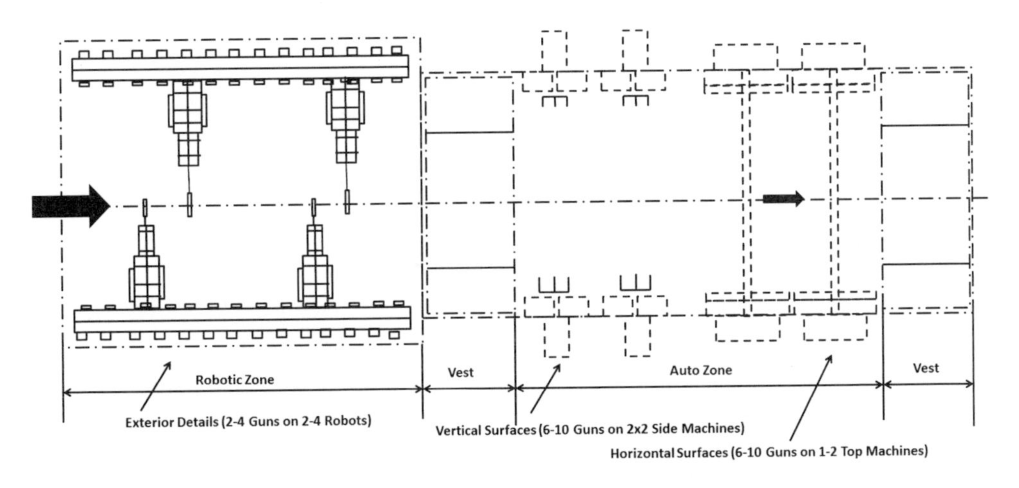

Figure 4.36 A typical paint spray booth robotic configuration

conclude that each spray applicator is designed to deliver a different paint footprint, the combination of the different paint footprints allows more flexibility in dealing with flat surfaces and the more complicated geometries. Larger footprints allow better paint productivity with fewer number of applicators.

The robotics control should also be able to adjust the painting applicator spray conditions, which include the flow rate, the flow pattern or fan (paint footprint). The flow fan is controlled by the shaping air and the paint flow rate. As the flow rate changes, the electro-static charge amount should be adjusted accordingly. All these adjustments are based on a specific paint viscosity, so the paint raw material batches should be carefully controlled, handled and continuously monitored.

The design of the over-lap is vital in ensuring a fine paint finish at the right desired thickness, yet done with the minimum paint waste or overspray. This requires the use of different paint footprints for each panel: flat area, area with features, and its edges. A large paint footprint is used for the flat area, a medium one for the edges and a small, narrow one for the intricate features. Changing the footprint on the fly is a very complicated process, because it requires changing the applicator flow rate, shaping the air pattern, the electro-static charge, the gun distance from target, and finally the robot painting speed.

The investment and the pay-back of robotic painters can be estimated following the formula proposed by [35] for industrial paint, as in Equation 4.16:

$$Y = P_{robot} /(W_s + P_s + R_d - [M_c + S_c]) \tag{4.16}$$

where Y is the number of years for the robot to pay its cost back, P_{robot} is the purchase price for a painting robot including installation cost (around ~\$500,000), W_s is the savings in terms of replaced labor costs (around ~\$40,000), P_s is the production savings due to increased productivity per year (around ~\$30,000), R_d is the robot depreciation per year (around ~\$70,000/yr), M_c is the robot maintenance cost per year (typically around \$6000/yr), and S_c is the additional staffing cost needed for the robots (~\$60,000). So a typical pay-back period will be around 5–7 years from installation, this number might be further reduced depending on the production volume or, in other words, the robot utilization ($m^2_{vehicle}$ / atomizer/hr).

4.8 Paint Quality Measurements

Achieving and controlling the paint quality is one of the most complicated steps within the automotive manufacturing processes, due to the large number of the controlling factors and the inter-relationship in between them. Such controlling factors include;

1. *The painting booth conditions*, in terms of the booth temperature and RH because both of these variables affect the solvent evaporation rate as the paint travels from

the paint applicator nozzle onto the BiW surface. The evaporation rate affects the solvent content and paint flow over the panels, which ultimately affects the paint color shade; less solvent in the paint (due to more evaporation) leads to a darker shade and vice versa. The waterborne formulations are even more sensitive to such conditions hence they require a tighter operating window within the psychometric chart, as discussed earlier.

2. *The paint flow characteristics*, in terms of the paint particles' size and impingement speed at the BiW different surfaces. More paint particles translates into more surface area for the same paint volume, hence more solvent evaporation, also the finer particles can lead to a cascading effect in charging neighboring particles in electro-static implementation. However, there is a limit on the particles' size, below which the particle can carry the charge and might even split in smaller particles; this limit is called the Raleigh limit. The impingement speed affects the metallic flakes' surface alignment, the higher the speed, the more parallel the alignment. The flakes are typically made of either aluminum or mica flakes to add more shine to the paint through its high reflectivity. However, if the flakes are not aligned parallel to the surface, then the paint will reflect the incident light differently from different angles, leading to mismatched color perception. Additional metallic flakes' deviation is related to the flakes penetrating the film different layers leading to a defect called mottling.
3. *The paint material formulation*, in terms of the paint viscosity and surface tension attributes which affect the paint atomization into ligaments then into particles. The paint viscosity affects the way the paint flows over the painted surface and its curing characteristics. The percentage of solids within the paint chemistry should be considered due to their effect on the evaporation rates. The concentration of the metallic flakes and their distribution within the paint medium affect the paint color perception also.
4. *The paint applicator conditions*, which can be described in the applicator flow rate, the shaping air flow, the air flow in air-assisted guns and the rotational speed of the rotary bell applicators. Other effects arise from the variations in the applicator distance from the surface and the electro-static charging mode (internal vs. external electrodes). Furthermore, the charge amount versus the paint thickness can play a role in discharging paint particles to its electric charge which might lead to crater formations; this is evident at the panels' edges where thinner paint films are typically applied.
5. *The production-related variables* that include the production rate, and the types of the robotic painters and their painting speeds (gun speed across the surface). Other effects can come from the BiW design in terms of surface geometries that might lead to changes in the paint flow, in addition to the fact that some areas within the BiW might trap some of the conversion coating contents, such as silicones in the spray booths, leading to further crater formations. For each BiW body style, a detailed study of the gap filling and drain-off should be conducted.

All these variables are considered to be independent variables that govern the paint appearance and integrity. The paint appearance can be quantified based on the final film, geometric, color and gloss attributes.

The paint film color can be quantified by considering its color as a composite of a three-dimensional nature, that is a lightness attribute and two chromatic attributes that are the hue and saturation. The lightness is used to distinguish white objects from gray ones and dark from light colored objects; the hue, on the other hand, is defined as a color attribute to distinguish if the color is red, yellow, green, blue, etc. The saturation describes the color departure from that of gray color having the same lightness [39]. Thus the paint color can be mapped into a color scale with the coordinates of *L*, *a,* and *b* corresponding to its lightness, hue and saturation respectively.

In addition to other geometric attributes which describe the way the incident light interacts with the paint film profile; this interaction is also quantified through three attributes: the surface gloss, the haze and the translucency. The geometric attributes become more apparent when dealing with the metallic automotive paint formulations, with the gloss, hue, and saturation being the most important. On the other hand, translucency is more apparent when dealing with automotive clear coats. Geometric attributes are not evident when dealing with automotive surfaces that diffuse incident light. This is due to the fact that incident light interacts with automotive panels in one or a combination of following modes: specular reflection mode which can be described by the object gloss; a transmission mode which relates to clarity and translucency of the paint, and absorption mode which describes the light interaction with the color pigments within the paint structure. The incident light might get scattered due to micro-profiles that lead to multiple reflections in different angles, this can be evaluated and described as the paint film haze. One of the defects related to the film roughness and paint leveling is called orange peel, and this deviation will be discussed shortly in the coming text.

To evaluate the geometric and color composites, two sets of curves are required: the first is the spectrophotometric curve and the second is the goniophotometric curve. The spectrophotometry curves are produced through a spectrometer which is a light source capable of producing color of any selected wavelength and a photometer to measure the intensity of that light when projected on the surface under evaluation. The goniophotometric curves describe the variations in the light reflection as a function of the viewing angle, with the light source being fixed in location. Table 4.4 from [39] illustrates the different measurement standards for evaluating the surface gloss.

4.8.1 Paint Defects and Theory

This sub-section is intended to introduce the most common variations within the automotive paint applications that relate mainly to the film's final appearance. The

Table 4.4 The different measurement standards for evaluating the surface gloss

Gloss measurement	Standards	Application	Test Specifications, field angles	
			Source	Receiver
Specular gloss 20 deg.	ASTM D523	High gloss plastic films	0.75×3.0	1.8×3.6
	ASTM D2457	Appliances finishes		
	FTM 141A/6104	Automotive finishes		
Specular gloss 30 deg. Narrow angle haze 32 deg. Wide angle haze 35 deg.	ASTM E430	High gloss, non-diffuse metal	0.44×7.0	0.4×4.0
Specular gloss 60 deg.	ASTM D523	Differentiate between high, medium or low	0.75×3.0	4.4×11.7
	ASTM D2457	gloss for paints, plastics		
	ASTM D1455			
	ASTM C584			

theory behind the paint flow and leveling phenomena will be discussed in relation to the paint viscosity and the gravitational effects.

4.8.1.1 Theoretical Background

Once the paint is deposited on the BiW's different panels, the shear and gravitational forces start to affect the liquid paint flow and leveling characteristics. The balance and control of these forces lead to optimized film spread and flow across the surface. The shear forces acting on a liquid paint film of thickness t, and viscosity ϑ, can be described by the balance of the shear forces acting on the interface of film to substrate S_{int} and these forces acting on the film's external surface S_{sur} or the air-to-paint interface; the external shear forces are these forces exerted by the paint applicator, in other words, it is sensitive to the gun distance, flow rate, and speed. These forces can be described in Equations 4.17 and 4.18 respectively. Hence the flow of the paint film F_{paint} can be estimated based on the formula in Equation 4.19.

$$S_{int} = V_{int} \times \vartheta / t \tag{4.17}$$

$$S_{sur} = \vartheta \times \frac{dV}{dy} \tag{4.18}$$

$$F_{paint} = \frac{1}{2} \times \frac{S_{int}}{\vartheta} t^2 \tag{4.19}$$

The applied shear stress as evident from Equation 4.18 leads to a velocity within the paint film that controls its spread movement, the spread rate will be at its highest at the air-interface and 0 at the substrate interface.

Other forces acting on the applied paint film are related to the gravitational effects; one should note that the paint film behaves differently when applied on panels with different orientations; in other words, the paint spreads at different rates when applied on a flat panel on the hood when compared with the case when it is applied on a door. The gravity-induced paint flow might lead to paint defects such as paint sagging. The schematic in Figure 4.37 shows the forces acting on a paint film applied on a vertical panel.

The gravity forces for the schematic in 4.37 can be explained through the following equations (Equations 4.20 through 4.22) that describe the paint flow, spread rate (velocity), and gravitational induced force.

$$F = \rho g \times (t - y) \tag{4.20}$$

$$V_{int} = \frac{1}{2} \times \frac{\rho g}{\vartheta} t^2 \tag{4.21}$$

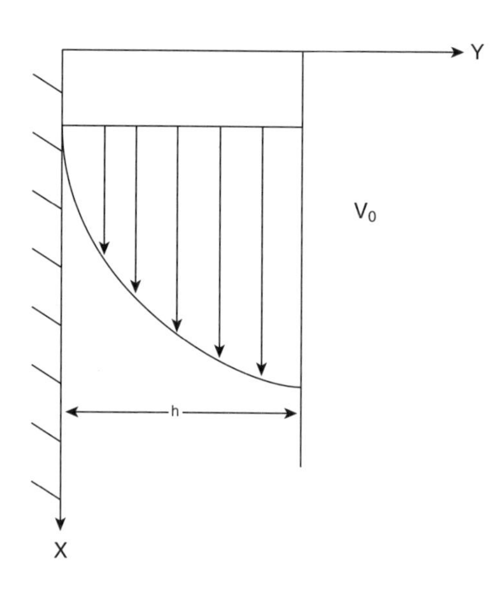

Figure 4.37 The forces acting on paint film when applied on a vertical surface

$$F_{paint} = \frac{1}{3} \times \frac{\rho g}{\vartheta} t^3 \tag{4.22}$$

Thus, from the above formulations, it is clear that the maximum sagging velocity will be at the paint-to-air interface, while the velocity profile across the paint film will be parabolic in nature, and can be controlled by manipulating the paint viscosity.

Additional forces that affect the paint spreading and leveling are related to the surface tension and energies, and its variations across the panels' surfaces. As described for the adhesive bonding case, any facial contaminants lead to variations in the surface tension; hence affecting the way the adhesive wets or spreads over the surface. This is also true for the automotive spayed paints. The surface tension forces can also be estimated and described mathematically in terms of the paint flow F_{paint} induced by a surface tension gradient across the panel (i.e. x-direction) $\frac{\Delta \gamma}{\Delta x}$ or $\dot{\gamma}$ as in Equation 4.23 and its gradient with time in the thickness direction $\frac{dt}{dtime}$ in Equation 4.24.

$$F_{paint} = \frac{1}{2} \times \frac{\dot{\gamma}}{\vartheta} t^2 \tag{4.23}$$

$$\frac{dt}{dtime} = \frac{t^2}{2\vartheta} \ddot{\gamma} \tag{4.24}$$

The schematic for a surface tension induced flow is given in Figure 4.38, showing the surface tension gradients across the panel.

From above, it is clear that the paint rheology controls its flow and leveling behavior. Rheology is defined as the science of flow and deformation of matter. So the knowledge of the paint rheological reactions due to induced shear/gravitational/surface tension forces is very important. Following rheological behaviors defines the way that paint responds to such forces;

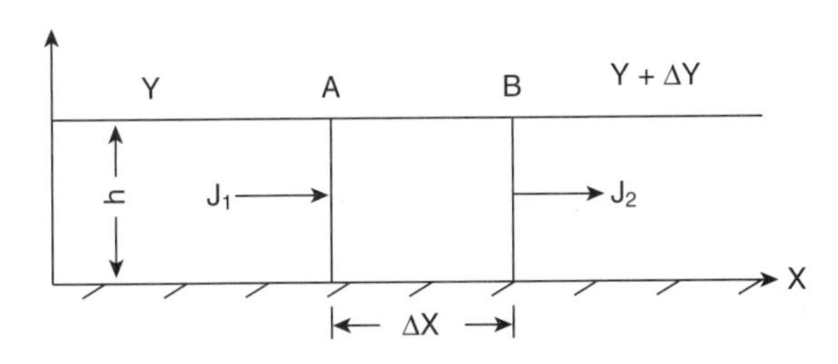

Figure 4.38 The forces affecting a paint film due to the difference in surface tension

- *Newtonian-based behavior:* is when the paint shear rate is proportional to the induced shear stress.
- *Non-Newtonian:* when the paint shear rate proportionality is not constant with the shear stress.
- *Dilatency:* when the paint volume (thickness) increases as the shear stress increases; i.e. shear thickening behavior.
- *Pseudo-plastic behavior:* when the paint film gets thinned by increased shear stress; shear thinning behavior.
- *Thixotropic behavior:* is the gel to liquid transformation induced by paint agitation, with time-dependent recovery to semi-solid state upon standing.

All the above behaviors motivate the addition of further controlling agents to modify the paint viscosity and its dynamic tendencies. Such controlling agents include clays and fumed silica to form hydrogen bonds within the paint structure.

One should note that the final paint roughness profile is an outcome of the substrate roughness, and the different layers' profiles in addition to the shrinkage rate while curing. The final roughness profile determines the paint's interaction with light and hence its color and geometric attributes. Automotive OEMs continue to work on improving their paint atomization processes and substrates' preparation steps to ensure finer paint depositions at smoother substrate profiles with minimum surface tension gradients.

The following describes the most common automotive paint defects and their main root causes and remedies. The main variations discussed here are: cratering, air entrapment, haze, orange peel, fish eyes, and solvent popping. Other defects do exist, however, these types are the most common.

- *Cratering:* is described as the formation of small bowl-shaped depression with circular edges, in the paint film. Craters might be caused by dirt or surface contaminants, or the discharging of the electro-static paint at the panel edges due to the thinner thicknesses. The main cause, however, is typically related to silicone or mineral oil contamination that has low surface and hence can spread over the panel surface easily. The low surface energy of silicone induces a surface tension gradient and consequently a surface tension-driven paint flow around the contaminants, creating the crater. The silicone is typically hidden within the panel seams or in between joints.
- *Fish-eye:* is a crater with a flat center; caters have a raised center with circular edges. The main cause of fish-eye is having undispersed paint ligaments on the surface; however, proper surface tension can spread such globules with time.
- *Haze:* is related to the light diffusion across the panel surface leading to a dull surface look, in other words, haze is the lack of surface gloss. Haze is mainly

related to a rougher paint film profile that might be due to poor coalescence of the paint, the accumulation of paint pigments, or due to micro-voids. Haze might be the result of uncontrolled atomization, leading to large paint particles. Haze is typically measured by evaluating the reflected component of an incident light aligned at 20 °, any reflections collected beyond 20 ° +/– 1 ° are considered to be indicative of haze.

- *Orange peel:* is described as the surface waviness that resembles the surface of an orange. Orange peel is mainly related to paint leveling and its shrinkage upon curing. Orange peel can be described in terms of its profile wavelength. Mainly orange peel can be classified further according to its impact and cause; so it can be categorized into:
    - *Dulling:* with wavelength around 10 μm and an amplitude of 0.1 μm or less, causing a lack of gloss while the waviness is not visible. The dulling is mainly related to the non-uniformities in the paint film caused by poor paint atomization and dispersion.
    - *Shrinkage:* with wavelength around 300–400 μm, with an amplitude of 0.4 μm. The shrinkage tends to cause spotty dull areas, its main root cause is believed to be related to the volumetric contraction upon curing and cooling.
    - *Wave:* with wavelengths around 35 μm, and an amplitude of 1 μm. This type is visible to a human observer as an orange-peel surface, and is mainly caused by the atomization process and the improper leveling of the paint once deposited on the surface. This can also be related to the paint viscosity variations.

So from the above discussion, the wave type can be controlled and quantified when the paint is being atomized and deposited on the surface. Its leveling and settling phases control the wave type. As the panels travel in the curing oven, the combina-

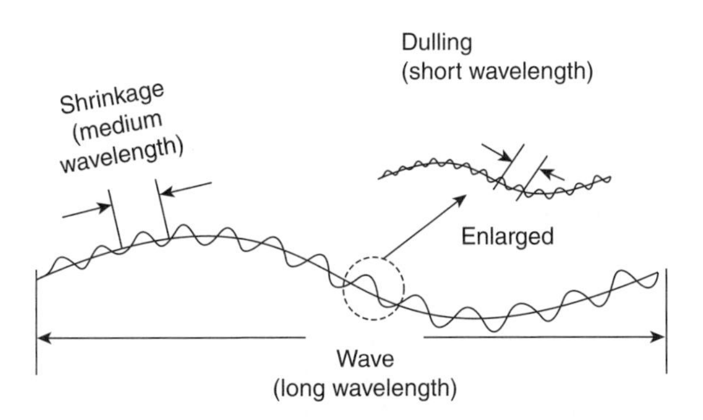

Figure 4.39 The orange peel different waviness

tion of the paint film and substrate go through volumetric contraction and shrinkages leading to the shrinkage-based orange peel. And finally as the paint cools in the final stages of the oven, it goes through the third type of the orange peel- i.e. the dulling. The schematic in Figure 4.39 depicts these different types.

- *Solvent pop-ups:* solvent popping is caused by the improper curing of the paint film, typically curing done at high ramp-up rates. The high ramp-up rate does not allow the solvent content to evaporate freely, leading to solvent entrapment, which then expands further leading to its pop-ups.
- *Corrosion:* all the above discussion is based on deviations that affect the paint look and appearance perceptions, on the other hand, the paint integrity characteristics relate to the paint film-build up in terms of its thickness and the interfacial adhesion between the different layers and the substrate. Also the paint chemical composition in terms of certain pigments' concentration is important in deciding on the paint sensitivity to certain environmental forces such as UV sunlight. However, the most important paint defect with relation to the paint integrity is corrosion because it leads to permanent undesired transformations not only to the paint film but also to the metallic substrate.

  Corrosion can come in different forms caused by different root causes, these types can be summarized into following types:
  - *Perforation-based corrosion:* where the salty electrolyte is hidden in interior panels and is in direct contact with the metallic substrate. Also, the plugging of drain-holes within the BiW might lead to this type of corrosion. This type is most evident in the lower interior of doors, rocker panels, and tailgates.
  - *Galvanic corrosion* is the corrosion resulting from having two different metals with different nobility in close/intimate contact, such as Al and steel. This type can also be called bimetallic corrosion or galvanic battery corrosion. The more active, or anodic, metal corrodes rapidly while the more noble, or cathodic, metal is not damaged. This means insulating all the joining points between Al and steel within the BiW using an adhesive bond layer.
  - *Crevice corrosion* remains a major problem because of current uni-body design, which includes a mass of box sections and joints. It is almost impossible to eliminate all the small cracks and openings between joined surfaces that are prime sites for crevice attack.
  - *Pitting corrosion* is a localized attack, usually caused by chlorides. The mechanism governing pit growth is similar to that of crevice corrosion.

Table 4.5 from [23] describes the different types of coatings applied to avoid and mitigate these corrosion variations. Figure 4.40 shows the different locations where these coatings are applied. The under-body sealer and wax material selection and their locations within the BiW shell are designed based on the different panels' sensitivities to the different types of corrosion described above.

Table 4.5 The different types of pre-coating formulations

Coating formulation	Process	Typical coating deposition	Panels
	Hot-dip based coating systems		
Hot-dip zinc	Galvanizing sheets using hot dip process	70–100 g/m^2	Inner and outer panels, applies to structural components
Hot-dip aluminum-silicon alloy	Hot-dip of Al sheets into Si enriched coating	38 and 60 g/m^2	Chassis components and exhaust systems
Hot-dip lead-tin alloy	Hot-dip process in 10% tin coating	40 g/m^2	Fuel system; tanks and lines
	Electro-deposition based coating systems		
Electro-plated zinc	Dip into electro depositions tanks	10–50 g/m^2	Inner and outer panels
Electro-plated zinc-iron alloy	Dip into electro depositions tanks, with 15% nickel enriched coating	30–50 g/m^2	Inner and outer panels
Electro-plated tin	Thin tin coatings on cold rolled sheets	10 g/m^2	Inner and outer panels

Reproduced by permission of American Iron and Steel Institute and Auto/Steel Partnership © 2010.

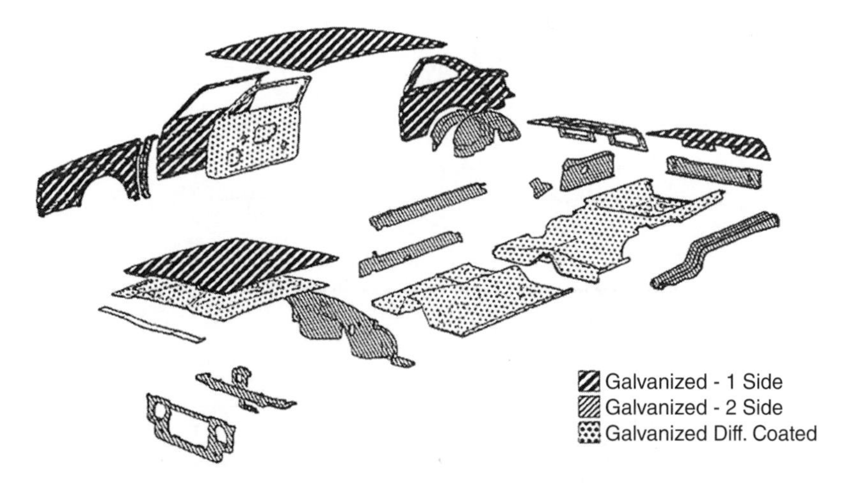

Figure 4.40 The different galvanized coatings applied and their locations in a BiW. Reproduced by permission of American Iron and Steel Institute and Auto/Steel Partnership © 2010

Exercises

Problem 1

Draw a complete painting booth layout; by filling in the boxes shown below; note: this booth employs a five-stage zinc conversion coating station, with wet on wet coating system.

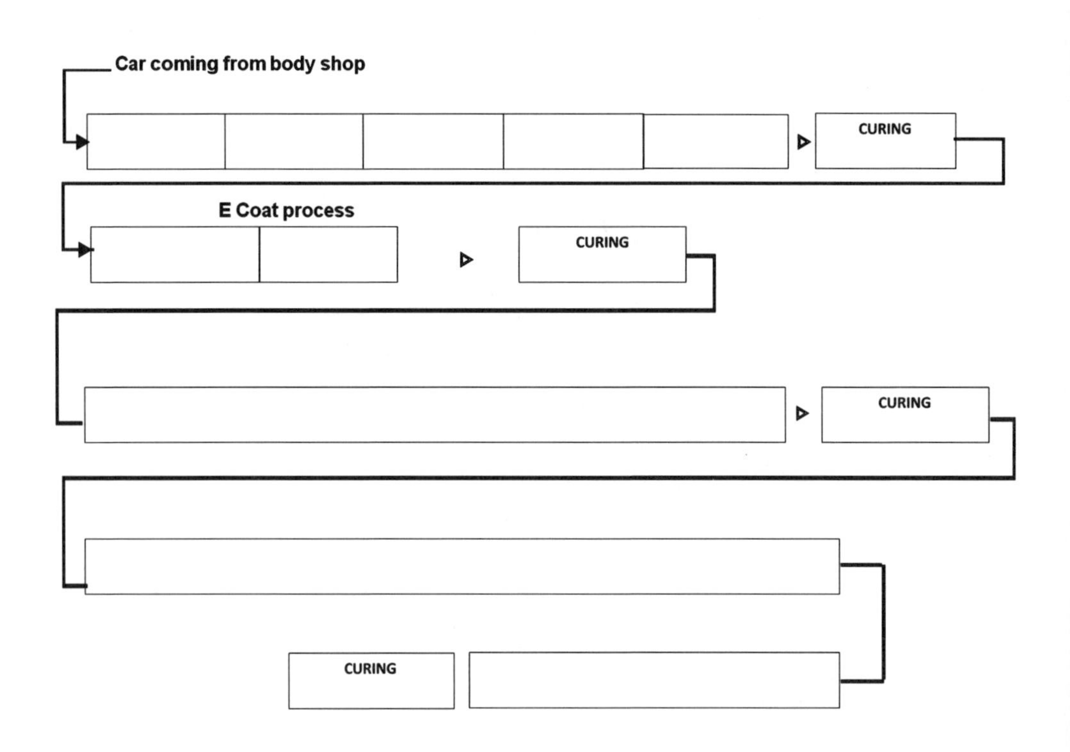

Problem 2

An automotive OEM "MEO automotive" is producing a new vehicle *CMW* on its production line that runs at 5 m/min, with 5.5 m pitch, using painting robotics with trigger on time of 75%, and a 50 m^2 booth. The *CMW* has a 12 m^2 of surface area, and typically coated with 25 micron of primer. Based on this information answer the following:

1. Decide on the paint applicator (from the options described below), to provide the *most efficient utilization of the primer painting atomizers*. You might assume a flush-time of 10 seconds.

   <u>Painting applicator types</u>

   Model AA40 Rotating cup capable of 300 cc maximum flow rate, of a 60% solid paint, TE = 80%.
   Model BB30 Airless gun capable of 600 cc maximum flow rate, of a 45% solid paint, TE = 85%.

2. Based on your selection from (1), compute the following (assuming 5000 working hours / year):
   i. the production volume of the *CMW* vehicle per year;
   ii. the paint requirements (gallons) for one month of production (note, 1 gallon = 3785 cc);
   iii. the paint booth space utilization and Productivity.

Problem 3

An automotive OEM "MEO automotive" is producing a new vehicle *CMW* on its production line that runs at 5 m/min, using painting robotics with trigger on time of 75%, and a 50 m^2 booth. The *CMW* has a 12 m^2 of surface area, and typically coated with 25 micron of primer. Based on this information answer the following:

1. Decide on the best combination of a paint applicator and a conveyor system (from the described below), to provide the most efficient utilization of the primer painting booths and its atomizers. You might assume a flush-time of 20 seconds.

   <u>Conveyor choices</u>

   Type (a) Continuous conveyor with 5.5 m pitch, Type (b) 2 minute Stop and Go conveyor.

Painting applicator types

Model AA40 Rotating cup capable of 450 cc maximum flow rate, of a 60% solid paint, TE = 80%.
Model BB30 Airless gun capable of 600 cc maximum flow rate, of a 45% solid paint, TE = 80%.

2. Based on your selection from (1), compute the following:
   i. the paint requirements (gallons) for one month of production;
   ii. the production volume of the *CMW* vehicle per year (assuming 5000 working hours / year).

5

Final Assembly

The final assembly area is responsible for assembling more than 3000 interior and exterior trim components in/on the painted vehicle shell. The assembly area features a variety of manual, automated and robotic stations, which are further aided by different types of autonomous guided vehicles (AGVs) and complex conveyor systems. The final assembly can be functionally divided into: trim line assembly, chassis line assembly, final assembly, and the testing area. In addition to some sub-assembly cells/lines dedicated to large components' sub-assembly such as the engine sub-assembly line, this is focused on installing the harnesses and the hoses, in addition to other miscellaneous items, into the engine; also joining the engine to the transmission and the torque converters.

5.1 Basics of Final Assembly Operations

The final assembly area is composed of a variety of sub-assemblies' cells in addition to a main line. The main line has a variety of stations that install and mount suppliers' parts and components into the vehicle shell. The final assembly area is considered a labor-driven process due to the high labor value-added work compared with other stations in the body assembly plant. For example, the stamping press line is considered a material- and machinery-driven activity, while the body-weld area is machinery-driven, and the paint-line is a material- and machinery-driven process. The greater contribution of the labor input requires further considerations with regard to human–machine interaction in terms of safety, ergonomics, and work standards and time studies. The work standards and time studies are needed to ensure that the process is done following the same procedure, consuming the same amount of time, and that any problems or defects are reported and integrated in the production plan. These issues are not evident in other areas in the automotive plant due to the dependencies on the automated, machinery-driven processes as in the robotics. At the same time, the final assembly area features a wide variety of sensors to recognize and track the

---

The Automotive Body Manufacturing Systems and Processes, First Edition. Mohammed A. Omar.

body-shell as it progresses through the assembly line, in addition to alignment sensors to ensure the accurate placement of the major components such as the power-train, the fuel tanks system, etc. The assembly area uses different types of automated fixtures to help carry the weight of the components in addition to facilitating its introduction to the production worker.

The sequence of the final assembly operations depends on different factors that include: the vehicles produced and their geometry, the special jigs and fixtures that might be required to install certain items, the production precedence relationship, and finally the automotive OEM's special style of manufacturing and assembly. The location of certain machines, special jigs or the conveyance method may limit the location and sequence of certain activities; the location of these special fixtures or jigs is typically called fixed points.

5.1.1 Installation of the Trim Assembly

This station is focused on installing the following main components: the electric wiring and harnesses, the shell insulation and radiator insulation, the air duct system, the headliner, the condenser sub-assembly, the pedal sub-assembly, and the fire wall insulation. Other parts are also installed in this station such as the wiper links and the washer tank and its hoses and connections.

To install the above described parts, a different conveyor system is adopted from the ones used in the paint area. So the shell is transferred from the paint-line conveyor system into typically a floor-mounted conveyor, the elevation and speed of the body-shell are adjusted so it suits the production workers and allow them spatial access and enough time to complete the job. Some OEMs have adopted a moving belt system for the line workers so that they can keep up with higher production rates. The first step in the trim assembly area is to remove the shell doors to facilitate the installation process, so large fixtures can be used inside the shell to aid the workers. The removed doors are hung on an overhead conveyor system that keeps running the doors through the plant without any value being added to them. A typical layout for the trim assembly area is shown in Figure 5.1.

Most if not all of the assembly steps require some sort of off-line support, which can be intermittent dependent on the need, or it can be continuous as is the case with the material replenishment and staging steps. The trim assembly area features a variety of robotic-aided operations to allow for faster, easier, and more accurate assembly. An example of this robotic aid is in the installation of the door liners, where a robot extracts a door liner from a magazine, passes it under an adhesive applicator, then attaches it to the door interior. Other automation examples include the assembly of the vehicle cockpit, where the cockpit module is pre-assembled in a different cell, then lowered into the body, where a robotic arm applies the adhesive to seal all the joints, and then bolts the cockpit to the body from the sides. Another

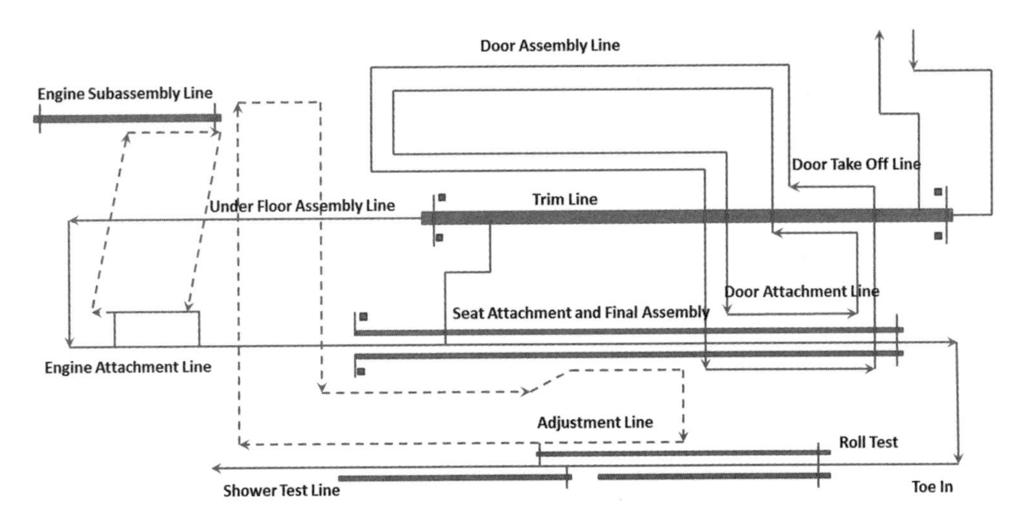

Figure 5.1 Typical layout for a trim assembly area

robotic-based assembly process is the assembly of the vehicle windshield. The windshields are first loaded onto pallets and centered in preparation for their installation. Then a robotic manipulator holds the windshield and passes it through an adhesive applicator, or in a different variation, one robot holds and manipulates the windshield while a second robot applies the adhesive film. Sensors mounted on the robotic arm ensure the accuracy of installation relative to the vehicle opening.

5.1.2 Installation of the Chassis

The chassis assembly area is also called the marriage area, where the power-train of the vehicle is coupled with the vehicle body-shell. To achieve this, the vehicle shells are transferred to an overhead conveyor system to permit the chassis installation from the bottom. The assembled power-train components are supplied from a sub-assembly area typically called the engine-line assembly area. The engine-line area features all the steps needed to install the different hoses, controllers and cables to the main engine body, in addition to coupling the engine to the transmission and the torque converters. The engine sub-assembly utilizes different types of conveyance depending on the accessibility needed, the station's configuration (left- and right-side workers), and the weight of the assembled power-train; typically a combination of an overhead system and an AGV is used. The final assembled power-train is then mounted onto an AGV or a trolley equipped with a hydraulic lift, and then shipped to the marriage area.

In the marriage area, the vehicle shell will be synchronized with the AGV so both meet at a specific location that features the power torque machine that will use bolts

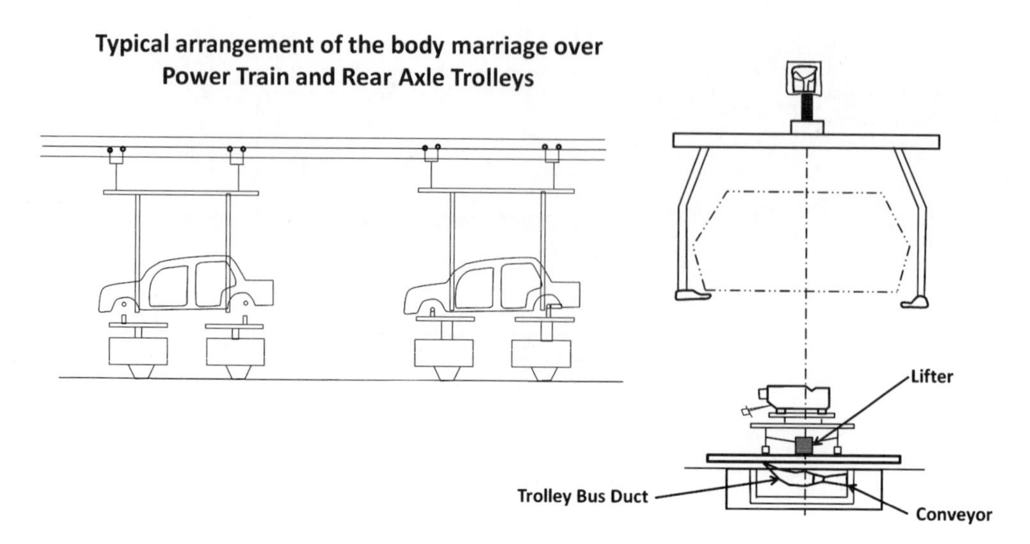

Figure 5.2 The body marriage area showing the rear axle trolley

and fasteners to join the power-train and shell together. Additionally, the elevation of the suspended shell is programmed according to specific settings to facilitate the mounting process. The power-train is lifted via the hydraulic lift to meet the stationary shell. This process sequence is shown in Figure 5.2. Several features within the vehicle shell are typically used to guide the power-train assembly to ensure its location and orientation within the vehicle; some laser projection-based sensors might also be used to ensure accurate placement.

Once the power-train is safely secured, the shell can move forward for further installations such as the exhaust system, also other components might have been already installed before the power-train, such as the fuel tank. At the end of the chassis line, other components are further installed and might include the wheels, and the tires. Due to the weight and shape of the tires, some fixtures are typically used to help better introduce the tire to the line worker. In addition, power-tools with built-in torque sensors are used to ensure fault-proof operation. Applying the right torque is essential in delivering mechanically sound joints that will not fail or fatigue due to under-torque or over-torque conditions. Forthcoming sections will discuss the assembly torque requirements and its relationship with the joint friction.

5.1.3 Final Assembly and Testing Area

The final assembly area features the vehicle shell on tires for the first time; the shell is then transferred from the overhead conveyor system into a double slat conveyor system. The final assembly area includes a single or a double slat conveyor system in addition to a variety of arm and drop lifters. Further component installations are

Figure 5.3 A chassis dynamometer

completed in this stage with the addition of providing the different fluids for the fuel-tank, transmission, engine, etc. to make the vehicle drivable. Also, the doors are coupled with the vehicle shells in this area.

After the assembly is complete, the completed vehicle starts the testing phase, where the following tests are conducted: alignment tests for wheels and turning radius, headlight test, side-slip test, engine drum test, and the brake test.

The alignment test focuses on adjusting the wheels through the manual adjustment of the top of the front and rear wheels and the camber of the front and rear wheels. Further alignments ensure that the steering wheel is positioned accurately. The alignment test is typically done in a chassis dynamometer chamber, shown in Figure 5.3. Additionally the turning radius can be tested and adjusted by measuring the right and left turning angles of the front wheels.

The headlight test evaluates the photometric axis of the headlamps by projecting them on a screen and then measuring any deviations.

The drum test is conducted by driving the vehicle into the chassis dynamometer to check the vehicle driving conditions. Also, the brake test evaluates the brakes' performance by applying and measuring the braking force of each wheel (drag, service brake, parking brake).

5.2 Ergonomics of the Final Assembly Area

The ergonomics of the assembly processes help the workers to conduct their tasks with ease and within the task time. The main ergonomics concerns in the assembly area are: installing heavy components, the frequent installation of medium to light-weight components, the installation posture, and the human hand utilization. All these aspects should be analyzed while considering that production workers are all different, they have physical and mental limitations, and humans have certain predefined reaction to certain scenarios.

Based on the above considerations, any component weight should be evaluated based on the frequency of the installation, for example, a tire weighing around 20 kg but handled around 1000 times per shift should be considered a heavy weight component $20 \times 1000 = 20$ tons. Additionally, monotonous work routines should be avoided by counting the number of times a certain operation or a repetitive motion is done per shift, for example, if it reaches 4000 times, then job rotation should be considered or an automation aid should be implemented.

The workers' posture and reach to equipment are typically analyzed while installation to make sure they comply with anthropometry principles. The basic anthropometric principles in regard to ergonomically safe work environment include:

- *The importance principle*, which states that the components necessary for safe and efficient operation should be placed in the most accessible positions.
- *The frequency of use principle*, which states that those components used most frequently should be placed in the most accessible positions or locations.
- *The function principle*, which states that components with closely related functions should be located close to each other.
- *The sequence of use principle*, which states that the components used in sequence should be located close to each other and in logical sequence of use.

In regard to the posture guidelines, the work environment should follow these guidelines:

- Workers should have an upright and forward-facing posture during work.
- The work points should be visible with head and trunk upright or head inclined slight forward.
- When the work requires the worker to stand, the weight of the worker's body should be distributed between the two feet.
- Where muscular force is needed, it should be by the largest appropriate muscle group.
- When the force has to be exerted repeatedly, it should be possible to exert it with either of the arms or legs, without any change to the equipment.
- Rest pauses should allow relief of all loads.

The typical practice in the automotive manufacturing is to conduct ergonomics audits for each step in the production sequence. The audit includes a variety of performance metrics such as lighting conditions, components distances and weights, etc.

5.3 Mechanical Fastening and Bolting

The automotive assembly processes rely heavily on the mechanical fastening and joining procedures to assemble and mount the different components and trim assemblies. The accurate mounting and placement of these assemblies are vital in ensuring the desired performance. The mechanical joining presents the ability to join modules to the vehicle shell with the possibility of removing the module in future application (non-permanent joints) for service or maintenance purposes. Mechanical joining can achieve an indirect way of joining dissimilar metals, hence avoiding the galvanic corrosion effects.

This section discusses the different variations of the mechanical joining technologies and their application to the automotive industry, in addition to the theoretical background required to understand the torque–tension relationship and their interactions with the interfacial friction. One should note that the threaded fasteners used for mechanical joining are normally loaded in tension so understanding the applied torque vs. the achieved tension relationship is vital.

The mechanical joining or fastening is meant to include the use of nuts, bolts, rivets and screws as a means to connect two components through a threaded element. The common element between the different screw fasteners is their helical thread, which can be a right- or left-hand thread, which leads the screw to progress or advance into a joint or a component or nut when rotated to achieve the connection. The ISO description of a thread is better described pictorially as in Figure 5.4. The thread terminology includes the following terms that should be properly defined to allow for better understanding of the thread action: (1) the pitch is defined as the distance between corresponding points on adjacent threads, with the measurements conducted parallel to the thread axis; (2) the thread lead is the axial movement (advance) of the screw once rotated one revolution; (3) the thread flank is the straight side of the thread between its root and its crest; (4) the effective diameter and the root diameter, with the effective diameter being the diameter on which the width of the spaces is equal to the width of the threads, measured vertically to the thread axis,

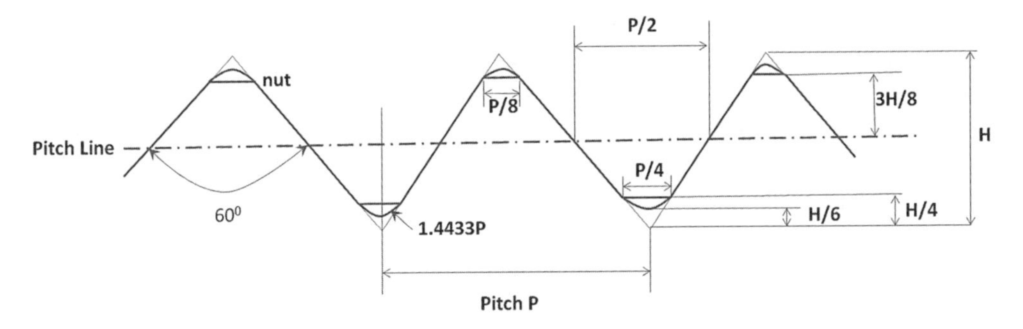

Figure 5.4 The ISO description of mechanical thread

Table 5.1 The thread designation based on ISO standards

Nominal size / Thread Dia.	Thread pitch	Width		Head height		Tapping drill	Clearance drill
		Max.	Min	Max.	Min		
M1.6	0.35	3.2	3.08	1.225	0.975	1.25	1.65
M2	0.4	4.0	3.88	1.525	1.275	1.6	2.05
M2.5	0.45	5.0	4.88	1.825	1.575	2.05	2.6
M3	0.5	5.5	5.38	2.125	1.875	2.5	3.10
M4	0.7	7.0	6.85	2.925	2.675	3.30	4.10
M5	0.8	8.0	7.85	3.65	3.35	4.20	5.10
M6	1.0	10.0	9.78	4.15	3.85	5.00	6.10
M8	1.25	13.0	12.73	5.65	5.35	6.80	8.20
M10	1.5	17.0	16.73	7.18	6.82	8.50	10.20

Note: All dimensions are in mm.

while the root, minor diameter is the smallest thread diameter when measured at right angles to the thread axis. ISO metric standards define the threads starting with the letter M, followed by the nominal diameter and the thread pitch, for example, an M6 × 1.5 is an ISO thread with diameter of 6 mm and a pitch of 1.5 mm.

The other system in defining threads for mechanical fastening application is proposed by the UNS (Unified National Standard series threads), and it defines the threads as Coarse C and Fine F, hence UNC and UNF threads. Tables 5.1 and 5.2 display some of available threads as designated by the ISO and the UNS systems.

Other mechanical elements are also used in conjunction with the mechanical fastener to aid in the joining process and in distributing the applied loads, such as the washers. For example, washers can be applied under the bolt head or under the nut or in both locations to help in distributing the clamping load over a wider area and also to supply a bearing surface for the nut rotation. Washers can vary in shape and configuration from the typical disc-shaped washer to lock washers with surface projections that tend to generate additional support forces on the assembly. Hence the load-bearing surface area controls the fastener performance. Typically, one can define the average area based on an average diameter from the minor and the pitch diameters, to be the load-bearing surface area for tension. This can be defined in Equation 5.1, while the minor d_r and pitch d_p diameters for ISO threads are defined in Equations 5.2, 5.3 and Equations 5.4 and 5.5 for the UNS threads.

$$A_{avg} = \frac{\pi}{16}(d_p - d_r) \tag{5.1}$$

$$d_p = d - 0.6495 \times p \tag{5.2}$$

Table 5.2 The thread designations based on Unified National Standard series threads

Size	Nominal major dia.	Threads per inch	Tensile stress area
UNC 0	0.06		
UNC 1	0.073	64	0.00263
UNC 2	0.086	56	0.0037
UNC 3	0.099	48	0.0487
UNC 4	0.112	40	0.00604
UNC 5	0.125	40	0.00796
UNF 0	0.06	80	0.0018
UNF 1	0.073	72	0.00278
UNF 2	0.086	64	0.00394
UNF 3	0.099	56	0.00523
UNF 4	0.112	48	0.00661
UNF 5	0.125	44	0.0083

Note: All dimensions are in inches, area in in^2.

$$d_r = d - 1.2268 \times p \tag{5.3}$$

$$d_p = d - \frac{0.6495}{N} \tag{5.4}$$

$$d_r = d - \frac{1.299}{N} \tag{5.5}$$

Then the applied load will be distributed among the first few threads due to the inaccuracies in the thread manufacturing. In the case of bolts, the tightening of the bolt head translates into an elongation in the bolt length, the tension that causes this elongation or stretching effect is typically called the bolt preload. Several correlations exist to estimate the required preload for permanent and non-permanent mechanical joints. One of these correlations is described in Equation 5.6 in the case of a non-permanent joint:

$$F_{pre} = 0.75 \times A_{avg} \sigma_p \tag{5.6}$$

σ_p is defined as the proof strength of the bolt and can be estimated as 85% of the bolt material yield strength. For the permanent joints, the 75% factor can be simply changed into 90% to obtain the preload needed. The next step would be to compute the required torque to achieve the desired amount of preload within the joint. The torque can be computed using Equation 5.7, which utilizes the constant K that is dependent on the bolt material and its size, typical values for K range from 0.16–0.3, also this equation uses the bolt nominal diameter d:

$$T = K \times F_{pre} d \tag{5.7}$$

Then, from the above equation sequence, one can compute the recommended torque for an M10 bolt with yield strength of 400 MPa to be around 14.78 kN of wrench torque.

Additionally as the mechanical fasteners are applied within a certain assembly, any further loadings on such assembly should consider the clamping load applied in each fastener. Based on the flexibility of the fastener relative to that of the joint, it can be classified as a soft or a hard joint, however, most of the mechanical fastened joints are somewhere between the soft and the hard joints, so the applied load on the assembly joints can be assumed to be distributed between the clamped parts and the bolt.

The load sharing between the clamped assembly and the bolt can be mathematically formulated to compute the final force applied on the bolt and that applied on the assembled parts. Equations 5.8 and 5.9 introduce these two computations for the final bolt load F_{bolt} and the final assembly load F_{ass}, respectively:

$$F_{bolt} = F_{pre} + \frac{k_{bolt}}{k_{bolt} + k_{ass}} F_{ext} \tag{5.8}$$

$$F_{ass} = F_{pre} - \frac{k_{bolt}}{k_{bolt} + k_{ass}} F_{ext} \tag{5.9}$$

where k_{bolt} corresponds to the stiffness of the bolt and k_{ass} is the stiffness for the assembled joint. Typically, one can assume that $k_{ass} = 3k_{bolt}$.

Even though the above mathematical formulation introduces an easy way to compute the wrench torque requirement for mechanical fasteners, one should note that there is a more dynamic behavior between the applied torque and the achieved tension in the fastener, such a relationship is typically termed the signature curve of torque–tension. A typical signature curve is displayed in Figure 5.5, showing the different phases that the fastener will go through as it is being tightened.

The first phase is the preloading phase that aims to achieve the target preload tension at the seating torque. Further phases describe the elongation of the fastener and the plastic deformation.

Also, one further investigates the torque–tension behavior by monitoring their evolution with time, as displayed in Figure 5.6. This figure describes four distinctive zones of the tightening process: (1) run-down, which is the process of driving the fastener into the joint without achieving any torque or tension, followed by (2) the snug-fit phase where the fastener starts to get tightened in the joint, and the achieved tension is still comparable to the applied torque; (3) the third phase is where the achieved tension starts to deviate from the applied torque, however, both are increased linearly, this phase is called the elastic clamping phase; and (4) the final phase is the

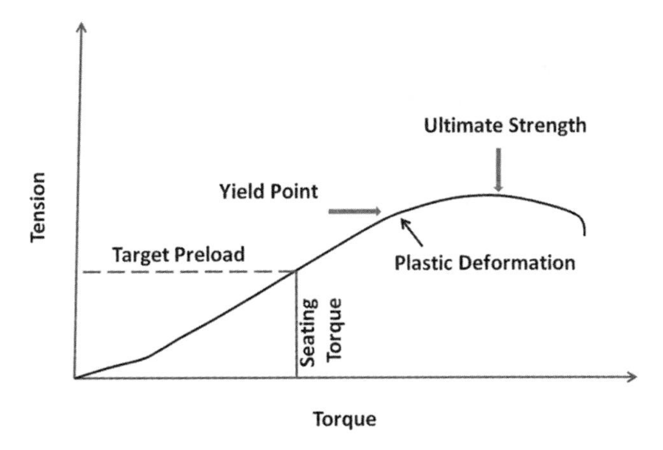

Figure 5.5 A typical torque-tension signature curve

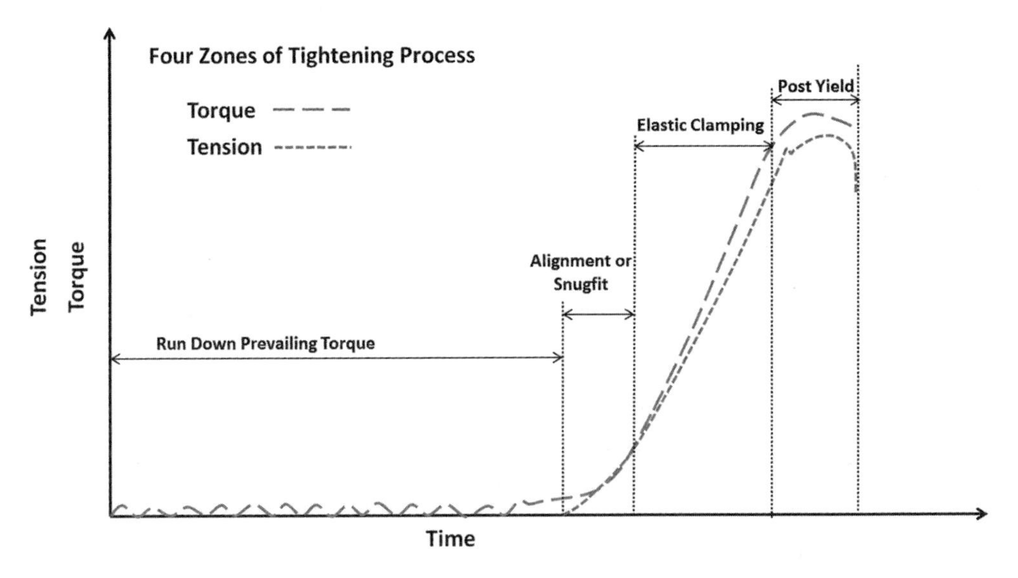

Figure 5.6 The torque–tension evolution with time inside a mechanical joint

post-yield, where the achieved tension reaches its maximum and then drops, corresponding to the applied torque signature. The main factor that affects the applied torque–achieved tension signature curve is the friction within the joint, because the amount of torque expended to resist the joint friction determines the remaining torque to tighten the fastener. Typically, friction consumes around 90% of the applied wrench torque in two areas: the fastener under-head friction (around 50%) and the thread friction (around 40%). The remaining 10% of the applied torque is consumed in creating the clamp load.

From the above discussion, changing the fastener under-head conditions, or manipulating the thread friction through lubrication, leads to drastic changes on the final clamp load. In one scenario, if the under-head friction is decreased by 20% (i.e. from 50% to 40% of applied torque), this leads to an increase in the clamp load from 10% to 20% of the wrench torque which is a 100% increase. So any minor changes in the washer conditions or contaminants on the fastener thread will lead to drastic impacts on final clamp load.

Other variables that affect the fastened joint behavior are dependent on the assembly method in terms of alignment, sequence of applying fasteners and torques, and the control method adopted to monitor the torque and achieved tension.

So establishing a correlation to describe the achieved clamping force (tension) with the applied torque enables better control of the assembly and fastening process. Re-inspecting Figure 5.6, one can correlate the torque and tension through a linear relationship in the first part of the curve. This relationship can be established using the friction and fastener attributes. One such correlation is given in Equation 5.10:

$$T = \left[\frac{P}{2\pi} + \frac{\mu_t r_t}{cos\alpha} + \mu_b r_b\right] \times F \qquad (5.10)$$

where T is the applied torque, and F is the achieved tension, μ_t is the thread friction coefficient, while μ_b is the under-head friction coefficient, α is the half of thread profile angle, and r_b, r_t are the effectives' radius between the interfaces of joint to bolt head or nut and the thread effective contact radius, respectively. The effect of friction coefficient can be highlighted as in Figure 5.7, by showing a group of lubricated bolts (solid) vs. unlubricated ones (dashed) on the preload force.

The automotive OEMs have developed different fastening strategies to ensure delivering the designed clamping force by measuring and controlling the assembly conditions, not only the applied torque. Ensuring the adequate clamp force is critical to joint performance characteristics; for example, if the delivered clamp force is more than the designed value, then the fastener will tend to get damaged and damage the mating surface, in addition, it will have high mean stresses leading to shorter fatigue cycle life. On the other hand, if the clamping load is below the designed value, then the joined components will slip, causing noise and rattle, and the final joint will be loose and not capable of holding the service loads. Additionally, the designed clamp force value should account for the different loosening mechanisms in terms of the thermal expansion differences, the hole interferences, and the elastic interaction effects. The elastic interactions are dependent on the joint configuration; for example, an average of an 18% loss in the clamping load is due to steel, a 30% average loss is typical for sheet gasket, while up to 40% loss in the clamping load is related to the spiral wound gaskets. Other factors include the assembly preload scatter accounting for about +/– 30%, and the surface conditions of the joint might lead to deviations within 10%.

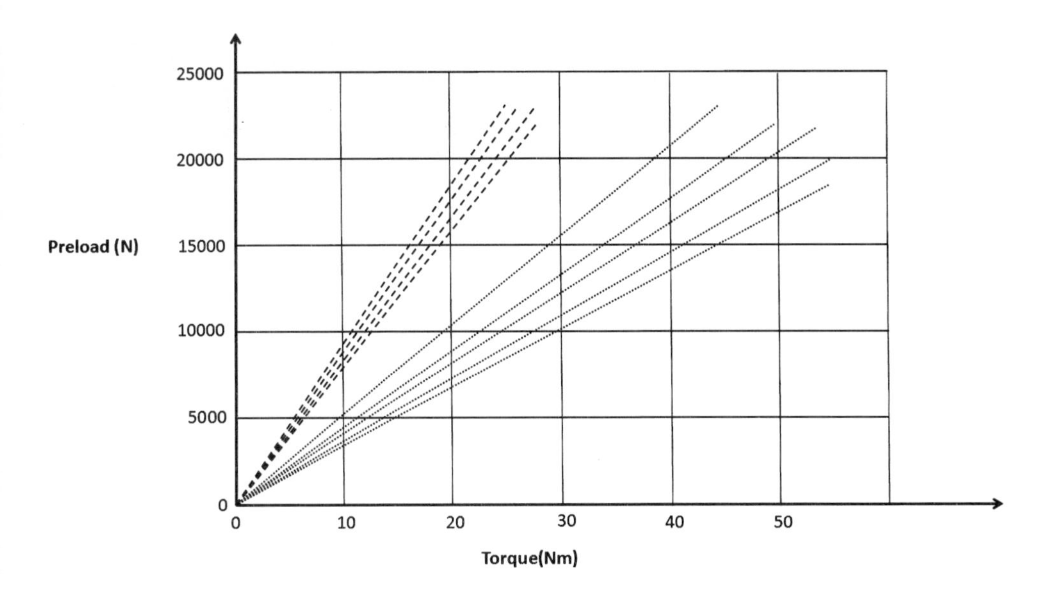

Figure 5.7 The torque–preload relationship

To accommodate all the above variations and the torque–tension interactions, the automotive industry have adopted several fastening strategies that include the following:

1. *Fastening with torque control*, which can be done in two variations: torque control only and torque control with angle monitoring too. The torque control, as the name implies, focuses on the continuous measurement of the applied torque, assuming a predefined torque target value where the target preload is achieved. This strategy leads to variations within the achieved clamp load within +/– 25–35% from the desired value. However, this technique is popular because it requires low cost equipment.

   The second variation of the torque control is complemented with angle measurement. This strategy is based on establishing a torque–angle window where the desired clamp load can be achieved. Figure 5.8 shows such a window.

   The evolution of the torque with the angle can also provide useful information about the joint conditions; indicating some of the assembly defects. Figure 5.9 illustrates some of the typical assembly defects signature in terms of the torque/angle combinations. Several pneumatic screwdriver clutch and pulse-based tools can be complemented with the angle control capability, however, this comes at a higher cost.
2. *Angle (rotation) control-based fastening strategy,* which is dependent on measuring and controlling the fastening based on the number of turns the screwdriver makes after achieving the target torque. This strategy results in +/– 15% preload

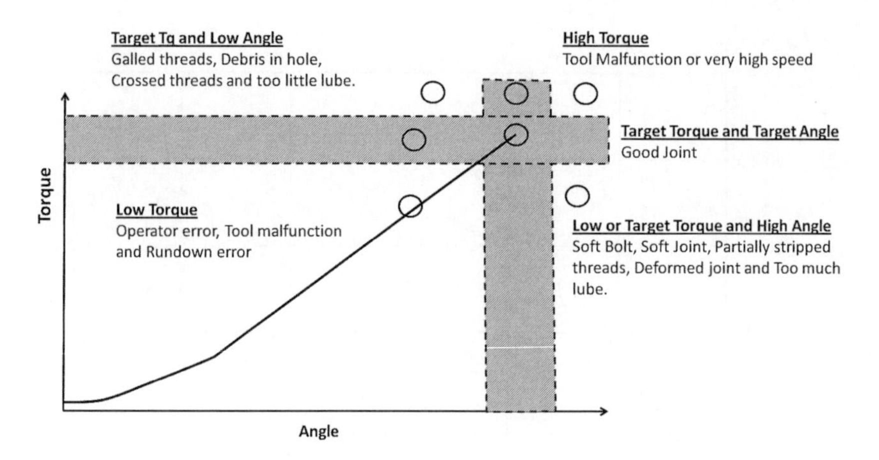

Figure 5.8 A typical torque, angle window

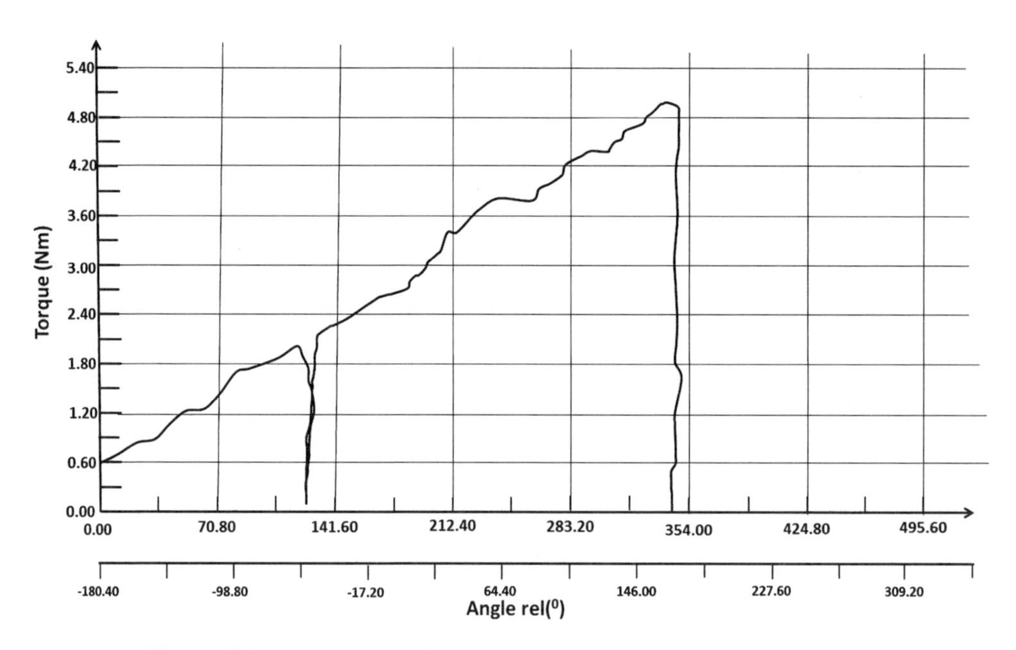

Figure 5.9 Assembly defects in terms of torque–angle combination

variations. This strategy is more accurate than the torque-based monitoring because it achieves the target torque (threshold), then rotates the fastener a controlled number of turns, and finally it checks the final cut-off torque to see if it is within the torque–angle window. These three steps are depicted in Figure 5.10.

The angle control strategy relies on the consistency in the load versus deflection relationship, which is of a linear nature corresponding to the fact that each 360°

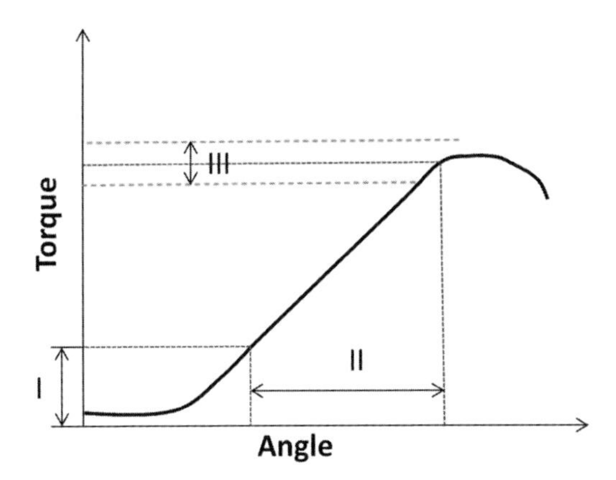

Figure 5.10 Angle rotation control steps

rotation translates into 1 pitch worth of length, that is, the 1 mm of fastener stretch in the joint. The fastener stretch δ can be directly related to the preload applied on it, through Equation 5.11 by knowing its stiffness k_{bolt}. Remember that the stiffness is defined as the resistance to deflection. Additionally, the stretch or elongation can be further described in terms of the number of turns or angle θ through Equation 5.12; thus one can combine these two relationships to obtain Equation 5.13, which directly relates the preload tension force with the rotational angle:

$$F_{pre} = \delta \times k_{bolt} \tag{5.11}$$

$$\delta = P \times \frac{\theta}{360} \tag{5.12}$$

$$F_{pre} = k_{bolt} \times P \frac{\theta}{360} \tag{5.13}$$

This also can be plotted as in Figure 5.11. Figure 5.12 shows the way to map the targeted preload into a specific angle range, then map the angle range along with the torque to establish the torque–angle window.

The *yield fastening strategy* is based on monitoring the amount of slope change as the torque progresses in terms of number of rotations. This procedure is similar to the torque–angle control strategy, however, it is more appropriate for hard joints because of its sensitivity to the joined material's behavior. The accuracy achieved through this strategy is estimated around +/– 5.8% of the desired preload.

The classification between hard and soft joints, as stated previously, depends on the relative flexibility of the fastener relative to the joint. For example, if the joint is made of a very stiff fastener (relative to the joint material) and is subjected to an external force higher than the clamping load, then the fastener will bear most

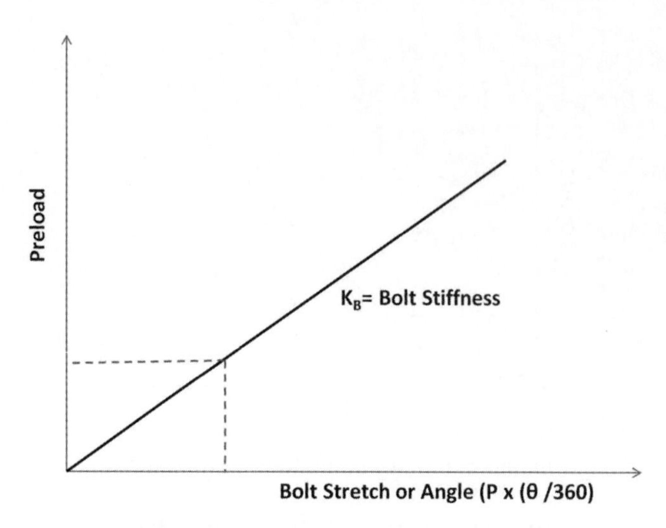

Figure 5.11 Bolt stretch versus preload

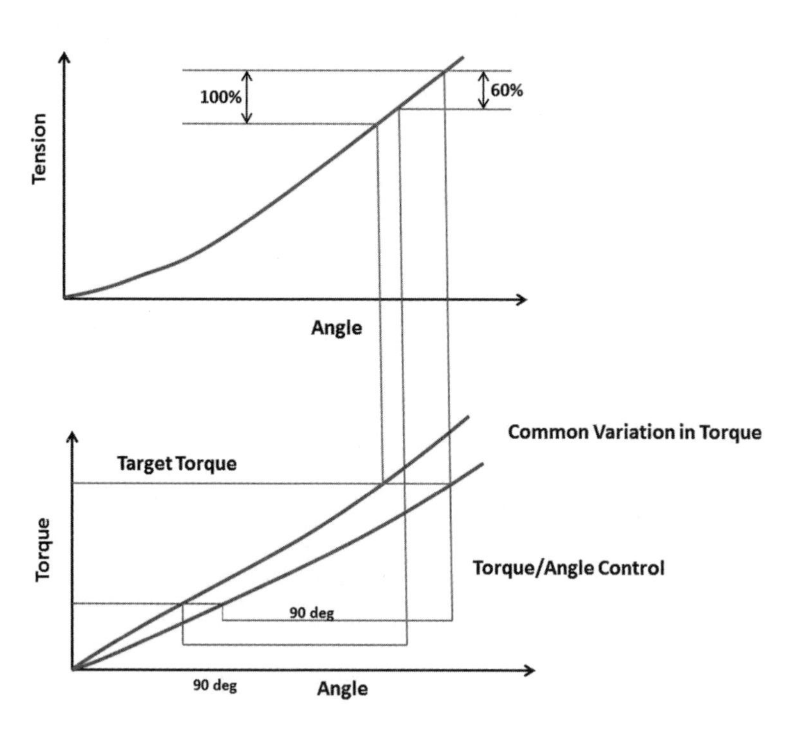

Figure 5.12 The preload mapped into torque, and angle windows

of the load in addition to the clamping force, this joint scenario is called a hard joint. On the other hand, if the fastener is less in stiffness from that of the joint, then the external load will go toward reducing the clamping force till the joined parts are separated, then the fastener will bear the external load, this joint is classified as a soft joint. The load distribution between the joint and the fastener is divided between them according to their relative stiffnesses, as described in Equations 5.8 and 5.9.

To illustrate with an example, if a joint is composed of a set of four M8 bolts, with a desired clamping force of 20 kN, knowing that the yield strength for the bolts material is around 400 MPa and that an external load of 10 kN has been applied on the joined assembly, after an initial 6 kN preload. Will the joint separate?

Using Equation 5.8 to compute the load carried by the bolt, one can find that $F_{bolt} = F_{pre} + \frac{1}{4} F_{ext} = 4.25\, kN$ (remember that the joint stiffness is typically three times that of the bolt) while that on the assembly $F_{ass} = F_{pre} - \frac{3}{4} F_{ext} = 6 - 1.8754 = 4.125\, kN$ because the F_{ass} yielded a positive value, this means the joint will not separate.

3. *Stretch control strategy*, which is based on monitoring the stretch or elongation within the fastener to estimate the clamp load. The elongation is typically measured using an ultrasonic transducer that sends a sonic pulse through the fastener and computes the time it took the pulse to reflect and travel back, hence estimating the elongation in the fastener.

Each of these fastening strategies requires the use of a certain type of screwdriver tooling that accommodates the required sensory and control capability. The following text introduces some of the variations in mechanical fasteners.

- *Screw joints*: the mechanical screws can be designed to be in the form of pierced, through or blind-hole joints, such configurations are shown in Figure 5.13. Because mechanical screws are used to join dissimilar metals, careful consideration should be given to the joining screws material type; the choice of stainless steel is typically adopted to avoid any galvanic corrosion issues. Additionally, because the automotive trim assemblies feature a lot of aluminum-based materials with lower compressive strengths, the joining interfaces should be protected through the use of washers.

  The screw joints should be designed based on the intended application (panels thickness), expected service life, the production cost, and the service environment in terms of thermal and chemical effects. This design is based on the proper selection for the screw joint type from the following: pressed nut, screw nut, flow hole forming, drilling screw, etc. However, one should note that in screw joining, there is a limited length for load bearing, which translates into weaker joints. Possible solutions to this problem are adopted through forming cylindrical collars for the screw joints, hence the load is distributed over more surface area.

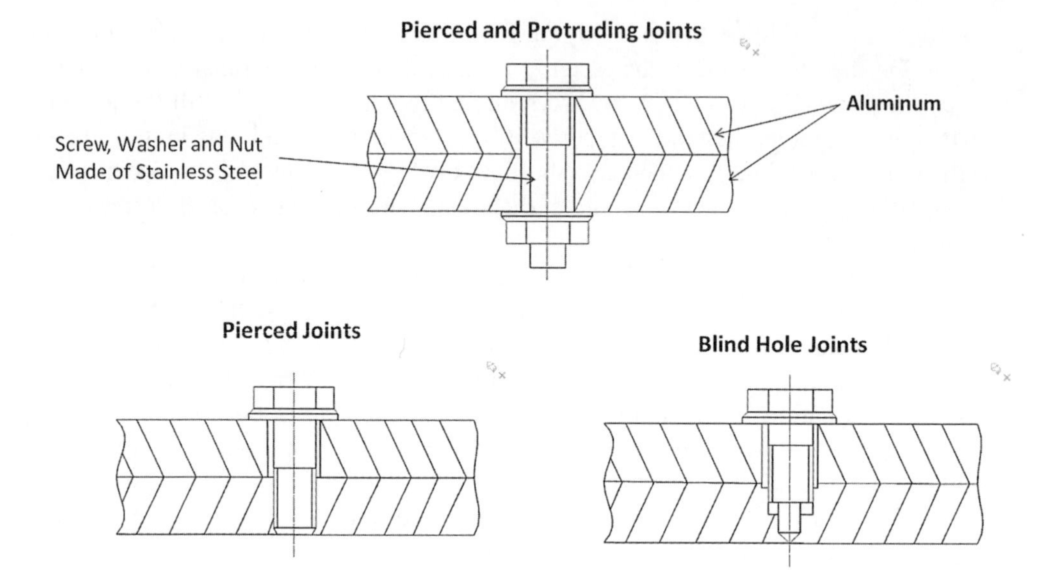

Figure 5.13 The different configurations of mechanical screws

- *Mechanical riveting*: new variations of the mechanical riveting process are becoming more accepted in the automotive and the aerospace industries, due to their indirect riveting or joining action that does not require intimate contact between the joined panels. This indirect riveting is achieved through auxiliary joining elements hence this process is suitable in joining any aluminum trim sheets to a steel shell. This indirect joining can be achieved by using one of the following different types of indirect rivets: the solid rivets, the blind rivets, the huck bolts or screw rivets, and the punch rivets.
- *Clinching*: describes the process by which a direct joint is achieved through localized plastic deformation with or without local incision. Clinching is typically classified based on the kinematic of its tooling into single and multi-step clinching, and based on the joint configuration into clinching with and without local incisions. A single step clinching process is displayed in Figure 5.14, while a multi-step clinching is shown in Figure 5.15. In clinching with local incision, a permanent joint is created using an applied shear force and a penetration force, to limit the joint region. Additionally, a compression force is applied to push the sheet material out of the plane and further compress it and flatten it to create a quasi-shaped locked joint.

A basic comparison between some of the different types of mechanical fastening is tabulated in Table 5.3 against the traditional spot-welding and adhesive bonding joining methods.

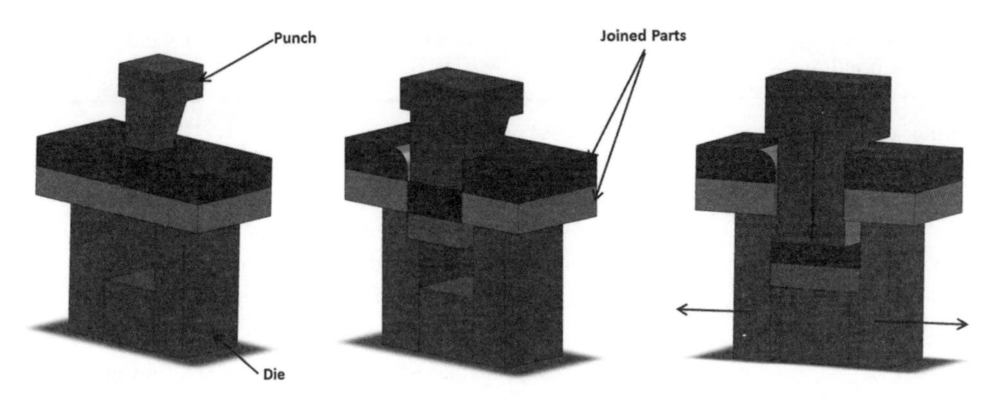

Figure 5.14 A single-step clinching process

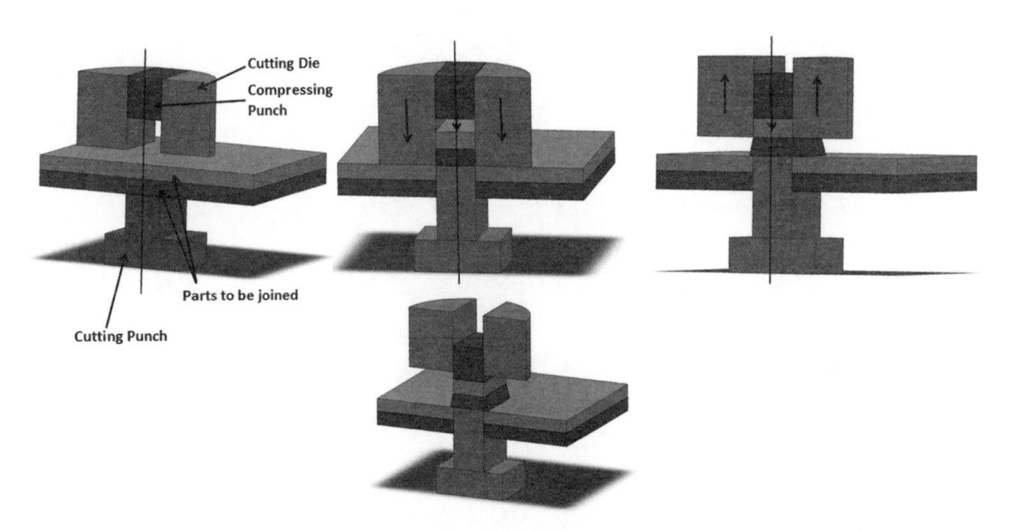

Figure 5.15 A multi-step clinching process

Table 5.3 Comparison between the mechanical fastening scheme and spot welding, and adhesive bonding

Joining technology	Application area
Spot welding	Stressed and unstressed joints Feasible for small to large production quantities Can be modified to join dissimilar metals Not applicable for Class A finishes
Adhesive bonding	Stressed and unstressed joints Applicable to small to large production quantities Can join dissimilar metals Requires surface preparations
Clinching	Stressed and unstressed joints Applicable to small to large production quantities Can be used to join dissimilar metals
Riveting	Stressed and unstressed joints Applicable to small to large production quantities Can join dissimilar metals

Exercises

Problem 1

What are the main operations done within the chassis installation area?

Problem 2

What are the basic ergonomic work principles that should be considered in final assembly?

Problem 3

What is the required wrench torque needed to install an M10 bolt with yield strength of 320 MPa?

Problem 4

What is the difference between the effective diameter and the root diameter of a screw?

Problem 5

Name and explain each of the different clamping force monitoring strategies used in automotive mechanical fastening.

Problem 6

Explain the relationship between the applied torque and the achieved clamp force in automotive fasteners, with the aid of a plot.

Problem 7

What effect will installing a washer in a mechanical joint have on the final clamp force, will it increase it or decrease it?

Problem 8

Explain the mathematical relationship between the bolt stretch and the preload force applied on it.

Problem 9

What is the difference between the mechanical *flexible* and *hard* joints?

Problem 10

If a mechanical assembly is made up of four M8 bolts with material yield strength of 320 MPa, the designed clamping force is 32 kN. Will the joint separate (fail) if an external load of 15 kN is applied to the joined assembly, after an initial 6 kN preload?

6

Ecology of Automotive Manufacturing

6.1 Introduction of Automotive Manufacturing Ecology

The most accepted definition for industrial ecology as proposed by R.H. Socolow, is

> Ecology is intended to mean both the interaction of global industrial civilization with the natural environment and the aggregate of opportunities for individual industries to transform their relationship with the natural environment. It is intended to embrace all industrial activity … ; both production and consumption; and national economies at all levels of industrialization.

However, this definition does not provide a specific platform on where the current or future states in terms of the ecological performance can be measured, or recorded so that improvements and changes can then be better quantified and evaluated in a systematic way.

Thus, this chapter treats and addresses the industrial ecology as a tool for the environmental accounting of the automotive manufacturing facilities and activities. This treatment is more in concert with the definition proposed by Graedel [40] Graedel, 1995] where he defined the industrial ecology as the accounting for the environmental impact through everyday activities within a defined carrying capacity. From this definition, any quantitative ecology analyses should have the following aspects:

1. a defined accounting tool for defined impacts, in other words a unified unit;
2. a set of defined activities;
3. a defined carrying capacity to establish thresholds for impacts.

The Automotive Body Manufacturing Systems and Processes, First Edition. Mohammed A. Omar.

The above three metrics will be used throughout this chapter to help discuss and evaluate the ecological aspect of the automotive manufacturing processes and the automotive industry in general. Even though the industrial ecology is getting greater emphasis nowadays, no established methodology exists for its quantitative analysis. Some formulae, like the one in Equation 6.1, have been proposed to describe the environmental impact caused on a national level:

$$Env.Impact = Pop \times \frac{GDP}{person} \times \frac{damage}{unit\ GDP} \tag{6.1}$$

where the *Env.Impact* is the environmental damage, the *Pop* is the number of the population in a specific country, the *GDP* is the gross domestic product of that country. However, for the automotive industry one cannot apply same formulation that assumes that environmental damage is like entropy, i.e. it can only increase with any human activity. Additionally, the concept of industrial sustainability will only be discussed in terms of a type 2 system that is partial sustainability (recovery) of the resources from the products. Type 1 system consumes new virgin resources as an input every single time without any recovery or recycling of the products, once it reaches its end of life, into new resources, while a type 2 system has partial recovery, and a type 3 system is a 100% self-sustained one, i.e. a sustainable system that can recover new resources from its own output.

This book analyzes the automotive manufacturing ecology based on the following defined impacts: the energy consumed and its utilization efficiency, the material usage and its processing efficiency, and the non-value indirect products as production emissions and scrap. Their impacts will be defined in monetary units to enable their accounting, using a standard, common unit. The monetary unit facilitates the computation of the impact of any alternatives on the company profits. One more reason to use a monetary unit is motivated by the fact that the government environmental fines are also described using a similar basis. This chapter will also focus on the painting booth activities due to its higher percentage of the energy consumed around 60–70% of the total plant energy, in addition to its impact on: water usage and treatment, paint emissions, both VOCs and HAPs, and finally, its air handling energy requirements.

The first analysis for the automotive manufacturing ecology will be from the energy expended perspective. The next section will first discuss the different energy models used to measure the total energy consumed within a manufacturing facility, then present a case study that aims to improve these models by developing a specific energy model. The developed model will then be used to assess the impact of changing the body panel materials.

6.2 Energy Consumption and Accounting

Recent changes in the operating environment of the automotive sector have forced the original equipment manufacturers (OEMs) to re-invent their products and production styles due to market over-capacity and product lack of differentiation. To provide a numeric example, in 2000, the automotive industry had the capacity to produce twenty-five million more vehicles than the world needed as reported by Clarke [41], and the South Korean OEMs produced five times their projected domestic market. Additionally, the push to *greening* the automobile has resulted also in hybridized body-in-white (BIW) panels made from steel, aluminum and magnesium. However, such hybridized structures have resulted in an added cost and energy requirement to produce such automobiles. Energy consumption cost American vehicle production facilities around $3.6 billion in 1999, however, it represents a small portion of total output, which was nearly $350 billion that same year as reported by [42]. Because of this small proportion, less emphasis has been placed on energy efficiency as a mean of cost reduction compared with other cost factors; however, there are benefits of increased energy efficiency, which other cost reduction techniques do not possess. One of the main benefits is that reducing energy consumption does not generally have a direct effect on product quality or yield. In addition to being quality neutral, reduced energy consumption can reduce uncertainty in production costs, resulting in more predictable estimates of earnings and more effective production planning.

Another benefit of reducing energy consumption stems from recent emphasis on reducing emissions from factories in the United States. The effects of carbon emissions on the global climate have attracted increased attention from nations around the globe in recent years. A number of regulatory policies have also been proposed to cap carbon emissions or fine companies with excess emissions. Any such policies, if adopted, will have a significant effect on the automotive assembly industry, which is a major consumer of fossil fuels and electricity. Also, any increase in the industrial electricity or fossil-fuel prices will increase the monetary impact of the manufacturing energy expenditures.

The discussion above motivates a strategy or systematic approach to assess and benchmark the energy expenditure trend between different automotive production facilities, to identify best practices and opportunities for energy savings. Studies by the American Council for an Energy Efficient Economy (ACEEE) by [43], classify the industrial benchmarking efforts for energy-efficiency into the Long-Term Energy Forecasting (LIEF) methods and the Environmental Protection Agency's (EPA) Energy Performance Indicators (EPI). The LIEF methods focus on market solutions to achieving "best practices" in energy consumption, by assuming that the energy intensity can be modeled as a function of the market energy prices as reported by Ross and Thimmapuran [44]; furthermore, the LIEF collects its data from entire industrial sectors. On the other hand, the EPI focuses on technical solutions and uses plant-level data, which is more useful in providing specific, plant-level energy consumptions. In essence, LIEF assumes that the difference between average performers and best practices is based on energy prices, while EPI assumes the difference to be of a technical nature.

The following text provides a brief introduction to the EPI development and its main resources and goals; the current case study utilizes the EPI to construct a process-specific energy model by complementing it with a stamping energy model and energy break-down for each process.

6.2.1 The EPA Energy Model

The EPA, in conjunction with the Department of Energy, began the *Energy Star* program in 1992 to promote more energy-efficient products and manufacturing processes, with the key element of successful energy management identified as the need to assess the current state of energy consumption in the manufacturing facility, because this will help in assessing the progress made towards energy management goals. The automotive Energy Performance Index (EPI) was developed through this effort, by the Lawrence Orlando Berkley National Lab (LBNL) in 2005. The development of the auto EPI focused on sources of energy consumption as identified by a 2003 study by Galitsky and Worrell using survey data from vehicle manufacturing focus participants that included GM, Ford, Honda, Toyota, and Subaru from 1998 to 2000. The data set included 35 facilities comprising 104 observations. However, the data set is limited to only the American assembly operations of these companies, where assembly operations were defined to include body weld, paint, and final assembly of light duty trucks, passenger cars, and sport utility vehicles. Also, the scope of the EPI is limited to plant-level performance rather than being proces-specific and it excludes the panel-forming energy consumption, which according to a study conducted by the Automotive Parts Manufacturers' Association (APMA), combines to account for nearly 19% of total energy consumption in the manufacturing of vehicles [45]. At the same time, the EPI stochastic model can provide actual energy consumption data for large number of assembly facilities in the US, in addition, the energy consumed by the heating, ventilation and air conditioning (HVAC) can also be extracted when analyzing the *Energy Star* zip-code database for each plant.

This case study presents the proposed specific electric (kWhr/vehicle) and fossil-fuel (MMBtu/vehicle) energy expenditures models when applied to two lightweight design criteria: the maximum weight saved per minimum cost added, and the maximum total weight saved. This study provides useful information about the impact of the lightweight engineering decisions on the manufacturing energy requirements, which might aid in highlighting the manufacturing variables involved in the total life-cycle of automobiles. This is becoming an important issue, because most of the automotive OEMs are focusing on lightweight engineering efforts to prepare for any increase in the Corporate Average Fuel Economy (CAFE) standards. Also, most of the lightweight engineering designs are based on replacing steel with Aluminum and other lightweight alloys and composites, which typically tend to have higher, but yet unquantified manufacturing energy signature. This trend is clear in the material break-down difference between 1950s and 1990s vehicles, as displayed in Table 1 [46]. Further, the interest in vehicle structures with Al percentages of >30% of the vehicle total weight is being renewed as in the study of aluminum intensive vehicles (AIV) by [47].

To assess the actual manufacturing energy requirement for each of the design criteria, the text constructs a model that is process-specific and includes the stamping and forming operations. This is done in two steps: the first extracts the actual energy requirement from the EPI index as (kWhr/vehicle), and (MMBtu/vehicle) for the welding, painting, final assembly and HVAC; while the second step computes the panel forming energy requirements for Steel and Aluminum panels. Additionally, the study provides the cost analysis for each of the two designs' major panels. The study starts by introducing the EPI stochastic and the look-up computation needed to extract the (kWhr/vehicle), (MMBtus/vehicle) in Section 6.2, while Section 6.3 addresses the correlations and assumptions used to predict the forming energy requirements of the reverse-engineered BIW in automotive stamping and blanking lines. Section 6.4 discusses the proposed requirements within the context of previous studies and proposals. Finally. the conclusion summarizes and discusses the findings, recommendations, and future work.

6.2.2 *Specific Energy Requirements from the EPI Model*

In brief, the EPI is designed as an assessment tool that provides a percentile ranking of an automotive manufacturing facility when compared with the best practices in the industry. The main inputs to the EPI code are: the annual energy expenses, the line speed, the vehicle wheelbase, and the plant location. The plant location is used in conjunction with data from *Energy Star*, which describes the total number of days and degrees a facility's HVAC systems will be required to heat or cool the building, the HDD and the CDD respectively, which are taken from the same data used by *Energy Star* to rate energy performance of commercial buildings.

For a statistical model to be useful as a benchmarking tool, it must be able to capture the behavior of the best performing facilities rather than the average performance. Most standard regression techniques capture the behavior of the average performers of a sample population rather than of "best practices" as required by the EPI. Stochastic frontier regression is one method which can capture the desired "best practice" behavior. The study by Greene [48] describes the stochastic frontier regression; whereas a standard regression curve of the form shown in Equation 6.2 would presume that any deviation from the curve is a result of random and normally distributed error, ε. However, stochastic frontier regression assumes that deviation from the "best practices" curve is primarily due to inefficiency, u, in the process being described. This principle illustrated by Figure 6.1 in which the solid line represents a standard least square regression curve while the dashed line represents a Corrected

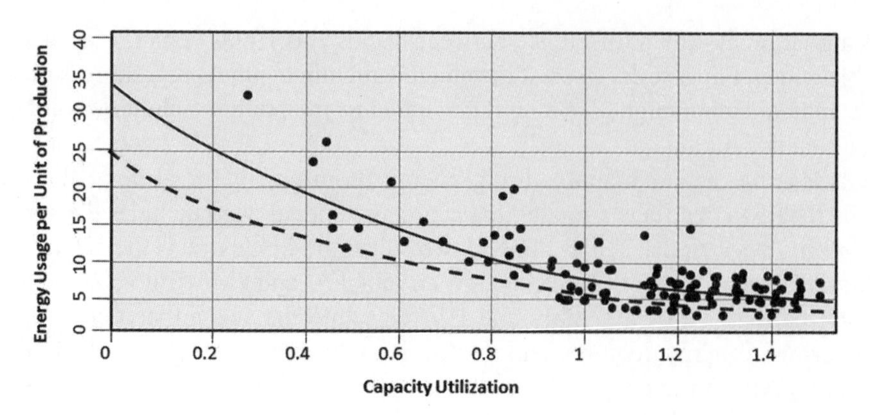

Figure 6.1 The Corrected Ordinary Least Squares (COLS) approach to stochastic frontier regression

Ordinary Least Squares COLS approach to stochastic frontier regression. Hicks in [49] used this approach in deciding on the average and the best practices for the milk and the beverage industry. Furthermore, using this approach the inefficiency of a manufacturing process can be described by Equation 6.3.

$$E_i = \alpha + \beta y_i + \varepsilon_i \tag{6.2}$$

$$u_i = E_i - Y_i \times (h(I_i, \beta) + \vartheta_i) \tag{6.3}$$

where, u_i is the process inefficiency, E_i is the energy consumed, Y_i being the annual production, h is frontier specific energy use (kWhr/vehicle) which is a function of I_i the system inputs and β a coefficient vector to be estimated, and finally ϑ_i is the noise in data set.

For the purpose of developing a percentile ranking, inefficiency must be assumed to follow some statistical distribution, which yields an EPI score given by Equation 6.4; where P is the probability that energy use is greater than the "best practice" level and F is the probability density function for some appropriate one-sided distribution for inefficiency.

$$P[u_i \geq E_i - Y_i[h(I_i, \beta) + \vartheta_i]] = 1 - F(E_i/Y_i - [h(I_i, \beta) + \vartheta_i]) \tag{6.4}$$

To predict the energy consumed per vehicle (kWhr/vehicle) from the EPI model, a gamma distribution might be used to describe the inefficiency. The result of this statistical analysis yielded two equations for electricity and fossil fuel consumption as described mathematically in Equations 6.5 and 6.6 respectively; with the *WBASE* being the wheel base of the vehicle, and the *Util* is the facility utilization.

$$E_i/Y_i = A + \beta_1(WBASE) + \beta_2(HDD_i) + \beta_3(HDD_i^2) + \beta_4(Util_i) + \beta_5(CDD_i) + \beta_6(CDD_i^2) + u_i - \vartheta_i \tag{6.5}$$

$$F_i/Y_i = A + \beta_1(WBASE_i) + \beta_2(HDD_i) + \beta_3(HDD_i^2) + \beta_4(Util_i) + \beta_5(Util^2) + u_i - \vartheta_i \quad (6.6)$$

The complete β_i coefficients for both equations can be found in Table 2 from [50], which displays the summary of statistics and derived values from the 1998–2000 automobile plants survey. Here, the utilization is simply defined based on a capacity to actual production volume for a 7-hour shift, 2 shifts per day, and 244 working days per year. Additionally, a look-up function identifies the energy inefficiency associated with the EPI rank value supplied by the user. The combination of EPI "best practice" energy use and inefficiency forms the specific energy consumptions; *Ei/Yi* (kWhr/car) and *Fi/Yi* (MMBtu/vehicle), for the process being evaluated.

The proposed look-up function in this study operates on the EPI database by scanning its inefficiency tables, searching for a match of the plant rank and zip code. Then, the code computes the specific electric and fossil fuel energy consumptions using Equations 6.5 and 6.6 respectively. The search subroutine also extracts the CDD and HDD from the *Energy Star* database and decides on the data error using a statistical model. The current look-up code runs as a Microsoft Excel macro. In essence, the macro operates using the EPI stochastic model in reverse, i.e. it requires the plant rank as input to produce the energy electric and fossil-fuel consumptions and the HVAC portion. This computation quantifies the EPI ranking in terms of (kWhr/vehicle) and (MMBtu/vehicle) making it easier for production plants to assess their current consumptions (rank) in terms of actual energy expenses and to further quantify and monitor their future progress and goals. Additionally, the HVAC consumption can further be analyzed to correlate the production data with the outdoor air temperature using current multi-variable tools such as the Lean Energy Analysis LEA from [51], and [52].

Also, the macro converts the fossil fuel consumption *Fi/Yi* (MMBtu/vehicle) into the total specific energy represented in (kWhr/vehicle), which is useful in translating the consumption into CO_2 emissions, as will be discussed in Section 6.4.

6.2.3 Panel-Forming Energy

The specific energy extracted from EPI does not include the panel-forming energy. This section addresses the panel-forming energy consumption for both Aluminum and Steel coils using current automotive press-lines, which include energy consumed in: blanking, bending, deep drawing, trimming and stripping, as described mathematically in Equation 6.7;

$$E = (1 + \alpha)\sum_{k=1}^{m} E_k \quad (6.7)$$

where E is the total energy consumed in (MJ), α being the rejection rate, m the number of processes (blanking, drawing, trimming, etc.) for each component and E_k is the energy consumed per process, which is further described in Equation 6.8;

$$E_k = n \cdot F_k \cdot \beta \cdot \frac{t}{a} \tag{6.8}$$

where n is the number of shots required in making the component; F_k the force required doing cutting, blanking, bending, deep drawing and stripping, β is press velocity (constant), t is the time taken for the operation, and a is the number of parts per shot or the number of die cavities.

Finally, the force required for each operation is computed using Equation 6.9;

$$F_k = L \cdot th \cdot s \tag{6.9}$$

where L is length of the cut (for trimming), or the length of draw (for drawing), th is the panel thickness, and s is the material strength (shear for cutting, yield for drawing).

To calculate the total forming energy for the BIW panels, a passenger vehicle structure is scanned using a Pro-T Zeiss Coordinate Measuring Machine CMM, to retrieve each of the major body panel dimensions. The reverse-engineered BIW point cloud is further rendered into a CAD model using Solid-works and is displayed in Figure 6.2. A typical BIW consists of around 400–500 stamped panels, however, the following panels govern the body design and consume the majority of the forming

Figure 6.2 The BiW used for the study

energy: body-side outer, hood inner and outer, trunk inner and outer, roof inner and outer, door inner and outer panels, under-body, and finally the fenders. The equation sequence (6.6) through (6.8) is run for each of the above panels considering; number of operations, panel dimensions and thickness, in addition to the actual die and press used for each one. A typical 1% rejection rate is set for Steel and 3% for Aluminum due to its lower n-value and narrower deformation window. The number of forming and piercing stations is set based on the part complexity and final shape, to accommodate any post-draw needed to counter any expected spring-back. Also, the forming model distributes the energy consumption of the press mechanical losses and non-value added work on the different panels; for example, the energy for the presses to reach its zero point. A Draw Quality (DQ) cold rolled carbon Steel and 5052 Aluminum are used for the panels. Figure 6.3 shows an example calculation for the DQ Steel grade, displaying the different components comprising the BIW. To validate the proposed model, its total energy computed to form an Al BIW is found to be 1220 MJ/vehicle and for a Steel BIW around 967 MJ/vehicle, which is compared with the results of the International Iron and Steel Institute energy audit in 1994 [53] for stamping Al and Steel BIWs, IISI results are: Al around 1200 MJ/vehicle and for Steel around 1000 MJ/vehicle. Even though Al has lower yield strength, its lower n-value increases the number of shots required to make the same shape when compared with Steel, in addition to a higher spring-back, which increases its forming energy requirements.

Combining the panel-forming model with the look-up macro into a unified code can provide useful insights on the automotive manufacturing energy break-down, as displayed in Figure 6.4 (a) and the energy type in Figure 6.4 (b). Such analysis reveals

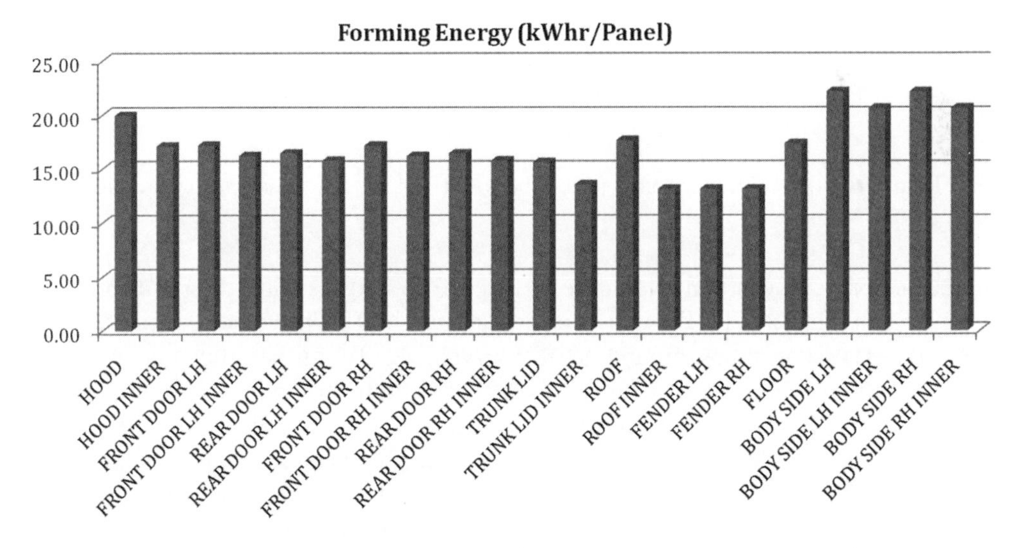

Figure 6.3 The forming energy for each of the BiW major panels

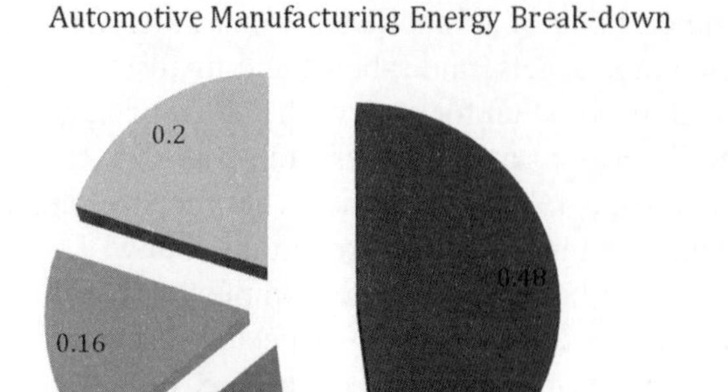

Figure 6.4 (a) The automotive manufacturing, assembly plants energy break-down

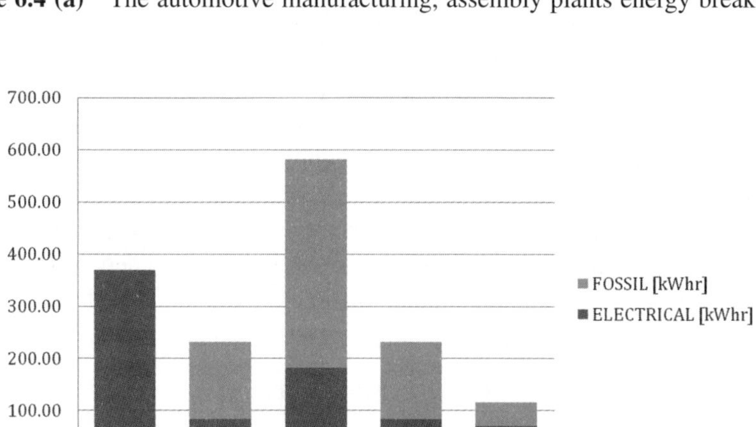

Figure 6.4 (b) Comparison between the fossil and electric energies consumption

processes or practices that should be targeted for energy-saving efforts. From this study, a complete Steel BIW will consume around 600 kWhr/vehicle, while a complete Al BIW consumes around 820 kWhr/vehicle. To further illustrate with an example, the forming model results presented can be further plotted as a sensitivity plot (of ±20% changes) as in Figure 6.5, to show the impact of the vehicle design options on the total forming energy. From Figure 6.5, one can see that from a product perspective, designing vehicles with smaller size (i.e. less surface area), thinner panel thickness, and less complicated features (less draw depth, and fewer number of shots) can result in less manufacturing energy consumption cost. While from a procedural perspective, optimizing the stamping process to reduce the scrap rate, or the number

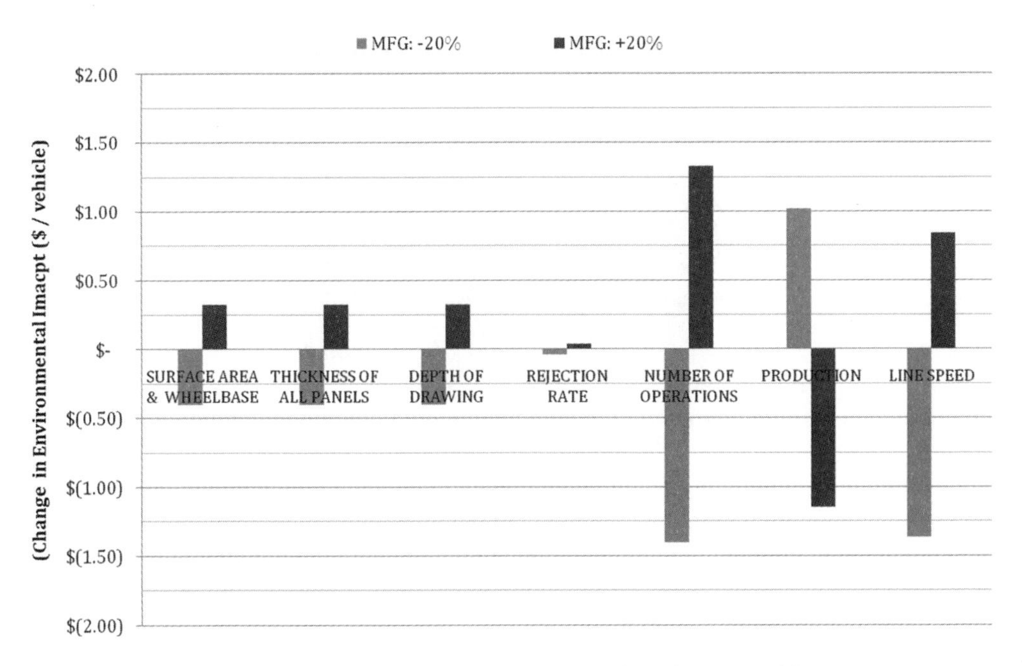

Figure 6.5 Sensitivity analysis in terms of environmental impact

of required operations, while increasing the process utilization (production rate), can all reduce the energy-forming requirements. For example, running the stamping line for 4 days/week with 10-hour shift is more efficient than running the typical 5 days/ week at with 8-hour shift. For the painting booth energy break-down, Klobucar in [54] proposed Figure 6.6 (a) to show the energy break-down for each operation within the booth. Also from our study, Figure 6.6 (b) can show the detailed energy type for the different painting operations. Such information is very useful to guide energy savings efforts and to help evaluate certain environmental decisions effect on the plant EPI ranking.

To provide a specific example, the case of replacing the solvent-borne paint with waterborne paint that has less VOC emissions but at the same time requires more HVAC energy to control the booth humidity and temperature. Additionally, as indicated above, the EPI rank can be quantified into actual energy consumption values and be further interpreted into an emission footprint for expended energy, which further can be quantified in monetary values. This simplifies the decision-making process and the return on investment calculations.

6.2.4 Hybridized Structures Selection and Energy Implications

In Sections 6.2 and 6.3 the total energy consumed per vehicle was computed for all assembly operations; this section decides on the energy consumed for each of the

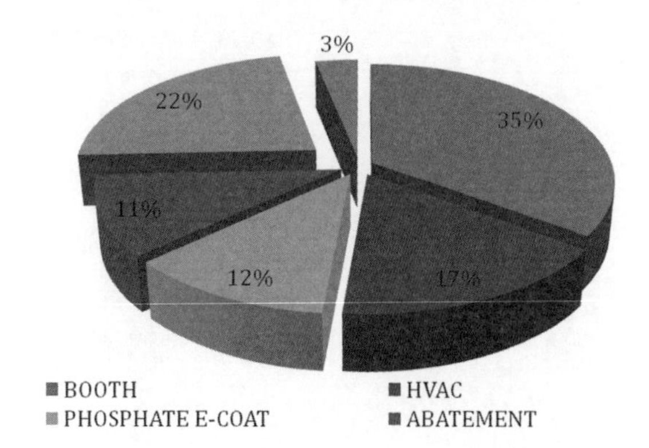

Figure 6.6 (a) Energy consumption in an automotive painting line

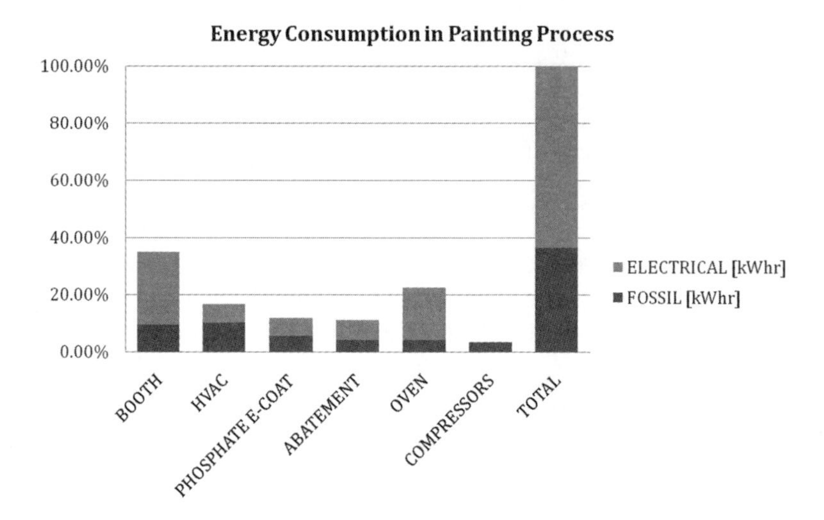

Figure 6.6 (b) The fossil and electric energies expenditures in automotive painting lines

two hybridized mix of the 5052 Aluminum and DQ Steel. In either design, the front module will be made of steel due to its functional requirement of supporting the engine cradle and other manufacturing complexities involved in hydro-forming Al. Also, the thickness of the new Aluminum panels is selected based on available coil thicknesses and the weight-to-strength ratio required for each panel function, mainly durability and crash-worthiness. For the maximum weight reduction, the Aluminum percentage should not exceed 30% of the total BIW weight, which is the typical starting point of Aluminum Intensive Vehicles AIV.

For the first criterion, the total cost associated with material purchase and fabrication for each panel is computed. The cost consists of the material cost, the fabrication

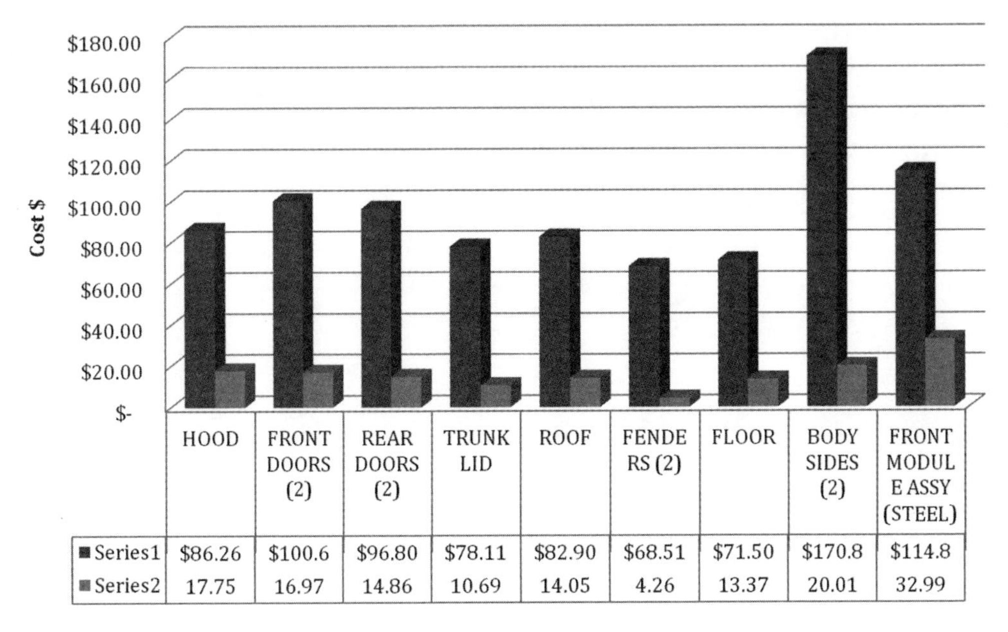

	HOOD	FRONT DOORS (2)	REAR DOORS (2)	TRUNK LID	ROOF	FENDE RS (2)	FLOOR	BODY SIDES (2)	FRONT MODUL E ASSY (STEEL)
Series1	$86.26	$100.6	$96.80	$78.11	$82.90	$68.51	$71.50	$170.8	$114.8
Series2	17.75	16.97	14.86	10.69	14.05	4.26	13.37	20.01	32.99

Figure 6.7 The cost and weight for each of the BiW major panels

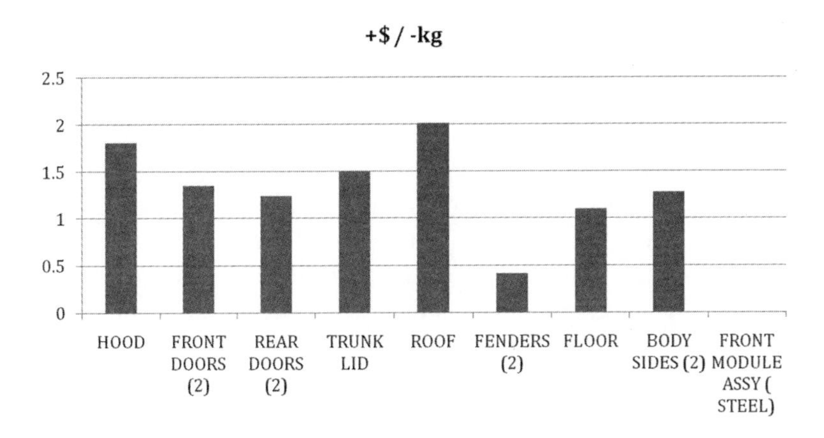

Figure 6.8 The dollars added per weight saved for each major body panel

cost, and the overhead cost. The material costing is based on each panel's blank dimension multiplied by its thickness and then its material density; this results in the stock weight needed for each panel. Such calculations are based on each panel's blank shape and size, while accounting for engineering scrap. The overhead cost includes all the fixed costs associated with labor and facility insurance, lighting, etc. Figure 6.7 combines the cost and weight information for each panel when made of DQ steel; combining the cost and weight information for Al results in Figure 6.8

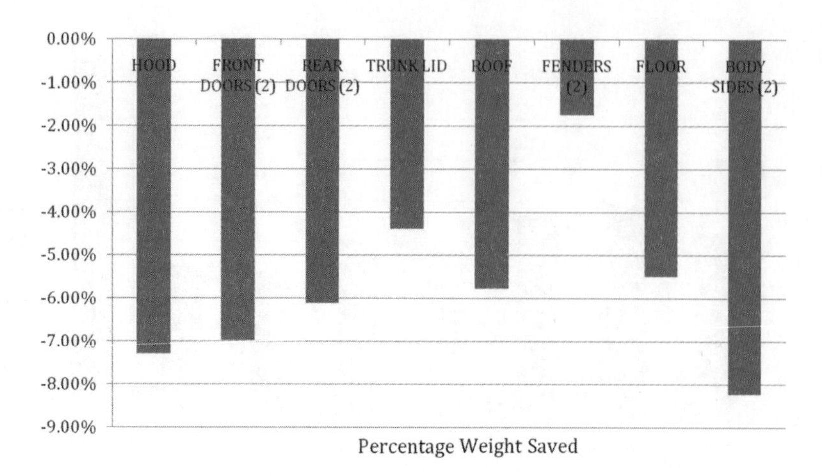

Figure 6.9 The percent weight saved, when replacing steel with Al

which shows the cost increase in $ per kg saved for each of the panels if made from Al. This figure indicates that replacing the fenders has the best savings. This is consistent with the current trend in the industry where OEMs have replaced steel fenders with plastic ones, as in the case of the BMW X-series. The cost calculations adopted 4.8:1.0 ratio between the Al and the DQ material prices, in addition to the added cost in spot welding and joining Al joints. For the second criterion, the weight of each panel, if made from Aluminum, is subtracted from its original steel weight and then divided by the total steel BIW weight, to generate Figure 6.9, which indicates that the highest weight savings around 8% will be by replacing the body-side outer. However, stamping the body-side out of Aluminum is not practical due to the different complicated geometries involved, which typically cause splits and tears. So, an Aluminum hood is selected for the second hybridized BIW. This selection also agrees with current commercial examples, as in the case of the Toyota Crown and the BMW X-series, with both having Al hoods.

The specific energy for each design; all steel BIW with Aluminum fenders and all steel BIW with Aluminum hood can be computed, to show the manufacturing energy implication. The Al fender design adds ~+10 kWhr/vehicle, while the Al hood design adds ~+15 kWhr/vehicle. Furthermore, a new design criteria might include the specific energy in the form of minimizing the (+kWhr/vehicle /-kg) for future designs. This analysis incorporates the manufacturing energy-efficiency in lightweight engineering decisions or criteria. In light of this criterion, the Al hood design allows for +15 kWhr/vehicle /–10.57 kg ~ 1.42 kWhr/vehicle/kg, while the Al fender design adds ~3.55 kWhr/vehicle/kg, which makes the Al hood design three times more energy-efficient.

To further describe the advantages of the new proposed criteria, one can study the emission implications of the added energy consumption, using such conversion tables from [55]. This table is included here as Table 6.3 on p. 284, which shows the conversion of the consumed energy into the emission species controlled and monitored by the EPA and termed "criteria air pollutant". In the case of CO_2, the results are in agreement with a CO_2 audit reported in [56]. Furthermore, each species' impact can be quantified in terms of dollar amount following the studies reported by [57, 58 and 59], which facilitates the conversion of (+kWhr/vehicle /-kg) into (+$/-kg) as shown in Table 6.4 on p. 285. Using the information from Tables 6.3 and 6.4 for each design option, shows the Al fender design to add $34,000 worth of environmental damage per year, while the Al hood option adds $45,000 per year; these values do not include the extra cost of energy. Even though the added environmental damage is not significant when compared with production sales, still it highlights each design monetary impact that might be implemented in the form of a government fine or added tax.

6.2.5 *Proposed Approach versus Previous Models*

The following paragraph discusses previous attempts to incorporate the energy efficiency in design decisions for the automobile structure.

The Swedish Environmental Institute (IVL) in collaboration with the Volvo car corporation have developed and proposed an analytical tool designated as the Environmental Priority Strategy (EPS) for Product Design in 1992. The EPS, as described by [60], designates an Environmental Index for each type of material used or to be used in the automotive industry to allow designers to select materials and components with low environmental impact. The index is computed based on the material impact during product manufacture, use and disposal, which is counted up as an Environmental Load Unit (ELU)/kg of material. An example of such ELUs is shown in Table 6.5 on p. 285. Computing and summing the ELUs for the different parts within an automobile allows the assessment of the complete vehicle impact. However, the EPS system is limited to its unit ELU which is not standardized internationally, thus limiting its acceptance. Additionally, the ELU were computed by ecologist, material scientists who assumed worst case scenarios of emission and focused on the raw material scarcity. Furthermore, they assumed the same ELU ratings for the manufacturing side if two materials are processed using the same technology and machine, without accounting for technical issues such as the difficulty in manufacturing. Additionally, the ELU rating does not provide expected energy expenditures, which further complicate the economic calculations for the different design options. Finally, the EPS system requires a complete life-cycle perspective of the material used, which requires extensive studies each time a new material is added and can be cost-prohibitive. In terms of materials scarcity, one can use a material flow diagram to track the material losses within the different extraction, fabrication and end-of-life phases, as in the diagrams proposed by [61] in Figures 6.10 (a) and 6.10 (b) for each steel and Aluminum application in the

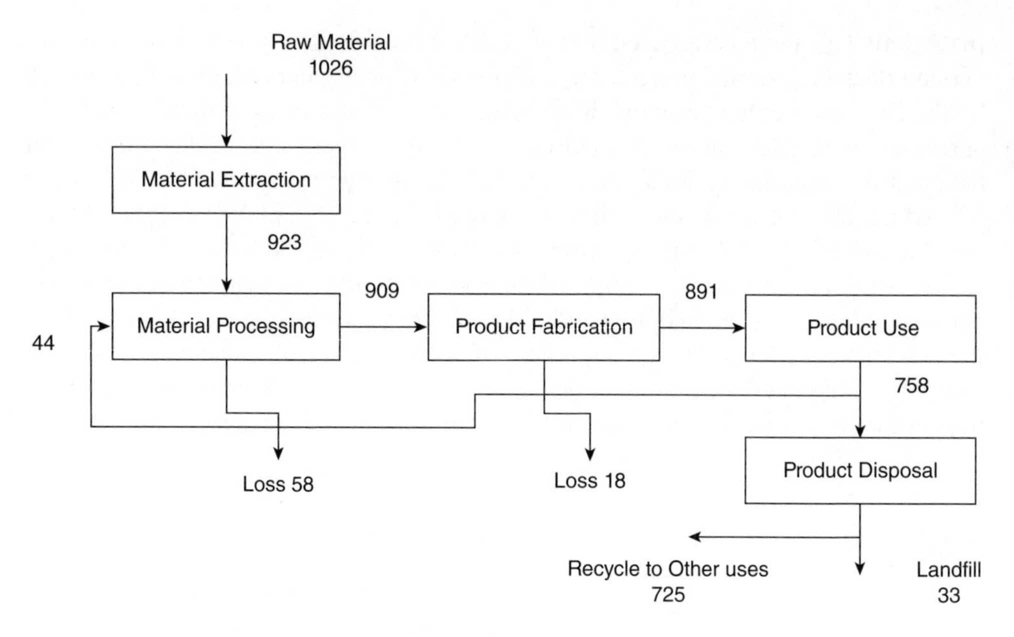

Figure 6.10a The steel material flow

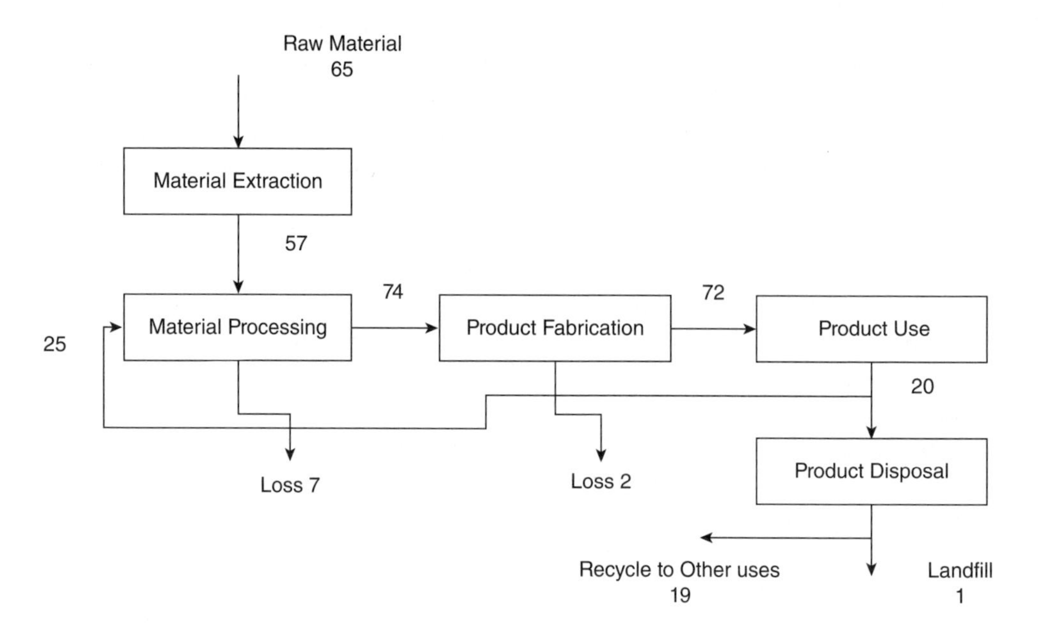

Figure 6.10b The Al material flow

automotive industry, respectively. However, using the material flows to evaluate design decisions for manufacturing energy-efficiency is complicated because of the difficulty in evaluating specific scrap rates in addition; the recycling modes and energy requirements are different for different materials and at different stages. For example, Al can be recycled horizontally (i.e. same class use or product) but requires higher energy for smelting, while steel is recycled vertically (i.e. into lower class products) but at lower energy requirements. Also, following a material flow approach might motivate the usage of large number of materials to replace the less efficient ones, which typically tend to reduce the scrap rate but at the same time it complicates the recycling because of sorting; in addition it necessitates different fabrication technologies which might challenge standardization or benchmarking efforts.

Even though presented study illustrates the energy computation within an automobile manufacturing phase, it does not distinguish between the different sources of expended energy, i.e. sustainable or renewable energy portion versus the non-sustainable. For example, some automotive facilities such as the BMW Spartanburg—SC and the Subaru plant in Indiana are replacing the natural gas used in the paint curing ovens, with methane from local landfills. This manuscript does not distinguish between natural gas and methane consumptions because both will be adding to the facility carbon footprint and emission counts, but further study is needed to address the issues of fossil fuel extraction and scarcity. Another example that challenges the direct specific energy consumption is from the Ford Motor Company, where solvent-borne paint is still being used with its VOC content collected and then burned as fuel in abatement units. The Ford Motor Company example requires multi-variable analysis to address all the added benefits versus the emissions and further implications. However, the main benefit from current computation is its process-specific nature, which benchmarks the facilities on the basis of direct (value-added) activities' energy expenditures to the product. To illustrate the importance of this nature, the Subaru plant in Indiana (SIA) has reached a zero-landfill status since 2007. This plant energy bill is higher than other typical plants of same size and product type, due to the energy consumed in recycling the plant's waste and packaging material. If the EPI or LIEF tools are used directly, it will indicate that this plant has lower rank than its peers, but if the current process-specific approach is used, the SIA value-added activities consumption is only computed.

Finally, the presented analysis of the emissions' economic impact is only intended to address the manufacturing emission footprint and not the complete life-cycle. Several published studies such as the IISI study in 1994 tackled the complete life-cycle emission analysis for steel and Al BIW designs. However, most of these studies do not include the manufacturing phase due to the highest emissions being in the refining, material extraction and product usage phases, in addition to their focus on revolutionary BIW designs i.e. Ultra-Light Steel Auto Body like the one reported in [62 Roth, 1998] and Aluminum Intensive Vehicles . Still, it is important to study the manufacturing footprint, because it aids in designing or selecting between transitional

BIW designs, i.e. traditional steel BIW with Al panels constituting a low percentage of total weight (<30%), which are most common in the automotive industry for mass production purposes. Also, identifying the specific manufacturing footprint is important for companies operating globally, under different governmental regulations and environmental taxation systems.

6.2.6 Conclusion and Comments on Specific Energy Modeling

The presented specific energy study analyzed the specific energy consumption in electric (kWhr/vehicle) and fossil-fuel (MMBTU/vehicle) for vehicular Body-In-White, for two typically used design criteria. The study analyzed the EPI stochastic model and complemented it with a gamma distribution function and a look-up macro, to retrieve the actual energy consumption data. Also, the macro operated on the *Energy Star* database to extract the HAVC portion of the facility total consumption. A new forming energy consumption model is also proposed and implemented for steel and Al panels and further verified against current stamping plant surveys for both Al and steel scenarios. The manuscript also presented the energy break-down for the stamping process from a product perspective through a sensitivity plot, to quantify the potentials for energy savings. The total energy break-down for an automotive assembly operation is also displayed to provide additional insights from a facility perspective. The manuscript has also translated the specific energy into a new design criteria (+kWhr/vehicle/ -kg) for lightweight engineering and in addition to emissions' species environmental impact. Future work will focus on incorporating a life-cycle perspective for each design by incorporating the product-usage (mile per gallon) and the Al and steel material extraction and recycling stages. Furthermore, other material options might be included; such as plastics, which require energy consumption modeling for an injection molding process.

6.3 The Automotive Materials' Ecological Impact

This section is intended to discuss another environmental impact for the automotive manufacturing activities, with focus on material extraction, processing, usage, and recycling. A typical automobile is composed of variety of materials to constitute its structure, aid its mobility function, e.g. tires and fluids, and provide comfort to its occupants, e.g. glass and refrigerants. A typical break-down of a typical automobile from material point of view is given in Figure 6.11.

Such materials pass through different phases during their lifetime; they are extracted, transported, processed into semi-finished shape, transported again, and then fabricated into final products. At the end of their service life, they might be recycled partially or completely, and vertically or horizontally. The extraction energy for the material listed in the pie chart in Figure 6.11 is displayed as a bar chart in

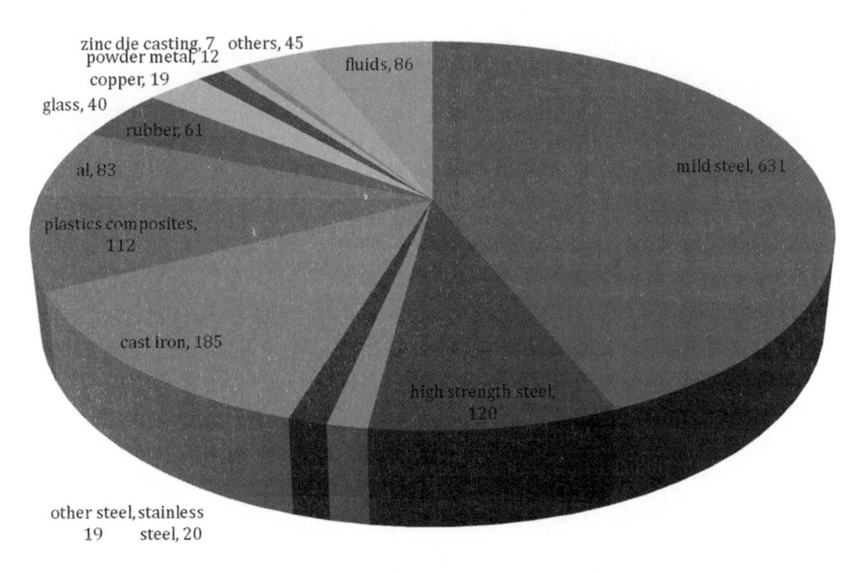

Figure 6.11 The materials and their percentages in a typical sedan

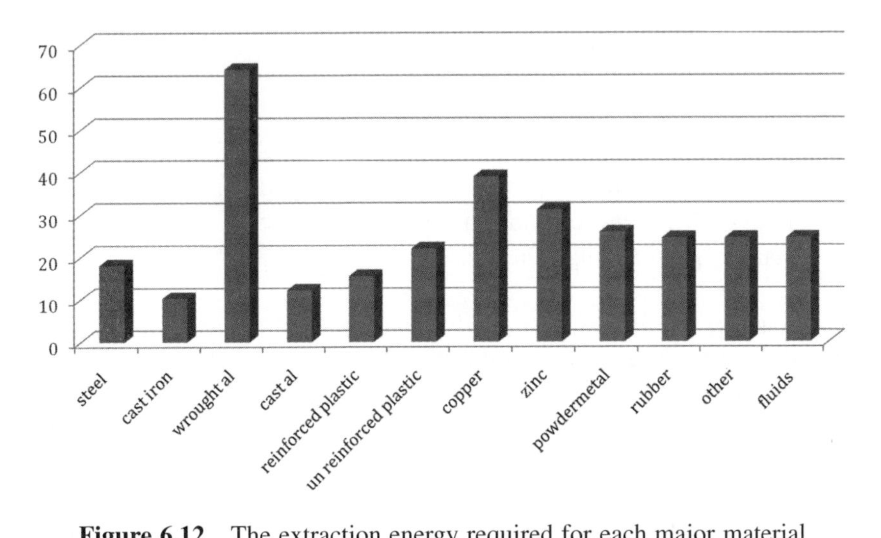

Figure 6.12 The extraction energy required for each major material

Figure 6.12. Additionally the changes in their use for the past thirty years are displayed in Figure 6.13. And their complete life cycle can be traced in Figure 6.14.

6.4 The Painting Process Ecology

The painting process consumes on average 50–60% of the total energy within an automotive production facility. In addition, it outputs HAPs in the form of NO_x, SO_x,

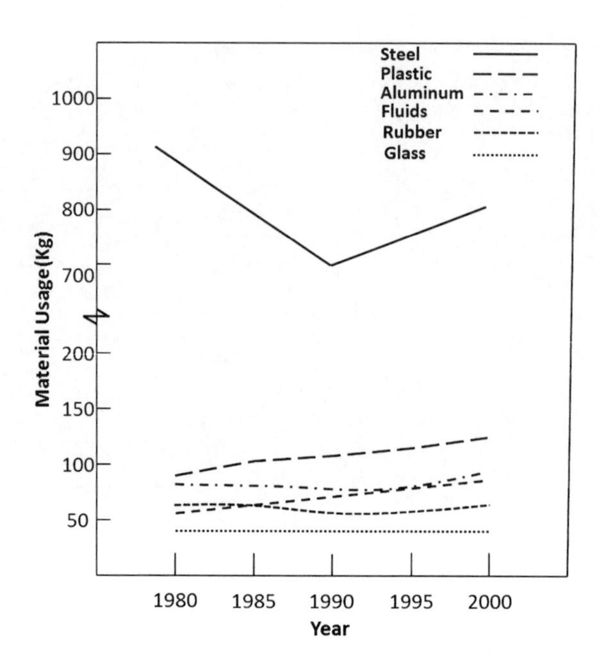

Figure 6.13 The material usage trend in automobile bodies

CO_2, CO, and VOC, in addition to waste water, and solid paint waste. So more and more automotive OEMs are considering new ways to reduce the painting process environmental footprint through the following:

1. *Adopting new paint formulations* with less to no VOC content; as a consequence of this, the waterborne paint technology is being widely used to reduce not only the VOC content but also the use of water to collect the over-spray paint. At the same time, due to the added challenges with this formulation in terms of the tighter air conditioning requirement, this translates into more energy usage in addition to the need for an intermediary curing step known as the flash-off. Other OEMs are still relying on high solvent content paints to avoid these issues, and further mitigate the VOC emissions from solvent-borne paints by implementing abatement units that incinerate the solvent emissions. The current governmental regulations on the VOC emissions are based on the Solvent Emission Directive SED of 1999, which limits the VOC emissions for vehicles produced in new installations to be 45 g/m^2 or 1.3 kg/vehicle at 33 g/m^2. And for vehicles made in existing facilities to be around 60 g/m^2 or 1.9 kg/vehicle at 41 g/m^2; this establishes a general threshold of 15 tonnes/year of solvent emission for OEMs. Some OEMs have resorted to abatement units to reduce their VOC emissions if they have very limited options in updating or modifying their paint shops.

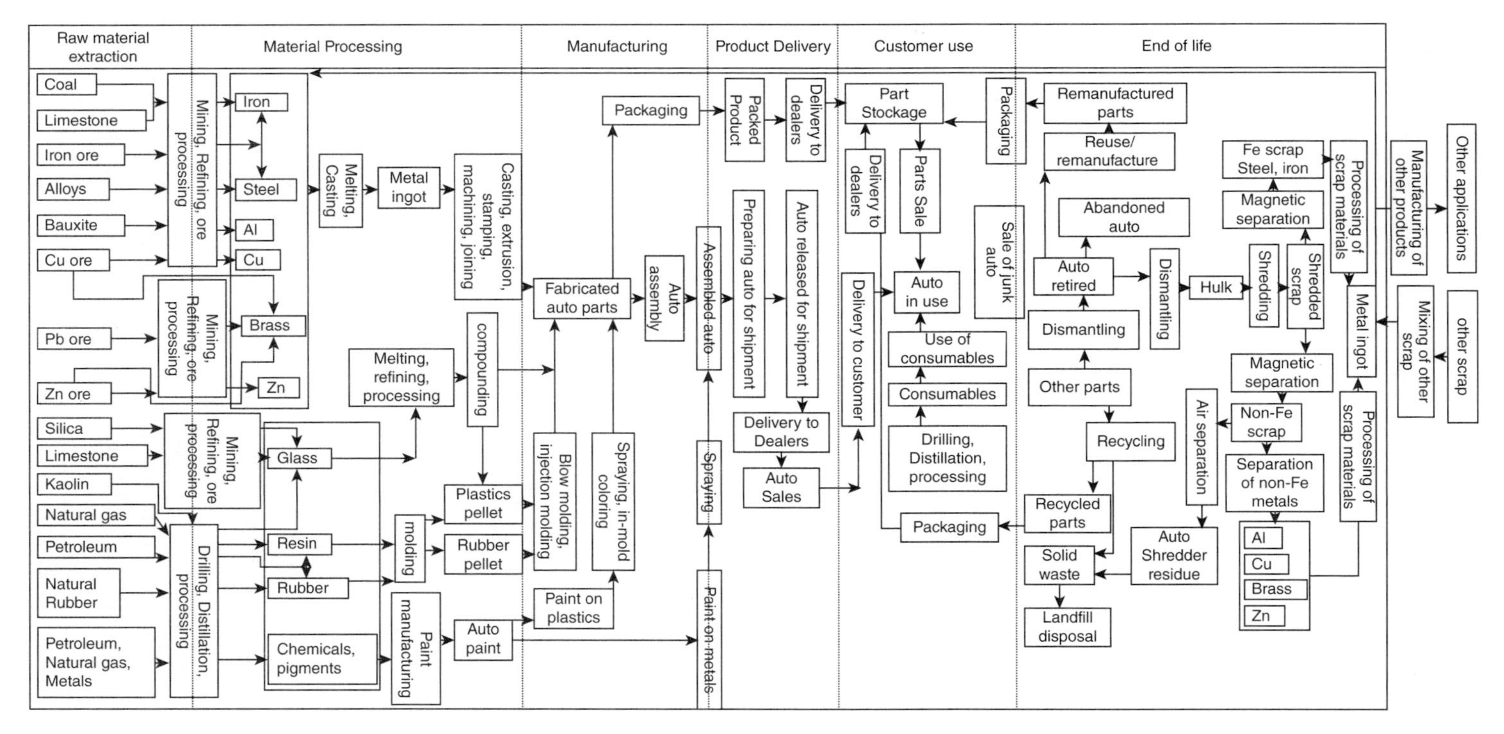

Figure 6.14 The overall material flow within the automotive manufacturing chain

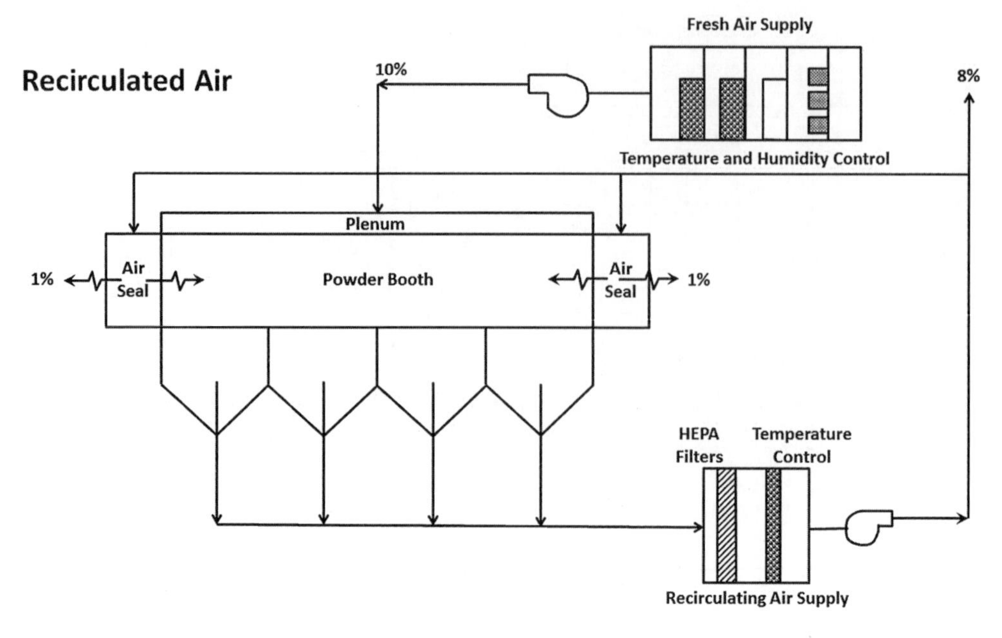

Figure 6.15 Recirculated air flow within an automotive painting booth

Another possible paint formulation with great environmental benefits is the powder coat technology, which reduces not only the booth water consumption used in capturing the over-spray paint but also the solid paint waste, because the powder coat can be collected and re-used directly within the booth without intermediate steps. However, the handling and application steps of the powder coating are more energy-intensive than those for the liquid paints. In addition, the power-coating atomization control and the surface final look are still being improved.

2. *New air circulating and management systems*, to reduce the energy consumption of the paint booth. The current state of spray booth consumes the biggest share of the paint shop energy, as evident in Figures 6.6 (a) and 6.6 (b). Its energy consumption fluctuates according to outside ambient conditions making it highly variable and difficult to control. A typical air reticulating system is shown in Figure 6.15. The air conditioning is done according to the psychometric chart to maintain a pre-determined booth temperature and RH levels. As discussed in Chapter 4, the waterborne paints require an even more limited operating window within the psychometric chart due to their fluidizing medium, that is, water. The typical air control for waterborne paints is set according to a set-point that should be continuously maintained through using heating and cooling elements to manipulate the air temperature in addition to a humidifier unit to change its RH. This procedure can be better shown in a psychometric chart as in Figure 6.16.

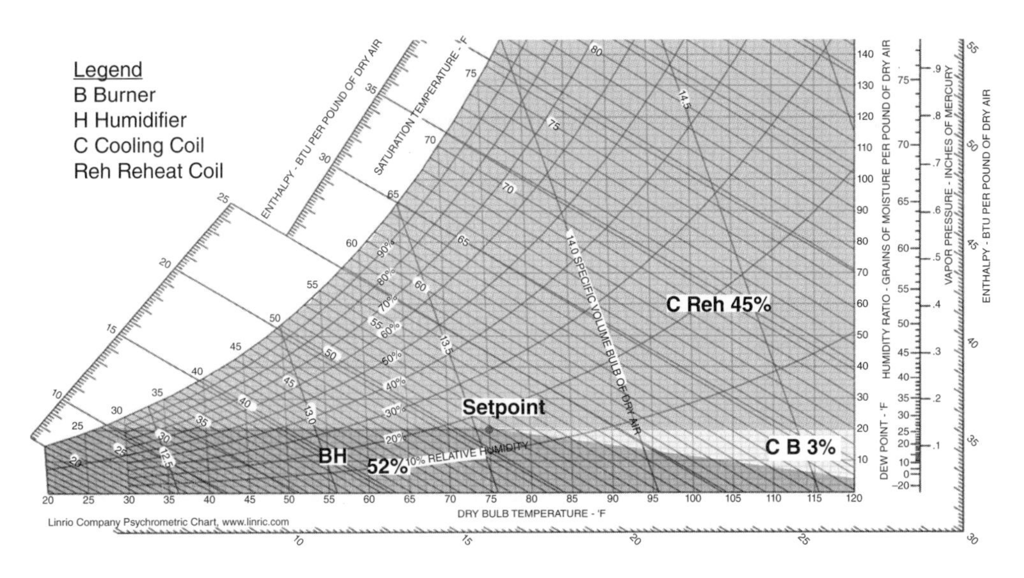

Figure 6.16 The set-point strategy for conditioning spray booths

A better control strategy for the air conditioning units can be done by eliminating the heating and cooling elements and focusing on a set-line or a variable set-point principle, which can be achieved by using the burner and humidifier units only. Again this control strategy is displayed on a psychometric chart in Figure 6.17. Additional adjustments can be done to reduce the volume of air being treated and fed into the booth by optimizing the air flow patterns in the booth chamber through computational fluid dynamic (CFD) studies; also, reducing the under-booth pressure drop results in lower flow rate requirements. Another effective air control strategy is based on increasing the recirculation of the exhaust air from the booth with the fresh air. This requires additional filtration steps with the air handling unit but still can be done as a retrofit to current booths. Figure 6.18 describes the air flow through the booth system and the current mix ratio. For any new technology in controlling and managing the spray booth air system, this new alternative should provide and not jeopardize the basic air systems' functionalities which mainly are a way to clear the evaporated solvents from the booth environment, a way to guide the flow of the over-spray toward a paint collection or spray apparatus, a way to ensure the equipment cleanliness by directing the paint particles away from it, and finally and most importantly to maintain the booth air conditions in terms of temperature and relative humidity. So, one can reduce the air flow requirement by reducing the booth volume which can be achieved either by making the booth shorter or narrower. Also, reducing the solvent evaporation and over-spray can play a major role in reducing the air flow requirements. One of the commercially (patented) available technologies that aim at reducing the air flow requirement is introduced by Duerr Gmbh that is based on splitting the air

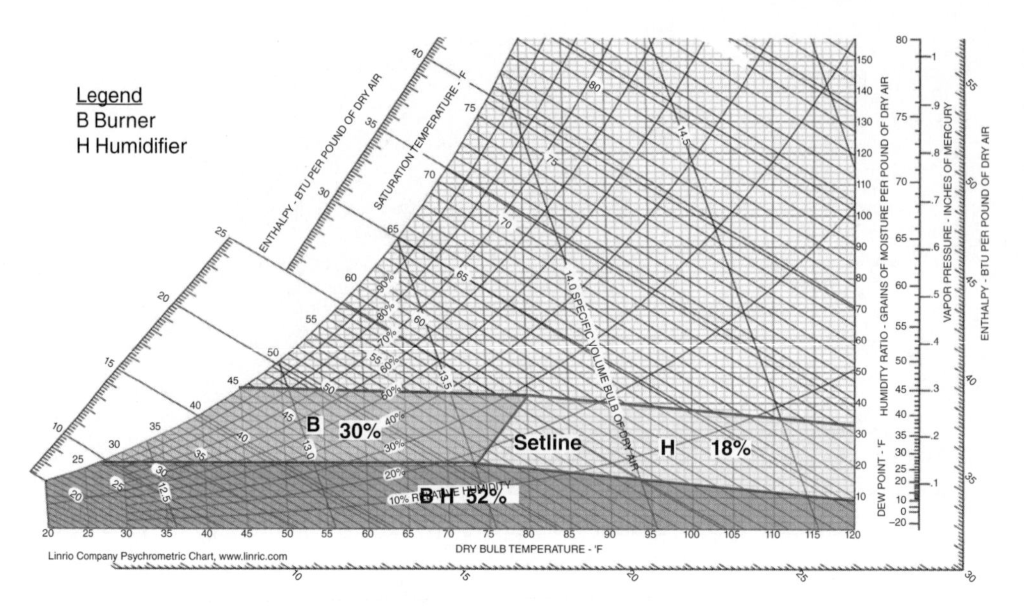

Figure 6.17 The set-line strategy

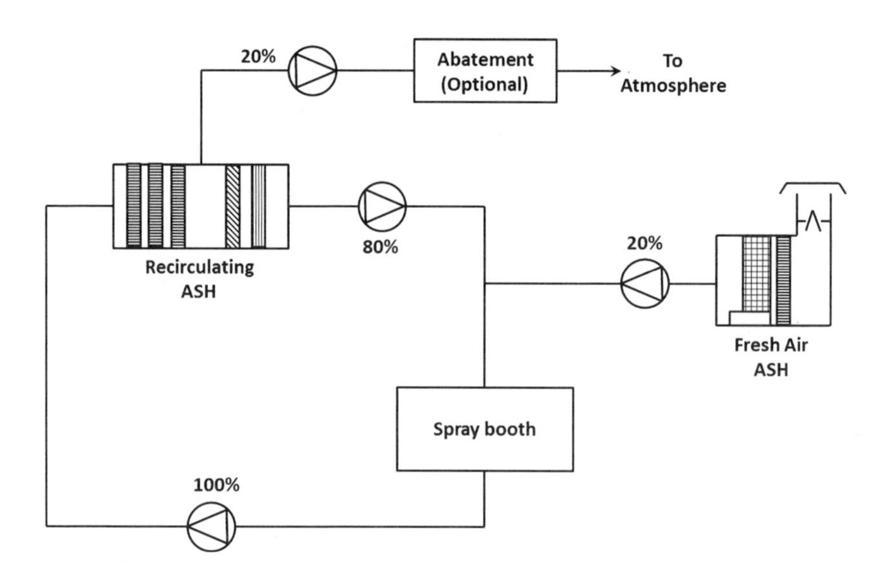

Figure 6.18 The typical air flow within the paint booth system

flow into different zones depending on its requirement, that is, the flow rate is higher at the equipment side and lower in the booth center; this system layout is shown in Figure 6.19. To highlight the role of technologies that aid in reducing the energy expended in conditioning the booth air, one should know that the total air flow within an automotive spray-booth can reach up to 800,000 m^3/hr.

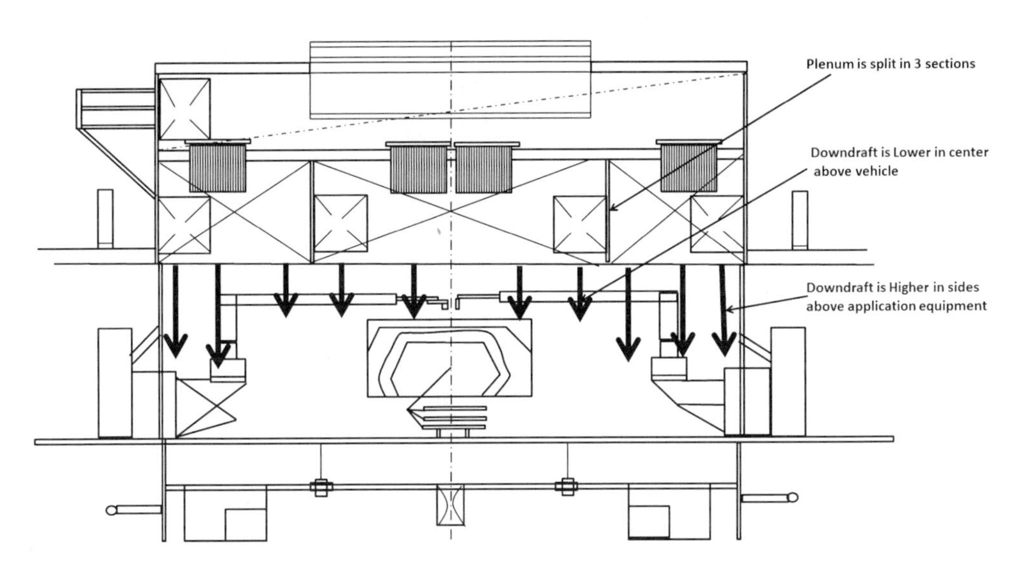

Figure 6.19 The split air flow system

Figure 6.20 The paint over-spray. Reproduced by permission of Duerr Systems, Inc © 2010

3. *Increase the paint transfer efficiency*, which has a direct effect on the total energy consumed in applying the paint, maintaining the booth air conditions and flow rate due to reduced over-spray, in addition to lower energy consumed in the over-spray paint collected by water which is then scrubbed and treated. Furthermore, increasing the transfer efficiency reduces the amount of VOC and HAP emissions and the amount of solid waste collected from the paint over-spray. The over-spray collection is shown in Figure 6.20. Automotive OEMs have adopted several

technologies to increase the transfer efficiency including but not limited to new paint applicator designs and innovations. Such efforts have resulted in the development of the rotary bell applicators and the continuous improvements in its design. First, the rotary bell applicator had major mechanical problems due to its high rotational speeds, which was resolved with the use of air bearings, in addition, several variations of the rotary applicator were introduced to paint the different required coats, for example, a rotary bell applicator was developed for metallic paints, and another variation for the cartridge-based color change system. Also, recent research is being conducted to study the effect of the rotary bell surface profile on the paint ligaments' stability and atomization; this principle is similar to the golf ball dents that aid in stabilizing the turbulent air flow around the ball while in flight.

Other technologies that led to better transfer efficiency include the use of the electro-static charging of the paint particles to create an attractive force between the paint and the vehicle shells. Further electrode designs are still under development to increase the charge efficiency.

Also, the way the painting robots are programmed has a direct effect on the paint transfer because of the precise control of the robot motion and its gun distance, in addition, the gun shaping air, all affect the paint flow characteristics and pattern. Examples of some of these advancements introduced in this area include the dual bell system from ABB, where two rotary bell applicators are mounted on one robotic manipulator to increase the number of paint passes at the same robot passes. Additionally this technology helps in controlling the film build-up through a better paint deposition routine achieved through the sequencing of the two guns in terms of pattern and flow rate to apply the required thickness at the right location with using spatial increments. This technology has been reported to decrease the applied paint from 1200 cc to 700 cc when painting a large surface area vehicle. Also, increasing the number of paint applicators mounted on a single robotic painter leads to more compact spray-booths with higher flexibility because these robotics will be able to adapt to different panels' orientations, eliminating the need for reciprocating painters. One implementation of this approach has been developed by ABB under the name QuadGun, which, as the name implies, consists of mounting four paint applicators on a single robot arm, as shown in Figure 6.21. The QuadGun has the potential to reduce the paint booth size by almost 30–40%. The fewer the number of robots, the narrower the spray booth and the more flexible the painting process; another robotic application that can also reduce the number of robots within a booth is the over-hanging robots, because this configuration allows them to have wider reach and accessibility, as shown in Figure 6.22.

Other robot programming methods that can reduce the over-spray paint are based on reducing the number of trigger-off time for a robot, by ensuring that the robot painting path is continuous, for example, the vehicle hood can be painted through different variations as displayed in Figure 6.23, however, the first design allows for continuous application with no trigger-off time.

Figure 6.21 The multi-head paint applicator system

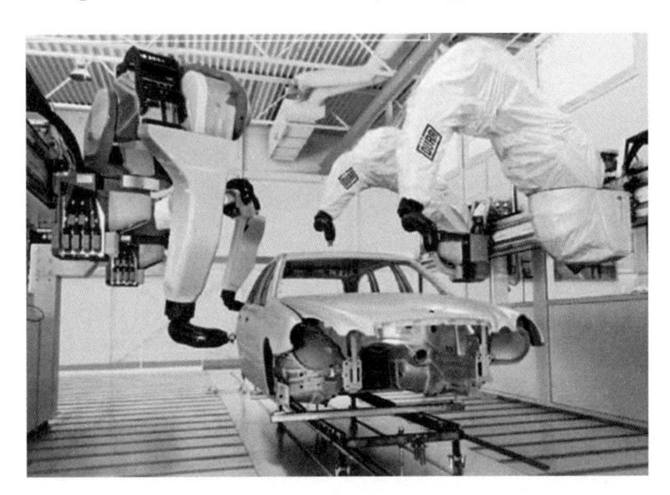

Figure 6.22 Overhanging robots. Reproduced by permission of Duerr Systems, Inc © 2010

4. *Improving the curing oven's efficiency.* As stated in Chapter 4, the paint shop area features different ovens dedicated to curing the different paint layers in addition to the applied under-body sealer material. These ovens operate by relying mainly on the heat convection to raise the temperature of the vehicle shell to be within the paint curing window. Typical values for these ovens' efficiency can be estimated to be around 10–20%, which translates into large thermal losses that lead to bad fossil fuel utilization. A more detailed analysis of the energy input and outputs of a typical top-coat painting ovens is given in Figure 6.24, which shows that only around 10% of the inputted thermal energy is consumed by the vehicle shell, while the exhaust consumes the remaining majority.

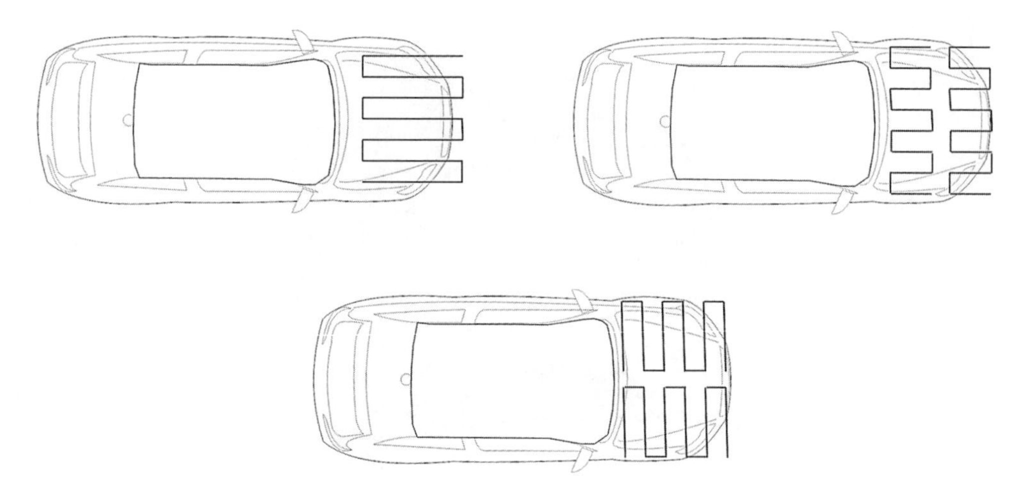

Figure 6.23 The paint sweep variations for a hood

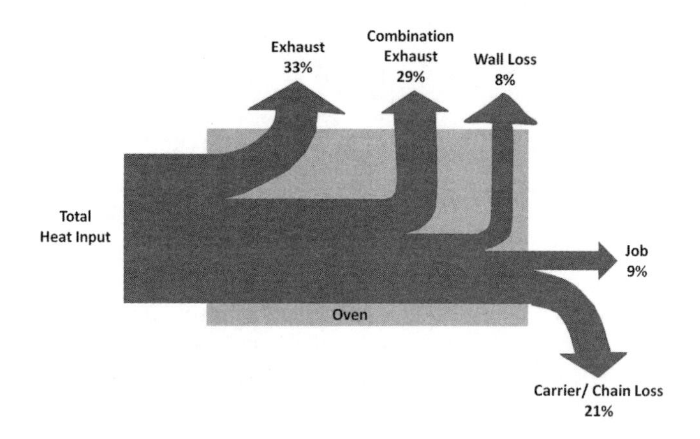

Figure 6.24 Energy distribution for a typical paint curing oven

Several automotive OEMs started to use the methane that is generated from local land-fills, as a fuel for the curing ovens to improve the ovens' environmental footprint and make it more reliant on sustainable energy sources. Additionally OEMs have started to rely on wet-on-wet paint formulations as in the case of the top-coat and clear-coat, where the clear-coat is applied without the need to cure the top-coat. Also, automotive OEMs and paint shop suppliers are still researching and improving radiation-based curing technologies such as the infrared-based and the UV-based curing schemes, where the paint film is targeted for heating without the need to heat the underlining substrate. This not only increases the curing

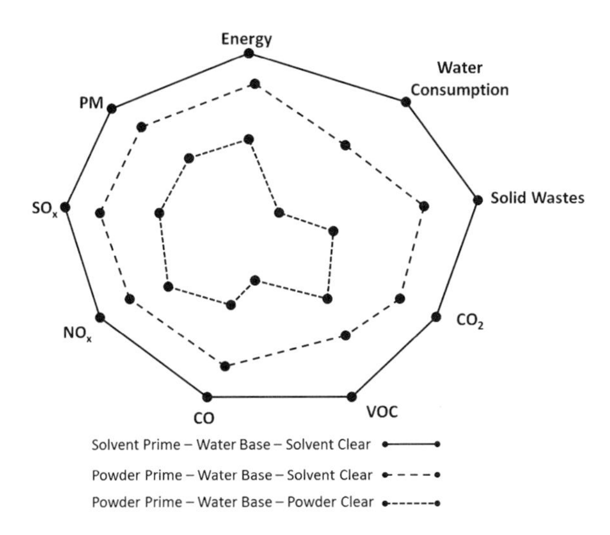

Figure 6.25 The evaluation of powder-coat formulation from a process point of view

energy utilization but also reduces the space required for the ovens (typically from 40–100 ft depending on type).

Other areas of potential savings include improving the paint film formulations so it can serve dual purposes, for example, eliminating the primer coating by having its role replaced by new modified E-Coat formulations that are able to provide the surface profile needed for the top-coat. This not only saves the energy consumed in applying the paint, but also the associated waste, water consumption and the curing effort. Additionally, some potential savings can result from using the ovens' exhaust thermal energy for other applications within the plant, such as for the E-Coat and phosphate baths or the plant's HVAC.

To conclude this section, one can establish a set of criteria to evaluate and qualify the paint shop practices, paint formulations, or any proposed technologies, based on their environmental impacts. These criteria should include emissions, solid waste, water consumption, and the energy expended. To provide a quantitative example, the plot in Figure 6.25 shows the comparison between using a powder-paint formulation for the primer film, against that of a solvent-based formulation. Figure 6.25 can then quantify the benefits from any changes or investments.

At the same time, one can further quantify any proposed changes from the monetary perspective too, by conducting the paint calculations introduced in Chapter 4, for example, the effects of increasing the transfer efficiency and the increasing the robotic painter trigger-on time percentage are shown in Figures 6.26 (a) and (b), respectively.

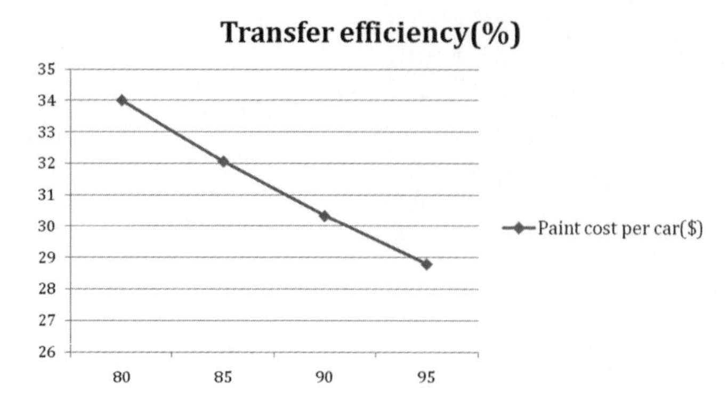

Figure 6.26 (a) Increasing the transfer efficiency effect on the final cost

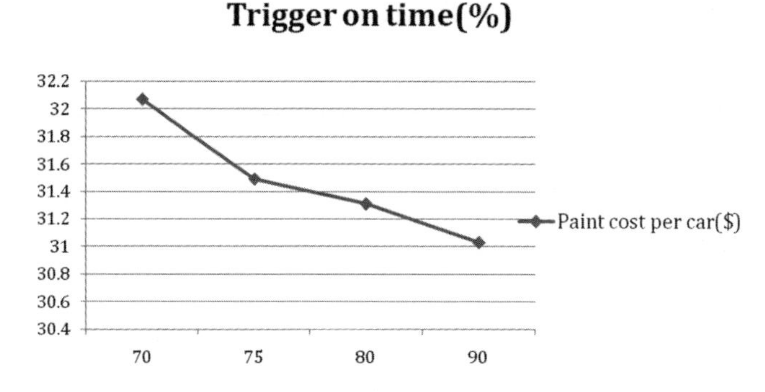

Figure 6.26 (b) Increasing the trigger-on time effect on the final cost

6.5 Ecology of the Automobile

The life-cycle analyses conducted in regard to the automotive industry have shown that most of the environmental impact is in the usage period of the vehicle, i.e. during the vehicle service life, because of the continuous generation of the carbon dioxide and other pollutants from the vehicle prime mover that is the internal combustion engine. To explain this further, one can analyze the environmental impact in terms of the energy expended during the product different life phases: raw material extraction, transportation, fabrication into semi-finished products, final production, transportation and distribution, and finally the usage and the end of life recycling. This approach is followed because the consumed energy can be converted into air emissions, water consumed, and other solid wastes, assuming a specific energy source for each stage and energy conversion efficiencies. This concept is similar to the embodied energy approach proposed by Mike Ashby in [63], in which the author proposed that each product consumes resources, which can be translated into energy in MJ, in each of its life phases.

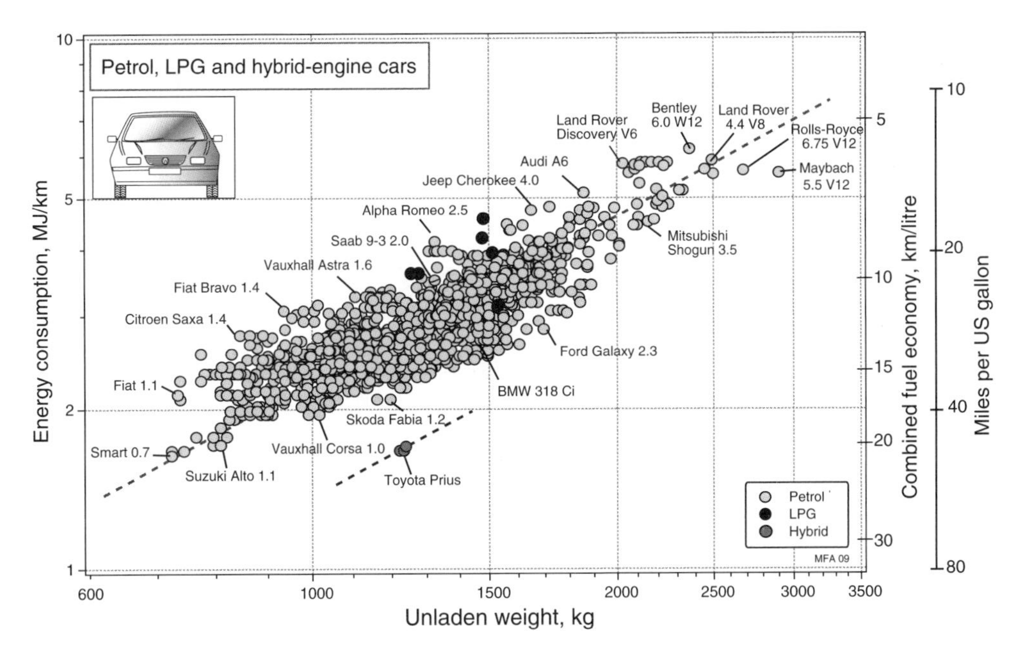

Figure 6.27 The energy consumed versus the vehicle weight, for different models and brands of petrol, LPG and hybrids. Reproduced by permission of Elsevier © 2009

Before introducing an environmental audit using the Eco-Selector software package [from Granta design] of the material, constituting a complete vehicle found in the pie chart in Figure 6.11, the environmental footprint of vehicle usage for variety of models and brands will first be discussed. This discussion is based on presenting a set of graphs showing the energy consumption for vehicles with different weights and with different engines using gasoline (petrol), Diesel and Liquefied Petroleum Gas (LPG) fuels, in addition to showing their CO_2 emission per distance traveled (in kilometer, 1 km = 0.62 mile) against the expended energy. The reasoning behind showing such information is to first highlight the different vehicles' CO_2 and consumed energy footprints against their weight, and, second, to correlate these two variables, i.e. establish a function to describe the CO_2 (g/km) in terms of the consumed energy (MJ). The energy consumed versus the vehicles' weight is given in Figures 6.27 and 6.28; while the CO_2 emissions versus the consumed energy is given in Figures 6.29 and 6.30 for gasoline and diesel engines, respectively.

The above figures display clear correlations between the energy consumed and the vehicles' weights and the CO_2 footprint versus their weight, for the vehicle models and brands included in the graphs. These two correlations have been computed by [63] and presented here in the equation sequence from 6.10 to 6.12 for the energy consumed per km E_{km} per the vehicle weight $W_{vehicle}$ and Equations 6.13–6.16 for the CO_2 emission per energy consumed.

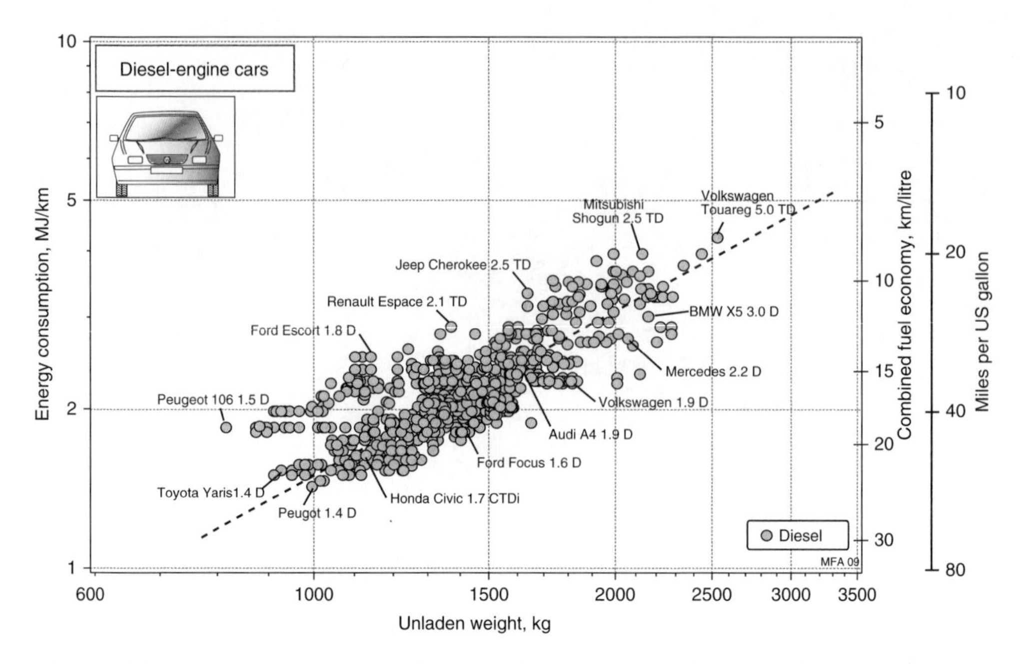

Figure 6.28 The energy consumed versus the vehicle weight, for different models and brands of diesel. Reproduced by permission of Elsevier © 2009

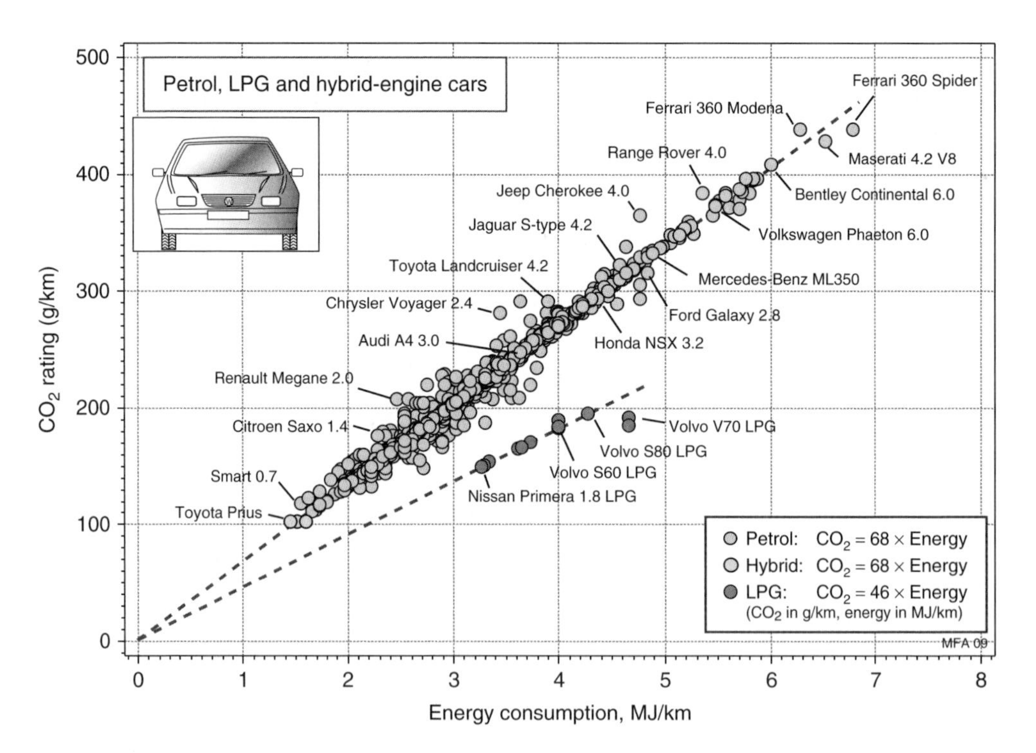

Figure 6.29 The CO_2 emissions versus energy consumed for LPG, hybrid vehicles. Reproduced by permission of Elsevier © 2009

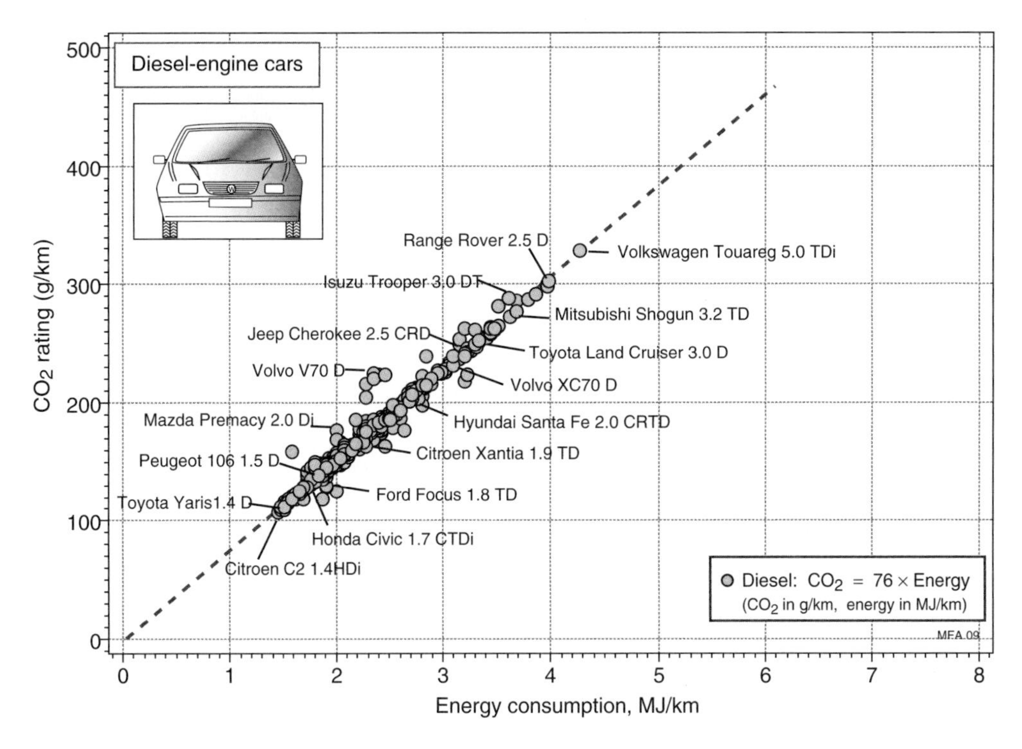

Figure 6.30 The CO_2 emissions versus energy consumed for diesel vehicles. Reproduced by permission of Elsevier © 2009

For vehicles with gasoline and LPG internal combustion engine, as in Equation 6.10:

$$E_{km} = (3.7 \times 10^{-3}) \times W_{vehicle}^{0.93} \tag{6.10}$$

For vehicles with diesel internal combustion engine, as in Equation 6.11:

$$E_{km} = (2.8 \times 10^{-3}) \times W_{vehicle}^{0.93} \tag{6.11}$$

For vehicles with hybrid assisted internal combustion engine, as in Equation 6.12:

$$E_{km} = (2.3 \times 10^{-3}) \times W_{vehicle}^{0.93} \tag{6.12}$$

And for the CO_2 emission per km from a vehicle with gasoline fuel versus its weight;

$$CO_{2/km} = 0.25 \times W_{vehicle}^{0.93} \tag{6.13}$$

And for the CO_2 emission per km from a vehicle with diesel fuel versus its weight;

$$CO_{2/km} = 0.21 \times W_{vehicle}^{0.93} \tag{6.14}$$

And for the CO_2 emission per km from a vehicle with an LPG fuel versus its weight;

$$CO_{2/km} = 0.17 \times W_{vehicle}^{0.93} \tag{6.15}$$

And for the CO_2 emission per km from a vehicle with hybrid assisted internal combustion engine, versus its weight;

$$CO_{2/km} = 0.16 \times W_{vehicle}^{0.93} \tag{6.16}$$

One can also investigate the rate of energy consumption per the distance traveled, which again from [63] is found to range from 1.3×10^{-3} (MJ/km.kg) for the hybrid assisted vehicles, to around 2.2×10^{-3} and 2.1×10^{-3} (MJ/km.kg); for the LPG and gasoline fueled vehicles, respectively. And to put things in the customer perspective, the data has been further mapped into two of the customer requirements: the cost of ownership and the 0–60 acceleration in Figures 6.31 and 6.32, respectively.

For the eco-audit, the embodied energy concept is applied to the different phases of the vehicle life, starting from raw material extraction (Table 6.1), the manufacturing phase (Table 6.2), the transportation of goods and materials (Tables 6.3, 6.4), and the usage (Tables 6.5, 6.6) to the end of life. The eco-audit is done using the Eco-Selector tool within the CES selector (product of Granta Design, UK). The details of the audit are tabulated for each phase and for each material type, stating the weight, manufacturing energy, and the transportation mode and distances traveled. The total energy expended at the end of life is shown in Figure 6.31, while the exact percentages are in Table 6.7. Figure 6.32 shows the CO_2 at the end of life, assuming 10 years service life with the percentages for each phase in Table 6.8. If the same auditing process is applied to a lightweight vehicle design, assuming a 30% reduction

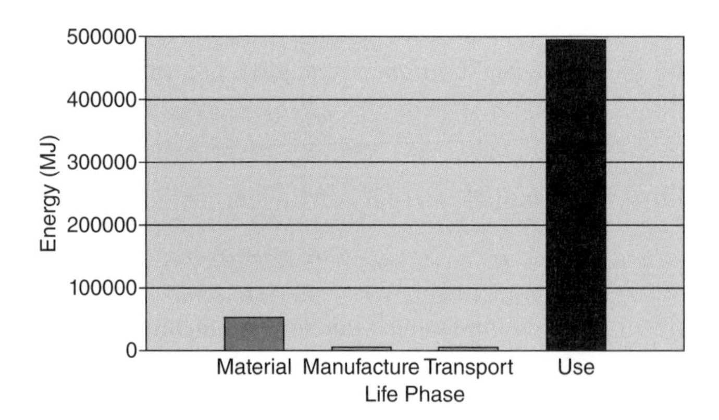

Figure 6.31 The energy consumed for each life phase, lightweight vehicle

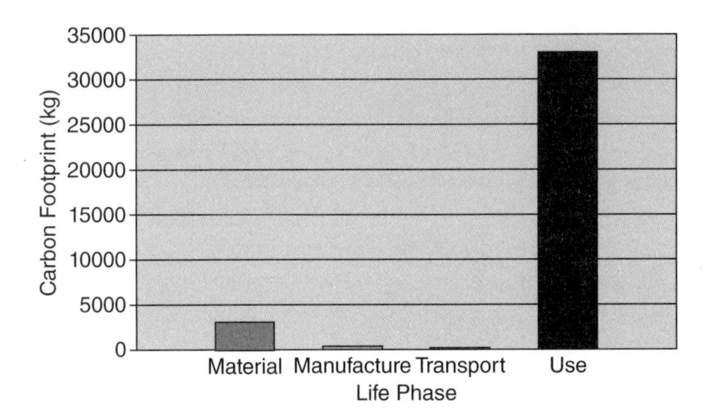

Figure 6.32 The CO_2 emissions for each life phase, lightweight vehicle

Table 6.1 Material extraction Eco-Audit

Component	Material	Primary	Mass (kg)	Energy (MJ)	%
Mild Steel body panels	Carbon steel, AISI 1015, as	22.243	589.670	13115.952	25.67
HSS Structure	Low alloy steel, AlSl 9255	24.357	113.398	2761.996	5.41
Powertrain	Cast iron, white, low alloy, BS	31.437	172.365	5418.694	10.61
Plastic Trim	PE (High Density, Low/ Medium)	88.815	108.862	9668.555	18.92
Al body panels	Aluminum, 5005, wrought, H6	126.987	86.183	10944.070	21.42
Wiring	Silver-bearing oxygen-free, h.c.	48.152	20.412	982.868	1.92
Windshield, windows	Soda Barium Glass	26.201	40.823	1069.595	2.09
Tires	Polychloroprene (CR, 17–50%)	112.253	63.503	7128.401	13.95
		0.000	0.000	0.000	0.00
Total			1195.216	51090.130	100

*When applicable, primary production values account for recycle fraction in supply.

Table 6.2 The manufacturing phase Eco-audit

Manufacture:

Component	Process	Processing	Mass (kg)	Energy (MJ)	%
Mild Steel body panels	Forging, rolling	2.342	589.670	1381.113	32.89
HSS Structure	Forging, rolling	3.436	113.398	389.613	9.28
Powertrain	Casting	3.760	172.365	648.171	15.44
Plastic Trim	Polymer molding	6.452	108.862	702.386	16.73
Al body panels	Forging, rolling	2.622	86.183	225.930	5.38
Wiring	Forging, rolling	1.838	20.412	37.512	0.89
Windshield, windows	Glass molding	7.930	40.823	323.726	7.71
Tires	Polymer molding	7.730	63.503	490.888	11.69
		0.000	0.000	0.000	0.00
Total			1195.216	4199.338	100

Table 6.3 The transportation Eco-audit

Transport:
Breakdown by transport Total product mass = 1.2e + 03 kg

Stage Name	Transport Type	Transport	Distance (km)	Energy (MJ)	%
Distribution OEM to Dealer	32 tonne truck	0.460	724.205	398.167	28.10
Supplier to OEM Mfg Facility	32 tonne truck	0.460	160.934	88.482	6.25
Inter-supplier stage 1	14 tonne truck	0.850	112.654	114.449	8.08
Extraction site to Supplier	Rail freight	0.310	724.205	268.330	18.94
Inter-supplier stage 2	14 tonne truck	0.850	160.000	162.549	11.47
OEM powertrain site to	32 tonne truck	0.460	700.000	384.860	27.16
Total			2581.998	1416.837	100

Table 6.4 The details of the transportation phase

Breakdown by	Total transport distance = 2.6e + 03 km		
Component	Mass (kg)	Energy (MJ)	%
Mild Steel body panels	589.670	699.009	49.34
HSS Structure	113.398	134.425	9.49
Powertrain	172.365	204.326	14.42
Plastic Trim	108.862	129.048	9.11
Al body panels	86.183	102.163	7.21
Wiring	20.412	24.196	1.71
Windshield, windows	40.823	48.393	3.42
Tires	63.503	75.278	5.31
	0.000	0.000	0.00
Total	1195.216	1416.837	100

Table 6.5 The usage phase Eco-audit

Mobile Mode	
Fuel and Mobility type	Gasoline—family car
Energy Consumption (MJ/tonne.km)	2.100
CO_2 Emission (kg/tonne.km)	0.140
Product Mass (kg)	1195.216
Distance (km per day)	56.327
Usage (days per year)	350.000
Product Life (years)	10.000
Total Life Distance (km)	197144.639

Table 6.6 The details of the usage phase

Breakdown of mobile mode by components			
Component	Mass (kg)	Energy (MJ)	%
Mild Steel body panels	589.670	244125.615	49.34
HSS Structure	113.398	46947.234	9.49
Powertrain	172.365	71359.795	14.42
Plastic Trim	108.862	45069.344	9.11
Al body panels	86.183	35679.898	7.21
Wiring	20.412	8450.502	1.71
Windshield, windows	40.823	16901.004	3.42
Tires	63.503	26290.451	5.31
	0.000	0.000	0.00
Total	1195.216	494823.843	100

Table 6.7 The complete audit showing all phases, from an energy point of view

Phase	Energy (MJ)	Energy (%)
Material	51090.130	9.26
Manufacture	4199.338	0.76
Transport	1416.837	0.26
Use	494823.843	89.72
Total	551530.148	100

Table 6.8 The complete audit showing all phases, from a carbon dioxide point of view

Phase	CO_2 (kg)	CO_2 (%)
Material	3111.244	8.52
Manufacture	322.984	0.88
Transport	101.117	0.28
Use	32988.256	90.32
Total	36523.601	100

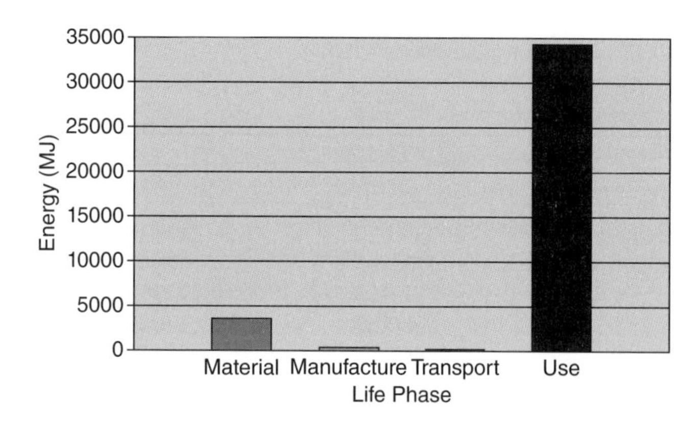

Figure 6.33 The energy consumed for each life phase, lightweight vehicle

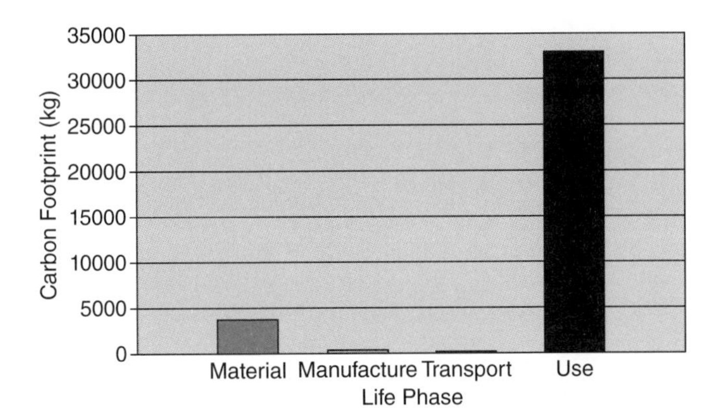

Figure 6.34 The CO_2 emissions for each life phase, lightweight vehicle

weight, which translates to around 360 kg savings, the final energy consumed MJ/km and CO_2 contributions will be that in Figures 6.33 and 6.34. Hence, inspecting the results from the two designs (lightweight vs. conventional) shows that 30% reduction in weight can lead to a 28% reduction in the CO_2 footprint and a 29% reduction in the energy consumed.

7

Static Aspects of the Automotive Manufacturing Processes

7.1 Introduction

The static aspect of any manufacturing plant constitutes its machinery allocations to the process layout, in addition to the geographical location of the plan within the manufacturing, supply, and the market network. So the static aspect focuses on optimizing the process layout and the resource allocation within the plant to achieve the production capacity and an optimized flow of material and value additions. This chapter addresses the static aspect of the automotive manufacturing from the different layout strategies applied in the industry in addition to the locational strategies that might be used to systemize the process of selection a production site. The book will introduce the different types of the automotive suppliers and their impact on the manufacturing process.

7.2 Layout Strategies

The layout strategies are used to provide a higher utilization of the production floor, equipment and the working force. In other words, these strategies control the fixed cost aspect of the automobile manufacturing, represented in the insurance cost, the HAVC, the lighting, and the equipment. They aim at providing an optimized spatial platform for material flow, people flow or movement, and the information transfer within the plant. This enables higher production rates, a flexible workforce where people can be exchanged inter and intra stations depending on production need or the product type. The optimized information flow helps in adjusting the production upstream and downstream instantly, which is important in stopping or adjusting the

The Automotive Body Manufacturing Systems and Processes, First Edition. Mohammed A. Omar.

production if a certain defect is detected while its root cause is spatially and/or temporally detached from the point of inspection.

Additional benefits from layout strategies were summarized by Francis in 1992 [1], as:

- Minimize the production time.
- Minimize the production cost.
- Minimize the material handling, in terms of time, cost and equipment type.
- Minimize the investment in the equipment.
- Utilize existing space more effectively.
- Maintain flexibility of arrangement and operations.

The manufacturing space layout strategies are typically classified into four main types, according to the nature of the production in terms of product variations and volume, these types are:

1. *The fixed position layout strategy*, which is applied to optimize the placement of bulky, large objects that would stationary though the production sequence. The fixed position strategy is typically used when the product variety and volume are low and is typically dedicated for production with small number of machines.
2. *The process-oriented layout strategy*, which is designed for production environments where low volume, high variety products are manufactured. The focus here is on the process. This strategy is typically used for workshop environments because their product changes every day. For example, a carpenter's workshop produces beds one day and doors the next, so the sequence of operations is important to the product flow because the product is well defined in terms of its frequency and volume. One of the main advantages of the process layout is that it allows for a high mix (variety) of products to be manufactured at high flexibility, hence it is more robust against product variations or fluctuations in product volume. At the same time, the process layout tends to increase the semi-finished products on the product floor, i.e. increase the Work In Process (WIP) levels, in addition, this layout might jeopardize the machinery utilization rates and complicate the material and information flows. To solve these issues, some optimization techniques can be adopted to have a process layout strategy with a single piece flow to reduce the WIP.
3. *The product-oriented layout strategy*, which is applied to production plants, where high volume, low variety product variation is manufactured. The focus here is on the product, because the process is repetitive and continuous, hence the machinery layout and material flow should be designed to optimize the manufacture of a well-defined product in type, frequency and demand. This strategy is appropriate for the automotive manufacturing facilities because the output is well defined in terms of specific (typically limited) number of vehicle models and stable produc-

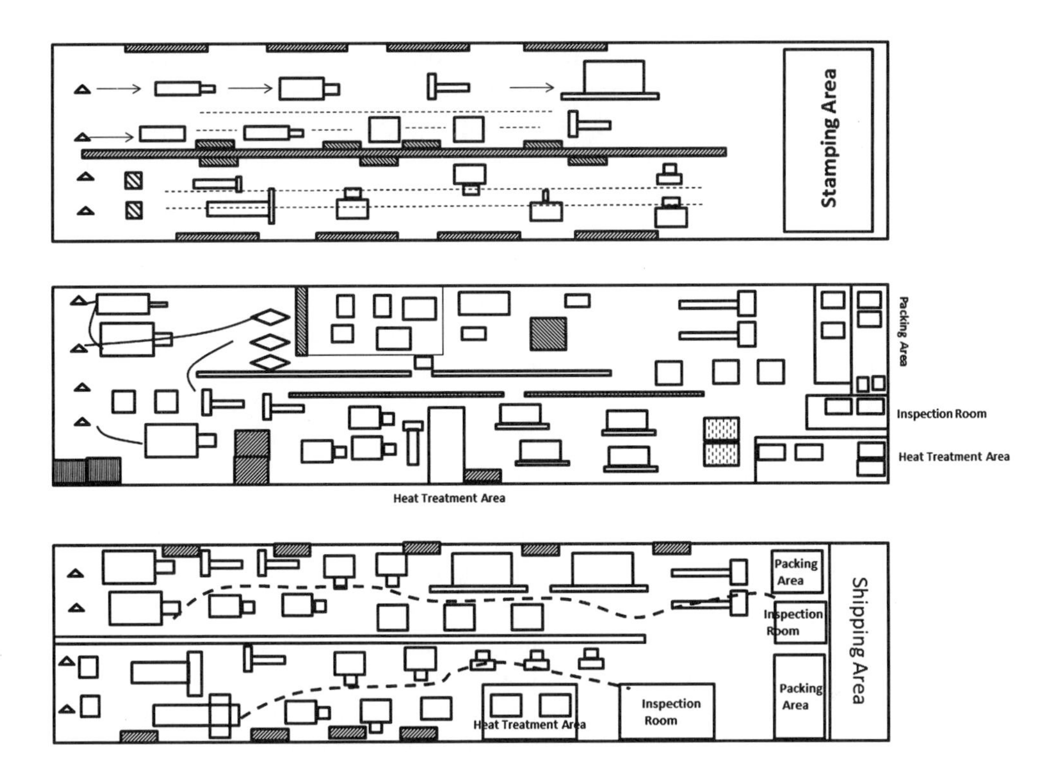

Figure 7.1 The three different layouts: process, cell and product

tion volumes. For this layout, the machinery utilization is specialized according to the product nature and the sequence of operations based on the product precedence relationships. However, once the layout is set according to the market forecast and the production plan, any changes or disruptions can affect the overall plant efficiency and operation. The automotive OEMs have adopted some complementary techniques along with the product layout to enhance the overall plant flexibility; most of these techniques are based on creating a single piece flow with a pull-based system; such techniques will be discussed later in the chapter within the context of Value Stream Mapping (VSM).

4. *The cellular layout strategies*, which are designed for medium volume production with medium to low product variety. The cell strategy allows for intermediate design between the product and the process layouts to accommodate production changes.

Figure 7.1 shows the floor space arrangements for the above strategies to highlight the difference in the equipment allocations.

From the above, the strategy selection process can be systemized based on the number of product variations P and its volume Q. so for large Q/P ratio a product-based layout should be adopted, while for a medium *Q/P* a cell layout is preferred, and finally the process layout should be used for a small *Q/P* ratio.

So from the above discussion, the first step in designing a plant layout will be to identify the following parameters: the number of product variations *P*, the product volume *Q*, the routing considerations *R*, and production task time or production rate represented by the time *t*, and any supporting activities that might be needed to aid the actual material conversion processes *S*. Such activities might include material replenishment or staging or any needed offline support such as frequent equipment maintenance, e.g. if dressing the spot-welding electrodes is done regularly at a specific number of welds applied, then the location of the welding robots should take into consideration the area where the dressing machine and the back-up electrodes are staged and prepared. After compiling these inputs, the open literature offers two methodologies to establish an optimized layout; the first is called the Systematic Layout Planning (SLP) and the second is the Systematic Handling Analysis (SHA). The SLP and SHA are similar in their overall structure, where the SHA analysis has three phases: the external integration plan, followed by the overall handling plan: and finally, the detailed planning phase. The SLP starts by defining the locations associated with the supporting processes and the main processes using a relationship chart, then an overall layout is defined, followed by a more optimized detailed layout. This chapter will focus on the use of the SLP method as proposed by Muther [64].

The SLP starts by inputting the parameters to decide on material flow route which is then modified per the activity relationship chart. The material flow is designed based on the operations required to make the product take into consideration the precedence relationships. The activity relationship chart then modifies the layout by describing the inter-relations between the different stations in terms of frequency or quantity of material or other services transport. After that, the modified layout is mapped onto the available space and any fixed points in the production floor. If the layout fits within the production floor, a space relationship chart is used to establish any main or auxiliary material flows between the stations. Final optimization is done to ensure the most proper placement of the machinery within the layout.

7.3 Process-oriented Layout

In a process-oriented layout, the product is variable in type and volume so the layout is dedicated to serving a sequence of processes for an unknown product, hence no precedence relationships are taken into consideration. The main examples of the process based layout are small workshops, other examples include emergency rooms, where different patients come for different reasons. The main goal for process layout is not based on meeting the production rate, in other words, it is not about creating the fastest route but it focuses on providing the highest flexibility to enable the handling of the different products at low cost.

The process layout strategy relies on minimizing the material handling effort to reduce the cost associated with processing different products. So from a mathematical point of view, the process layout is concerned about minimizing the number of material transfers and the quantities transported between the different stations; as in Equation 7.1, which describes the minimization of cost function $Cost_{min}$ between the different stations n, with the subscripts i,j being the divisions or the inter-departments, and X being the number of loads transported between station i to station j, with an associated cost C per unit load between i to j.

$$Cost_{min} = \sum_{i=1}^{n} \sum_{j=1}^{m} X_{ij} C_{ij} \tag{7.1}$$

So the process layout is applied with following guidelines and procedure in mind:

1. First, the engineer should collect and identify the work-stations and the relation between them in terms of the material transfer. This is followed by plotting a pictorial picture of the material transfer, which can be done through a simple schematic.
2. The schematic is modified to show the stations with the highest inter-material transfer to be close to each other. Then the cost information is added to each line showing a material transfer, at the same time, one should be careful about the transfer directionality and whether it has any effect on the cost, because products' weight and their content of value change as they are transferred between the stations.
3. The schematic is mapped into the actual available space to check for any constraints or machinery fixed points.
4. Finally, the schematic should be checked to make sure that the stations with the highest cost in material transfer are closest to each other within the space floor. Also, any applied modifications can be quantified by computing the initial cost and the cost after the modification is done.

To illustrate the above steps through an example, one can consider the case where five stations exist in a workshop, with inter-material exchange between them shown in Table 7.1. The first step is to ensure that the directionality of the transfer is important or not, if directionality is not important, then the table can be summarized as in Table 7.2.

Second, the cost information should be added to each material transfer activity, hence the cost information from Table 7.3 can be added to Table 7.2 to produce Table 7.4 which shows the actual cost exchange between each of the stations. Then the schematic is drawn in Figure 7.2 to show the transfer lines and the stations with the highest cost exchange.

Finally, the schematic is adjusted so that the stations with the highest cost exchange are the closest to each other, as in Figure 7.3. One can compute the initial and final

Table 7.1 Inter-material exchange

From/to Station	A	B	C	D	E
A	0	15	0	20	10
B	15	0	15	0	20
C	0	15	0	0	30
D	25	0	0	0	35
E	10	20	0	10	0

Table 7.2 The inter-material exchange without directionality

From/to Station	A	B	C	D	E
A	0	30	0	45	20
B	0	0	30	0	40
C	0	0	0	0	30
D	0	0	0	0	45
E	0	0	0	0	0

Table 7.3 The transportation cost

From/to Station	A	B	C	D	E
A	0	2	2	3	2
B	2	0	3	0	4
C	2	3	0	0	5
D	3	0	0	0	4
E	2	4	5	4	0

Table 7.4 The total cost for inter-material exchange

From/to Station	A	B	C	D	E
A	0	2	2	3	2
B	2	0	3	0	4
C	2	3	0	0	5
D	3	0	0	0	4
E	2	4	5	4	0

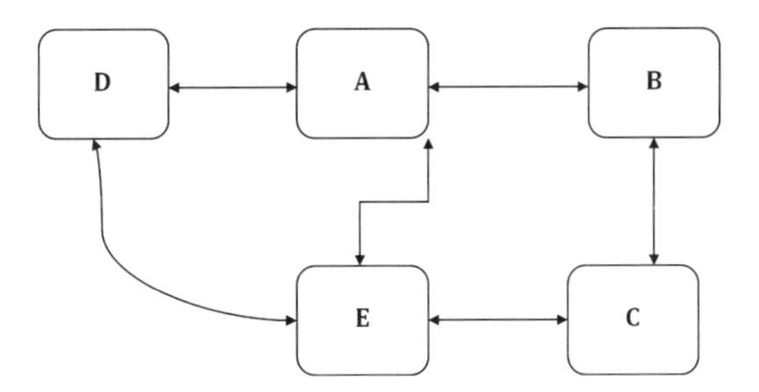

Figure 7.2 The stations' schematics and material exchange

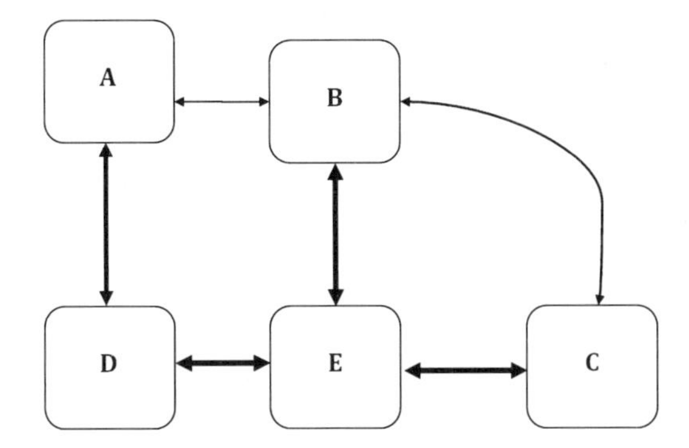

Figure 7.3 The improved layout

costs using Equation 7.1 to show the impact of the new layout design. The cost calculations can include the distance traveled through applying the cost as cost $ per unit distance traveled per m. And of course the proposed design should be fit to the available floor space, making sure it does not violate any space constraints or conflict with any fixed points.

The example discussed above is a simple case study that assumes no directionality or space constraints. Several software packages are commercially available for the design and optimization of the process-based layout; such packages include Computerized Relative Allocation of Facilities Technique (CRAFT), and the Computerized Relations Layout Planning (CORELAP).

7.4 Cell-based Layout Design

In a cellular-based layout, the goal is to facilitate the production and the material-handling efforts when dealing with a medium Q/P ratio, so that it meets the production rate needed. The cell layout focuses on grouping the products into families or platforms so that each family is handled by a separate cell. The grouping is done based on the resources' similarity required for each product type. Other grouping techniques are based on the product attributes in terms of shape, size and material type.

After the products are classified into families, the resources needed to make one family of products are grouped into one cell, which is specialized in performing certain activities or making one product family.

Again, the best way to explain the guidelines for establishing a cell based layout is through an example, if one considers the products from 1–8 from Table 7.5, and the processes (machines) required for each from machine 1 through machine 8; then three possible groups can be established as done in Table 7.6. These groups A through C are the proposed cells.

In most of the situations, the duplication of resources or the co-sharing, if possible, is needed in the cell-based layout due to the similarity between the product families. This is the case in the above example for machine 5 which is needed in cells A and B. so these two cells are located within close proximity so the products from both cells are processed through the same machine or machine 5 can be duplicated and placed in both cells.

Table 7.5 The product-machine manufacturing needs

	Product Models/Types						
Machines	1	2	3	4	5	6	7
I	X			X		X	
II		X			X		
III	X			X		X	
IV		X			X		
V			X			X	X
VI			X				X

Table 7.6 The cell-based grouping of products and machines

	Product Models/Types						
Machines	1	4	6	3	7	2	5
I							
III							
V			SHARED				
VI							
II							
IV							

7.5 Product-based Layout

In a product-based layout, the product is predictable in terms of its volume and variations, and so the layout is optimized to manufacture this product in a consistent and a repeatable manner. This layout is hence dedicated to a standardized product manufactured through a known sequence of tasks. This layout strategy is the most suited to the automotive manufacturing lines, because the product variations are typically limited to few vehicle models and the master production schedule defines the quantities from each variation on weekly basis. At the same time, this layout strategy is the most sensitive to any disruptions in the product demand and/or quantity. So, the main goal from a product-based layout is to maximize the resources' utilization and reduce the number of stations needed to manufacture the product. However, the consolidation of processes is controlled mainly by the product design, so to reduce the processing required, one needs to consolidate the parts within the product.

The product-based layout is typically done through following steps:

1. The processing effort is determined and divided into tasks, with their cycle time (needed processing time) computed.
2. The product precedence relationship should be consulted to determine the sequence of operations or tasks.
3. Using the takt time (the frequency of how often the customer pulls a product), the number of stations is established from the known cycle times and the takt time.
4. Finally, the tasks are assigned to the stations and the layout efficiency is calculated. Additional steps might be needed to help make sure that the stations are balanced and the idle time is minimized.

As before, illustrating the above steps through an example, the different concepts of cycle time and efficiency should be defined; the cycle time defines the processing time within a station and if it is not known (establishing stations phase) then it can be assumed to be the same as the takt time. The takt time is defined as the available production time per day divided by the actual customer demand of the product per day; mathematically as in Equation 7.2; also the numerical definition of the cycle time can described in Equation 7.3. With the difference being that the takt time assumes perfect synchronicity between the customer demand and the production output and rate, which is typically called the pull system-based production. On the other hand, the cycle time is dealing with a customer demand from the forecast data which is then translated into a master production schedule (MPS).

$$t_{takt} = \frac{available\ production\ time\,/\,day}{actual\ customer\ demand\,/\,day} \tag{7.2}$$

$$t_{cycle} = \frac{available\ production\ time\,/\,day}{planned\ customer\ demand\,/\,day} \tag{7.3}$$

Meanwhile, the layout efficiency is meant to describe the utilization of each work station, with the inefficiencies caused by the idle time, i.e. the time duration in which the machines or resources are not being utilized. The layout efficiency η_{layout} can then be described in Equation 7.4 as the summation of the task times from the different stations, divided by the total production time, which can be further computed as the number of stations n multiplied by the cycle time per station t_{cycle}.

$$\eta_{layout} = \frac{\sum_{i=1}^{n} t_{taski}}{n \times t_{cycle}} \tag{7.4}$$

One can consider the case of a production facility supplying 5000 units per week, composed of processing steps from A through I, with the cycle time for each task shown in Figure 7.4 along with the precedence sequence also depicted using the arrows.

The first step in establishing a layout for operations shown in Figure 7.4 would be to compute the number of stations required, which is done by computing the cycle time and the total production time. The cycle time is computed to be: the available time (40 hrs/wk x 60 min/hr) / the planned production (530 units), which is equal to 4.53 min/unit.

Then the number of stations can be computed through: the total work content ($\sum_{i=1}^{n} t_{taski} = 16$ min)/ by the cycle time (4.53), which is equal to 3.53 stations. Of course one cannot create half a stations, so 4 stations are established.

The four stations should follow the precedence relationships in grouping the tasks; so one suggested layout is shown in Figure 7.5.

So, from the suggested layout in Figure 7.5, one can estimate its efficiency using Equation (7.5) to be: total tasks' time (16 min) / number of stations (4) multiplied by the cycle time (4.53 min), which is equal to 88%. The inefficiency in this proposed layout can be further highlighted through Figure 7.6 which shows the utilization of each station with relation to the cycle time.

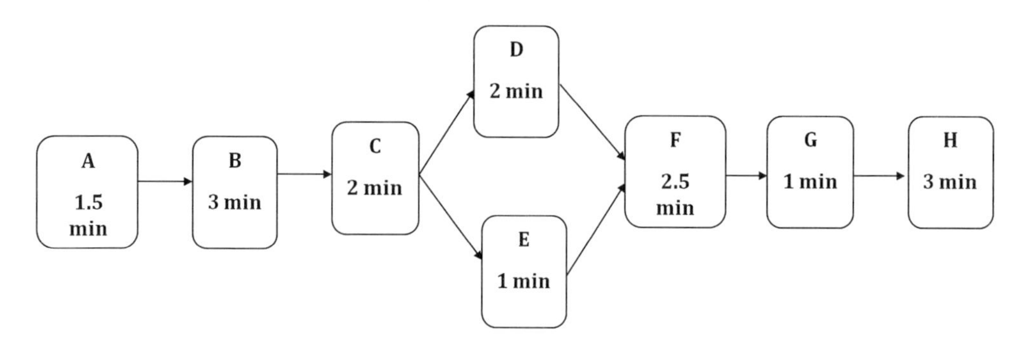

Figure 7.4 The schematic of the different stations and the material flow

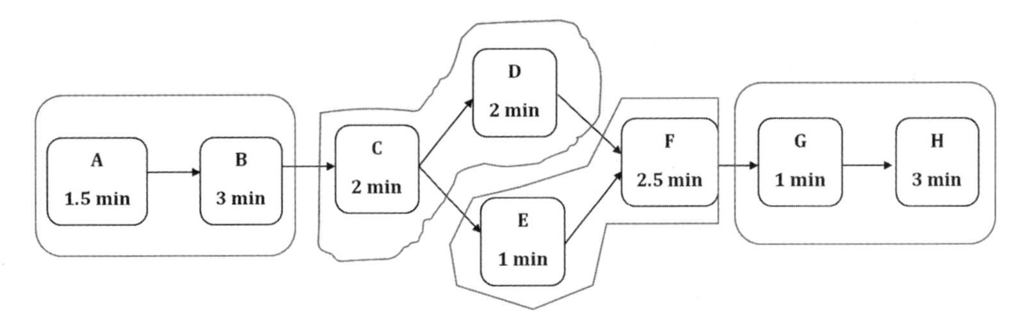

Figure 7.5 The improved layout showing the grouping of the different sub-stations

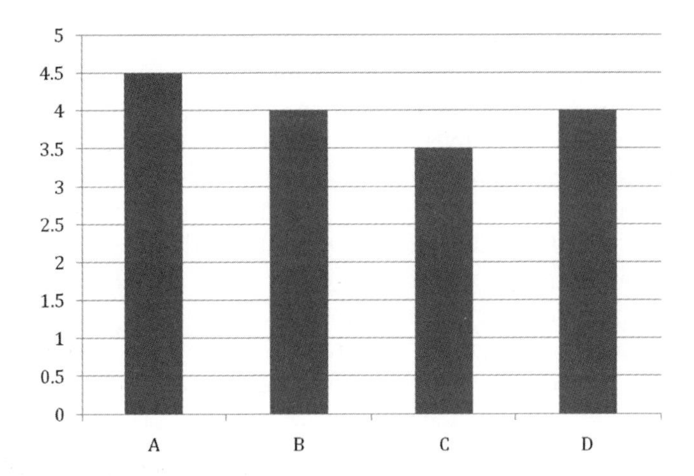

Figure 7.6 Utilization study of each station

The utilization in Figure 7.6 shows the imbalance between each station utilization, in other words, some stations are well utilized (100% utilization) and others are under-utilized as in station 2 which has an idle time of 4.53 – 4.0 = 0.53 min. every time a product is processed through it. This imbalance should be fixed, otherwise the production rate is not harmonized between the different stations, in addition, some stations might tend to produce more than their production share in their idle times; this extra production generates extra WIP levels within the production lines leading to over-produced units that are not instantly sold, i.e. converted from goods into cash.

The following section discusses the VSM as a tool to evaluate and optimize the product-based layouts. In addition to other lean manufacturing tools, that helps in dealing with the balancing issues between the different stations as well as creating a production rate (beat) synchronicity between the different production stations; this will be disused through the use of the Kanban system.

7.6 Lean Manufacturing Tools for Layout Design and Optimization

The general topic of the Lean Manufacturing (LM) or the Toyota Production System (TPS) has been treated and covered comprehensively in several books, so this section is only intended to describe some of the LM tools that are useful in the layout planning and optimization especially for the automotive production lines. The first tool covered here is the value stream mapping (VSM), which is used to help evaluate the current layout design typically called the current state, and the proposed value and time analysis for each station, working backwards to the material flow. The analysis should reveal some opportunities for improvement and help decide on a more optimized layout design, typically called the future state.

The VSM starts by establishing a schematic of the current layout, which is called the current state map. The map described each of the manufacturing steps in terms of its cycle time, its inventory level before and after the process, in addition to the time needed to consume this inventory. It also includes any down-time related to change-over time, as the case with the stamping die, which needs to be changed with a different die in order to switch between the manufacturing of one body style to the other. The change-over frequency is included in the station description; this change-over (CO) frequency can be every product every week, i.e. EPE 1 week, the longer the change-over frequency, the less flexible the station is. To show an example of a single process schematic, Figure 7.7 illustrates the case for a stamping station with 2 seconds needed to add value to the product, i.e. form the metal sheet cycle time

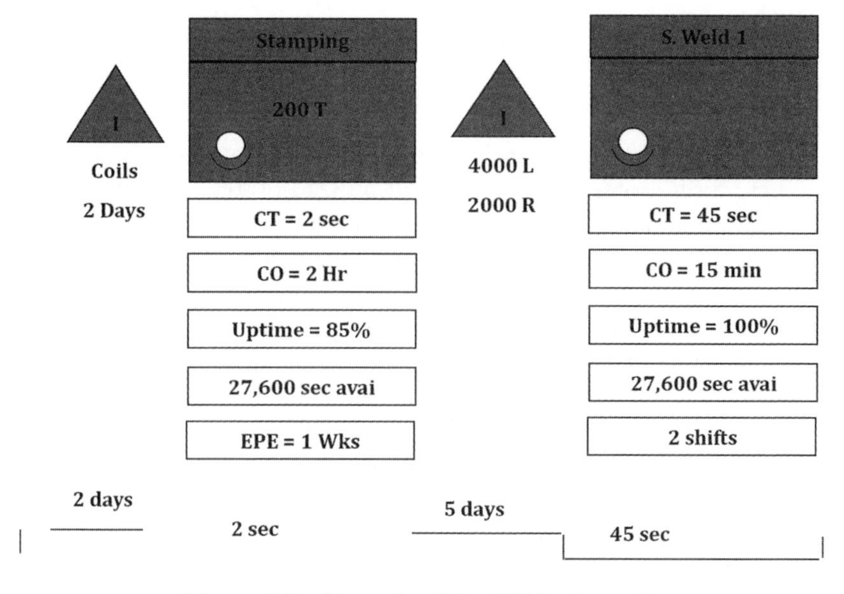

Figure 7.7 Example of the VSM schematic

(CT), with change-over happening every week, the production time available is the actual production time minus any breaks or lunch stops. Figure 7.7 shows a welding station, with different information, i.e. cycle time of 45 seconds; the horizontal line indicates the value-added time and the time needed to consume the inventory levels indicated as a triangle. The 2 days indicate 2 days worth of inventory while the 4000, 2000 indicate the number of units, while the L and R stands for left and right sub-assemblies for two variations of the product.

Additionally the VSM describes the value-added time versus the non-value-added time (e.g. time needed to consume the inventory levels) on a horizontal scale to show the total lead of the product. The lead time is the total production time, or the time it takes the raw material to go through the production processes, get transformed into semi-finished product, and then on to a complete product at the end. The lead time describes the time delay between the OEMs paying for the raw materials and parts and getting their money back; from the customer's point of view it describes the time delay needed to deliver a new product if the OEM is following a make-to-order strategy, which is the case for some of the luxury car brands. However, the lead time does not include the design and the development time for new models. For the two stations illustrated in Figure 7.7, the lead time shows that one part will go through 7 days and 47 seconds before it completes its journey through these two stations, out of which only 47 seconds are value-added and the remaining time is just inventory delays. This fact motivated automotive OEMs to reduce the inventory levels or the WIP, through applying a continuous one-piece flow.

However, the continuous one-piece flow cannot be applied in all occasions, due to the difference in the cycle times between the different stations. From the example above, one can observe that the stamping station will produce ~ 22 parts before the welding is able to finish one part. So the continuous one-piece flow should be used to couple processes or stations that have comparable cycle times. For the current case, a controlled inventory buffer system should be established to couple the stamping and the welding processes, such controlled inventory is typically called a supermarket, which, as the name implies, mean that stamping will only produce to replenish what's pulled by the welding station; in other words, in an actual supermarket when someone buys a specific item, the supermarket employee will only replenish what is bought from the shelves. Additionally, the addition of the supermarket between the stamping and welding stations balances the work-content between these two stations and synchronizes their production, according to the planned production volume and product type.

To synchronize and control the production relationship between the stamping and the welding, a kanban system is established. The kanban means a card in Japanese, and in the manufacturing context it means authorization card to make or acquire raw materials. The supermarket will utilize two kanbans, the first is called a production kanban and it authorizes the stamping station to make what is pulled by welding, and the second is the withdrawal kanban which authorizes the welding station to acquire the required sub-assemblies to make its product, through material handlers. This means that the stamping pace is controlled by the production kanban to avoid over-producing, and the supermarket guarantees that stamping makes what is needed by the welding, to avoid making the wrong product variation. On the other hand, the withdrawal kanban governs the welding station's supplies of materials and sub-assemblies so it can only manufacture what is needed, when it is needed. The production exchange process within the supermarket is shown in Figure 7.8.

Another LM tool that is typically used to complement the product-based layout strategies, is the leveling of the production (or Heijunka in Japanese) to ensure that at any given time the layout is manufacturing all the different product variations. Even though this translates into continuous change-over of the stamping dies and the robotic programs to suit the production of the different body styles, the leveling is vital to ensure that if the production stops at any time, the OEM is able to satisfy the customer demand from the different product variations.

To provide an example of the leveling principle, if a manufacturing company is making four body styles from A through D, with a customer demand for each style as shown in Table 7.7, then the best leveled daily sequence can be computed based on the following procedure.

First, the daily demand is computed as in the last column of Table 7.7, by dividing the monthly production by the number of working days per month, 20 in this case. Then each model demand fraction from the total demand is computed, e.g. model A is 100/200 = 1/2, while model D is 25/200 = 1/8. Then the highest common denominator is used as the minimum production batch where all these models should be produced, in this case it is 8 units. So the production should follow a production sequence of 4 vehicles of model A, 2 B, 1 C, 1 D; which can follow: AAAABBCD

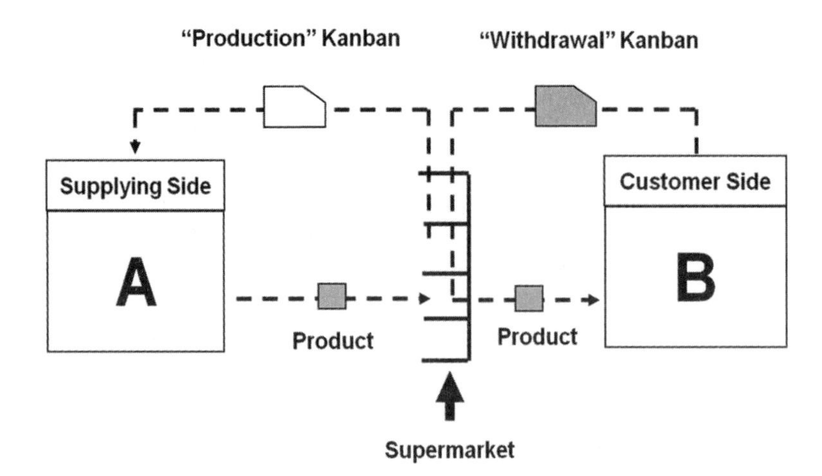

Figure 7.8 The supermarket concept between two stations and the flow of the kanban cards

Table 7.7 Customer demand for each product type per month

Model	Volume/mth	Volume/day
A	2000	100
B	1000	50
C	500	25
D	500	25
Total	4000	200

or better, ABACABAD. This will ensure that even if the production stops every 8 units, the OEM will be able to satisfy all his customer demands.

Another important LM tool for the product-based layout is the balancing, which aims at grouping the stations with low work-load (low utilization, or high idle time) together to create a better utilized bigger station.

The combination of the VSM schematic, the balancing, the leveling, and the reduction of the WIP through one-piece flow or supermarkets controlled by kanbans, all aim at optimizing the current state map into a future state map where the layout is improved based on the material processing (cycle time) and handling (storage, transportation) steps. However, to achieve some of the recommended modifications requires parallel improvement to the transformational and operational sides of the manufacturing plant. For example, to achieve a leveled production requires a drastic reduction in the change-over time; this has been evident for the stamping die where the change-over used to take hours and then improved through LM into less than 10 minutes; this has been achieved through the single minute exchange of die practices.

7.7 Locational Strategies

The location strategies present a systematic method to decide on the manufacturing facility location relative to the enterprise's other facilities, suppliers, and markets, and the benefits achieved from this location. Additionally, it helps the city's authorities in preparing their proposals to attract automotive OEMs' facilities. The locational strategies aim at maximizing: (a) the cost reductions through reducing the transportation cost from shipping materials and suppliers' parts into the facility and from the distribution of the final products to the dealerships or market locations. Additional cost reductions can also be achieved through the difference in the currency exchange rates between the manufacturing location and that of the market. Also, (b) it aims to increase the flexibility of the manufacturing facility by reducing the lead time and the movement of the specialized teams between the different enterprise or OEM facilities. Such teams include the pilot team that helps in the launching of new vehicle models or other suppliers' teams that aid in installing and running new manufacturing equipment. (c) The location can also help the OEM to acquire a skilled labor force, or new innovations if it's located near strong research and development centers or universities. The location selection process should consider the level of protection for the intellectual property in each location, because each OEM applies their own innovation and work practices that are considered confidential.

Different companies rely on different geographical allocations for their overall manufacturing network; these can be generally classified in four main types, as proposed by [65]:

- The centralized production, where one manufacturing location is chosen to make all the different product variations for the different market. This strategy helps the OEM to have centralized operations in one location, with their research and development, and supplier network being in close proximity. However, this strategy does not allow the OEM to reduce the shipping cost to their markets. Also, the OEM applying this practice might be subject to additional taxes and fees in foreign markets.
- The regional production is the opposite of the centralized production strategy where the OEM has a manufacturing facility in each of its main markets, dedicated to that market product variance and demand. This allow the OEM to avoid the additional taxes and fees, and shipping costs, but at the same time the OEM will have to work with his supplier network to co-locate on those markets. Additional challenges include the training required to ensure that the labor force in these markets are practicing the OEM's unique work standards. This trend of manufacturing can be further illustrated by the fact that foreign OEMs produce around 4 million vehicles in the US for the US market. Also, the Toyota Motor Corporation have established the Toyota Supplier Support Center to help train and qualify their supplier network in the US, so they are more adapted to the Toyota Production System and their practices, such as the Just In Time system, which requires the suppliers to provide lower quantities of their parts to Toyota but at higher frequencies.
- The regional specialization strategy locates the manufacturing facilities according to each location specialization, for example, the power-train is located in a country where the power-train manufacturing and testing infrastructure are high, while the body assembly plants are located where the stamping and body-weld skill set is high. Also, the location of the different manufacturing steps might be based on cost perspectives; in other words, the stamping facilities are located in countries where the equipment (presses) cost or material (steel) cost are low. In regional specialization, each country location exchanges their products with the other facilities so each location is able to assemble a final product, made in different countries, this is done to take advantage of the skills in each country and also avoid any fees or taxes.

- The vertical integration strategy locates the different manufacturing processes in different countries based on specialization but the different components will be integrated as a final product in one location. This is done to take advantage of the skills in the different locations and from the centralized location supplier network and established infrastructure. However, this mode requires the OEM to ship the final product from one location into the different markets.

These four modes are displayed in Figure 7.9.

To select the manufacturing location for any of the above four modes, three typical steps are followed; the first is the screening process which generates a list of some candidate locations, based on specific criteria that are unique for each OEM, some might be based on the wage differentials, the regulatory environment, proximity to the market locations, labor attitudes, the transportation infrastructure (ports, railroad network), and the suppliers' network in that area. The candidates selected in the screening phase are typically done at the country level not a well-defined city or regional location. The next step is based on the regional selection from the candidate list, using more specific criteria and tools, the criteria typically include estimates for

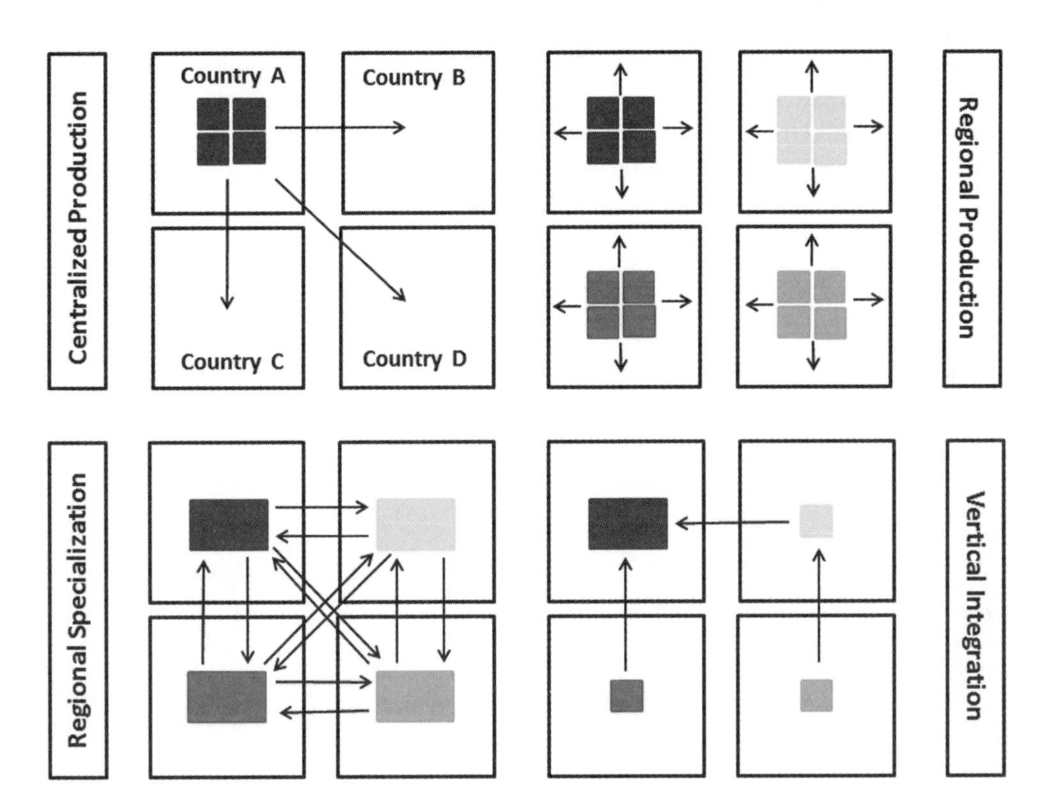

Figure 7.9 The different production strategies from locational point of view

the operating cost (material cost, utility cost, etc.), the incentive package from the community or state, and the labor force. The incentive package can play a major role in the decision-making process, to provide an actual example; in 1993, the state of Alabama offered Mercedes around $253 million worth of incentives to locate their US production facility in the state, this translates into around $169,000 / for each work opportunity created by Mercedes, which is around 1,500 jobs within the plant. However, this number is not descriptive of the actual benefits for the state of Alabama, because, having the Mercedes plant in Alabama generated around 84,000 jobs, due to the supplier network co locating in the area. Further benefits came with the Mercedes plant expansion and the co-location of the Honda plant in 2001 and the Toyota plant in 2002. Additionally, the Hyundai Montgomery, Al opened in 2005.

The tools used for this stage are presented in the next section in detail. The final phase of the selection decides on the site location, where the manufacturing facility is to be built. The criteria for this phase include the quality of life metrics such as the quality of education, the crime rate, and the housing. To provide an example, Table 7.8 shows the geographical separation (distance in miles) between the suppliers' network and the OEM assembly plant, for group of Japanese OEMs. Additionally Figure 7.10 shows the integrated BMW production network in Germany.

The labor force considerations should consider not only the wage differential but also the labor productivity and flexibility (learning ability), commitment (attitude, absenteeism) and . The labor wages should be normalized using the number of products per unit time. For example, the labor wage/productivity can be presented as the labor cost per unit produced as in Equation 7.5:

$$C_{unit} = \frac{labor\ wage / day}{units\ produced / day} \tag{7.5}$$

Table 7.8 The geographical separation (distance in miles) between the suppliers' network and the OEM assembly plant, for a group of Japanese OEMs

Assembler	Mean	Median	Std. Dev.	Minimum	Maximum	n
Autoalliance	366.47	245.00	453.83	5.00	2013.00	59.00
Diamond-Star	316.78	265.77	312.61	0.00	1726.29	50.00
Honda	262.47	152.00	379.63	0.00	2022.00	117.00
Nissan	343.21	251.20	347.35	6.96	1801.80	64.00
NUMMI	1749.85	1959.24	651.28	33.05	2191.87	25.00
Saturn	310.68	294.34	243.14	27.28	922.54	17.00
Subaru-Isazu	309.94	193.37	411.14	19.60	1837.73	53.00
Toyota	335.30	174.19	481.32	0.00	1897.00	67.00

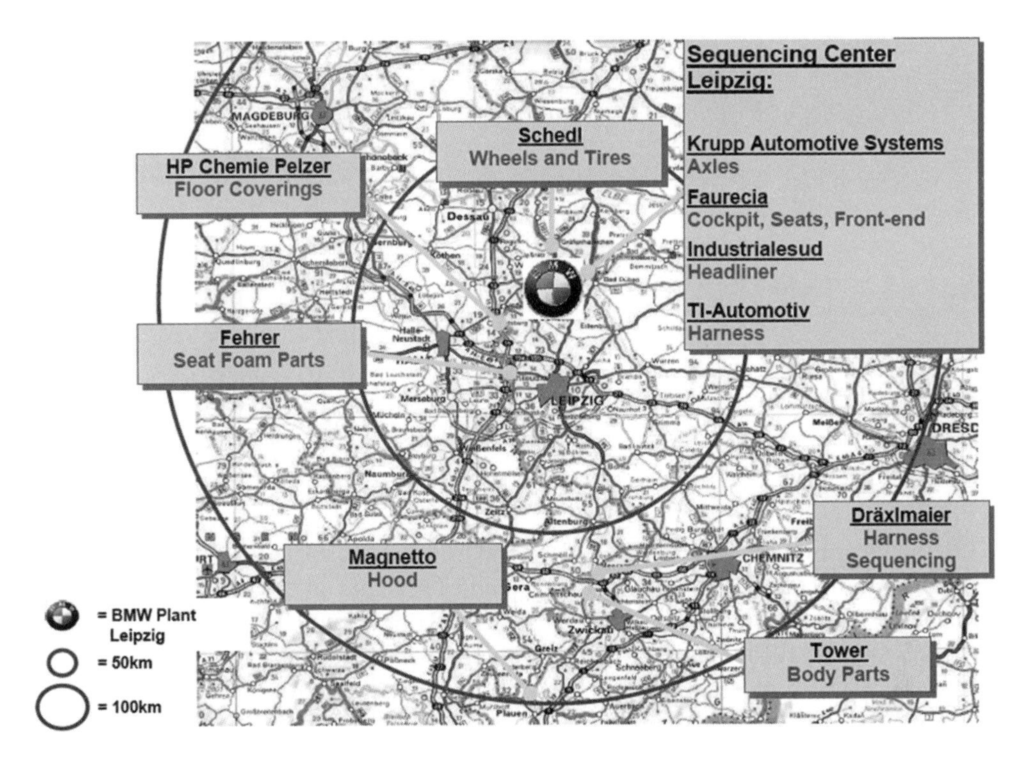

Figure 7.10 The BMW production Germany

Labor flexibility is typically measured in the workforce's ability to adapt to new products, new manufacturing practices, and new work assignments. These aspects can be measured by the learning curve of the labor force. The learning curves are generated based on the following main assumptions: the time requirements to produce or manufacture the first unit will decrease as additional products or units are made. Different learning curves formulations differ in describing the time decrease behavior, some describe it as a negative exponential function while others assume it to be logarithmic in nature.

For the learning curves, one can present its three main types to include: learning curves based on an arithmetic approach which is most useful if the production quantities double. The second type is based on logarithmic analysis, which assumes that the time needed t_n to make the product number n, is related to the time spent on the first product t_1, or product number 1, as the relation in Equation 7.6, where l is dependent on the product type and complexity, and is typically describe as the learning curve slope presented as a percentage. For example, the Ford model T production had an l value of 86% acquired between the first unit produced in 1910 and that produced in stable production.

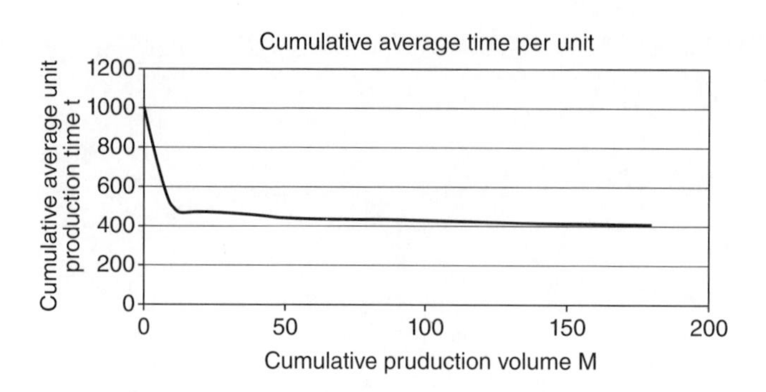

Figure 7.11 The learning curve

$$t_n = t_1 \times n^{\log(l)/\log(2)} \tag{7.6}$$

The third learning curve formulation is based on the use of tables specified for the product type, these tables are typically made based on previous experiences and knowledge base. A typical learning curve is shown in Figure 7.11 as the decrease in production time, till it reaches a steady value.

7.7.1 Locational Strategies Tools

This section discusses a set of quantitative tools that might be used in evaluating the different locations from the list of candidates. These tools are:

- The *factor rating method* is based on specifying a set of weighted criteria for each location. This method is typically used in the last stages of the process because it contains a lot of data with qualitative nature, such as the quality of education and housing. This method starts by specifying the criteria and their weights (i.e. their importance to the OEM), then each location is scored for each criteria using a predefined scale, which is typically normalized. Then the scores are multiplied by the weights for each criterion and summed to provide a final score for each location. This process can be done using a simple table or it can be more complicated using a Quality Function Deployment matrix, to allow for the interactions between the different criteria to be included. An example of the factor rating method showing the different factors is shown in Table 7.9, which is the result of information compiled by the state of Tennessee to attract automotive OEMs and suppliers.
- *The location break-even analysis* compares between the different locations based on the operating cost of the plant against the expected production volume. This requires that the fixed and variable costs are both identified and then the total cost is plotted for the number of produced units. Typically some locations might have

Table 7.9 The state of TN survey results for automotive OEMs and suppliers

Site Attribute	Excellent	Adequate	Inadequate
Access to market for your product	44.9%	54.4%	0.7%
Proximity to market for final product	29.7%	68.8%	1.4%
Access to marketing and advertising services	14.7%	79.1%	6.2%
Access to financial, accounting, and legal services	26.7%	67.2%	6.1%
Access to engineering and research & development services	12.7%	70.6%	16.7%
Access to raw materials	15.7%	78.4%	6.0%
Available supply of workers	13.4%	30.6%	56.0%
Skill level of available workforce	8.2%	37.3%	54.5%
Worker productivity	15.4%	66.9%	17.6%
Quality/adequacy of workforce training and development	9.7%	58.2%	32.1%
Wage rates compared to other potential sites	16.5%	78.9%	4.5%
Other labor costs relative to other potential sites	10.8%	82.3%	6.9%
Labor/management relations	39.4%	59.1%	1.5%
Cost of land	24.2%	69.5%	6.3%
Availability of land	24.3%	67.1%	8.6%
Availability of financial capital in Tennessee	19.8%	66.1%	14.0%
State taxes on businesses, individuals (franchise/ excise, sales)	8.5%	65.4%	26.2%
Local taxes on businesses (property, sales)	9.9%	71.0%	19.1%
Zoning and building regulations	15.6%	79.3%	5.2%
Environmental regulations and requirements	8.8%	83.8%	7.4%
Quality of interstate highways	43.4%	52.9%	3.7%
Quality of state highways	32.3%	59.4%	8.3%
Quality of local highways and roads	27.2%	55.1%	17.6%
Availability of quality rail service	9.8%	67.9%	22.3%
Availability of air transportation services	19.8%	58.8%	21.4%
Quality of electric power service	2.1%	60.6%	7.3%
Access to natural gas	27.8%	66.9%	5.3%
Price of natural gas	10.6%	82.6%	6.8%
Available water supply	28.3%	61.2%	10.5%
Adequacy of waste disposal	11.6%	78.3%	10.1%
Availability of high speed telecommunications services	18.0%	70.7%	11.3%
General business climate in Tennessee	33.8%	63.9%	2.3%
Quality of life	65.7%	32.8%	1.5%
Availability of affordable housing	36.0%	55.1%	8.8%
Low crime rate	19.3%	67.4%	13.3%
Quality of public schools	12.5%	55.9%	31.6%
Quality of private schools	44.2%	45.3%	10.5%
Higher education and research facilities	25.6%	60.9%	13.5%
Community recreation resources	25.0%	64.7%	10.3%
Right-to-work laws	41.6%	56.2%	2.2%
Cooperation of local governments	28.9%	60.0%	11.1%
Cooperation of state officials	20.2%	69.0%	10.9%

high fixed costs in terms of the tax structure but have a lower variable cost in terms of the utilities and health insurance costs. Additionally, big cities tend to be more expensive in terms of the fixed costs, however, due to their centralized locations which is near major ports, railway systems, or major airports, so the shipping cost to and from them is lower.

- The *center of gravity method* is typically used to study the proposed locations relative to other OEMs' facilities, intended markets, and the suppliers' locations. This method is focused on optimizing the geographical location of the new plant based on minimizing the transportation cost in terms of distances and the quantities of materials and finished products shipped to and from the location. This method can be mathematically described using the weighted coordinates of the plant based on the shipping cost/quantities, as in Equations 7.7 and 7.8 for the x and y coordinates, respectively.

$$x = \frac{\sum_i d_{ix} Q_i}{\sum_i Q_i} \tag{7.7}$$

$$y = \frac{\sum_i d_{iy} Q_i}{\sum_i Q_i} \tag{7.8}$$

Where, d_{ix} is the distance of exchanged materials form the x-axis, while the d_{iy} is the y axis distance, and Q is the quantity of material shipped to or from the location.

One of the main advantages of the Center of Gravity method, is that it can be mapped on an actual map from the Geographical Information System (GIS), which can provide additional dimensions to the selection. The GIS system provides the following information;

- Census data in terms of the city, county, zip code
- Maps of streets, highways, bridges, rail-way network
- Utilities (electric power, water, gas lines)
- All rivers, mountains, forests
- Major airports, colleges, hospitals, ports.

Additionally, some new software packages (MapInfo from Microsoft) can provide information about any competitor OEMs in the area.

- The *Transportation table method* which is based on optimizing the plant location with respect to the shipment (material, parts, finished products, etc.) from and to the plant. This optimization is done in the form of a transportation table. The transportation table combines the information about the sources' production capacities (other manufacturing facilities) and sinks' needs (markets or inventory locations) so that the plant location fits with the overall OEM network. The plant location is selected so that it reduces the cost of shipment. The transportation table is best used, when the need is to integrate the new plant with the existing OEM network to supply similar product type (vehicle models) to that of other OEM facilities. In other words, the goal is not to produce specialized products for specific markets.

The transportation table is generated by first preparing the information (production capacities) about the existing production facilities within the region where the new plant is to be located, in addition to the markets' needs. Also, the cost associated with shipping the products from each of the sources to each of the sinks is incorporated. Table 7.10 shows a sample transportation table, where the sources are compiled in the table rows and the markets or destinations are inputted. The summation across one row is the capacity for each facility, where the summation along one column represents that destination's total demand. In addition, the cost associated with the shipping is shown in the sub-boxes within the table cells.

After all the information is inserted in the transportation table, optimization should be done to better allocate the markets' needs from the available production

Table 7.10 Sample transportation table

Sources	Destinations/Markets						Total Capacity
	D_1	D_2	D_3	D_4	D_5	D_m	
Factory 1	C_{11} X_{11}	C_{12} X_{12}	C_{13} X_{13}	C_{14} X_{14}	C_{15} X_{15}	C_{1n} X_{1n}	a_1
Factory 2	C_{21} X_{21}	C_{22} X_{22}	C_{23} X_{23}	C_{24} X_{24}	C_{25} X_{25}	C_{2n} X_{2n}	a_2
Factory 3	C_{31} X_{31}	C_{32} X_{32}	C_{33} X_{33}	C_{34} X_{34}	C_{35} X_{35}	C_{3n} X_{3n}	a_3
Factory n	C_{n1} X_{n1}	C_{n2} X_{n2}	C_{n3} X_{n3}	C_{n4} X_{n4}	C_{n5} X_{n5}	C_{nn} X_{nn}	a_n
Total Requi.	b_1	b_2	b_3	b_4	b_5	b_m	

Key:
m: number of destinations
n: number of sources
ai: capacity of *i-th* source (in tons, pounds, liters, etc)
bj: demand of *j-th* destination (in tons, pounds, liters, etc.)
cij: unit material shipping cost between *i*-th source and *j-th* destination (in $ or as a distance in kilometers, miles, etc.)
xij: amount of material shipped between *i*-th source and *j-th* destination (in tons, pounds, liters, etc.)

capacities. This is typically done using an optimization code, however, this section presents one method that can be used iteratively to optimize the table after an initial solution is established as a starting point, this method is called *the stepping stone* method. The initial solution is initiated using what's called the Northwest Corner Rule, this starts from the north-west cell (and hence the name), and allocates the maximum production capacity to that cell without violating that source capacity or the market demand. When the first cell is completed, the subsequent vertical cells are filled and then the horizontal ones. Then the optimization is done using the stepping stone method, which starts by selecting an empty cell as a starting point, where one shipment is added to test the effect on the table, however, to avoid violating the production capacities and market demand, the added shipment should be shifted from another location. So, one should place a +1 at the first empty cell and then move in straight lines (up or down, left or right, no diagonal movement is allowed) to trace a close path, that starts and ends by the first empty cell. Also, all the closed path corners should be cells with shipment in them, i.e. not empty cells. For each movement, one should alternate the addition of a + and – signs representing the shifting of shipments; in other words the first cell is the empty cell that had +1 added, then the next corner (non-empty) should have a -1 placed in it, if it already had e.g. 10, then it becomes 9, and so on. After the first path is done, one should inspect the effect on the total cost, by tracing the same path and computing the cost through multiplying the cost box by the shipment quantity, if this yields an increase in the cost from the initial case, then the addition of +1 in the empty cell should not be done and that cell should be left blank. However, if the total cost comes to a lesser amount than the initial case, then more shipments should be added to that cell. After the cell is done, other empty cells are treated the same way. This is best done by a computer program or using the

Table 7.11 The stepping stone method

	Alpha	Beta	Gamma	Totals
A	+500 $5	$4	+ $3	**500**
B	+200 $8	−100 $4	$3	**300**
C	$9	+100 $7	−200 $5	**300**
Totals	**700**	**200**	**200**	**1100**

Microsoft Excel Solver option, where one can specify the goal of the optimization that minimizes the cost, defines the cost function as the cost per shipment per location multiplied by the shipment quantity, and finally defines the constraints that no marker is provided more than its total demand and no production capacity per facility is violated.

An example of the stepping stone method is shown in Table 7.11.

Exercises

Problem 1

CEGC Incorporated has three production facilities in: Greenville, Anderson and Clemson, to supply four different markets located in: Easley, Central, Pendleton and Seneca. The production capacity and market demand in addition to the associated cost for each location are shown in the tables below. (a) Provide the estimated shipping cost by formulating a transportation table for two scenarios: cost based on the North-West corner rule, and cost estimate based on the North-East corner rule. (b) Modify the production level at one facility to have the total capacity equal to total demand with least shipping cost for both scenarios.

	Easley	Central	Pendleton	Seneca
Greenville	$2.00	$3.00	$3.50	$1.50
Anderson	$5.00	$1.75	$2.25	$4.00
Clemson	$1.00	$2.50	$1.00	$3.00

Production Facility	Unit produced (Capacity)
Greenville	15,000
Anderson	40,000
Clemson	30,000
TOTAL	85,000

Market location	Units required (Demand)
Easley	27,800
Central	8,000
Pendleton	13,500
Seneca	33,000
TOTAL	82,300

Problem 2

CEGC Incorporated is establishing the layout for one of their facilities based on the information in the table below that describes the required tasks, time required and the precedence relationships. Provide (a) the flow diagram for this facility, (b) group the stations into cells based on cycle time, and (c) compute the idle time and efficiency. Note: CEGC working day consists of 8 complete hrs.

Station	Time required in seconds	Immediate predecessors
A	15	
B	26	A
C	15	A
D	32	B,C
E	25	D
F	15	E
G	18	E
H	10	E
I	22	F,G,H
J	24	I
TOTAL	202	

Problem 3

A firm must produce 40 units/day during an 8-hour workday. Tasks, times, and predecessor activities are given below. Determine the cycle time and the appropriate number of workstations to produce the 40 units per day.

Task	Time (Minutes)	Predecessor(s)
A	2	—
B	2	A
C	8	—
D	6	C
E	3	B
F	10	D, E
G	4	F
H	3	G
Total	38 minutes	

Problem 4

ABC Motors is considering three sites Alpha, Bravo, Tango to locate a new factory, to build their new SUV XL 500. the goal is to locate at a minimum cost site, with production expectation of 0–60,000 SUVs. If ABC Motors prefer site Tango, for which production volume will Tango be an acceptable choice? ABC Motors team gathered following information:

Site	Annual Fixed Cost	Variable Cost / SUV
Alpha	$10,000,000	$2,500
Bravo	$20,000,000	$2,000
Tango	$25,000,000	$1,000

Problem 5

Build a Microsoft Excel active model to evaluate the production location alternatives based on; the locational breakeven analysis, the center of gravity (markets, supplies, people), and the factor rating method. Then analyze the BMW production locations in Germany using your developed active Excel model.

Problem 6

Solve the following transportation table using the North-West corner rule and the stepping stone method for one iteration, then use the Microsoft Excel Solver to find the optimum allocations.

	Destinations/Markets				
Sources	D1	D2	D3	D4	Total Capacity
Factory 1	10	30	25	15	**14**
Factory 2	20	15	20	10	**10**
Factory 3	10	30	20	20	**15**
Factory 4	30	40	35	45	**12**
Dummy	0	0	0	0	**1**
Total Requi.	10	15	12	15	

Problem 7

An automotive OEM, have the capacities and demands shown below from its different factories and markets, respectively. Configure and solve a transportation table to optimize the capacity distribution for the different demands, using Excel Solver.

	Capacity
warehouses/factory	
A	15000
B	40000
C	30000
TOTAL	85000

Demand	
warehouse/markets	
E	27800
F	8000
G	13500
H	33000
TOTAL	82300

Cost matrix				
	E	F	G	H
A	$2.00	$3.00	$3.50	$1.50
B	$5.00	$1.75	$2.25	$4.00
C	$1.00	$2.50	$1.00	$3.00

8

Operational Aspects of the Automotive Manufacturing Processes

8.1 Introduction

This section introduces the different types of the procedural activities or tasks associated with the management of the automotive manufacturing and its planning. These procedural tasks can be usefully classified into two main types; the first is based on strategic planning, while the second is operational aspects. The strategic aspect is concerned with the high-level, long-term decisions on the vehicle models to be manufactured, and volumes for each market location. These decisions are typically advised based on market surveys and past experience. In addition, these decisions can control the location of the production facility if a new plant is to be built for new models or a new emerging market. On the other hand, the operational aspects are dedicated to translating the strategic goals into production realities through converting the production forecast into aggregate plans which is then further turned into the manufacturing production schedule (MPS). Also the operational activities include all the control and monitoring steps needed to ensure the delivery of the strategic visions, which include high-level monitoring activities such as evaluating the production capacities at the production locations and distributing the different markets' demands from the different production locations, for example, using the transportation matrix method discussed in Chapter 7.

Other control activities include online production control through procedural and hardware approaches. Additionally, the operational aspects further schedule the material flow and suppliers' parts into the facility, which is typically done through the material requirement or requisition planning (MRP) or enterprise requisition

The Automotive Body Manufacturing Systems and Processes, First Edition. Mohammed A. Omar.

planning (ERP) systems. This chapter will focus on the operational side because the strategic planning is typically handled by the upper management in the automotive OEMs and might be governed by engineering and non-engineering methods and drivers such as market economics, government regulations, and the operation of management style. So the next section discusses the aggregate planning activities and procedures, because having a good understanding of aggregate planning can assist in drawing a better overall picture of the vehicle manufacturing processes. The output from the aggregate plan is the input along with the vehicle bill of material (BoM) to the master production scheduling and MRP activities.

Other aspects of the operational side that are discussed include the standardized work schedules and procedures.

8.2 Aggregate Production Planning

Aggregate planning is an intermediate range (<18 months before production) resource planning tool that helps the automotive OEMs increase their confidence in the demand forecast by using an aggregate unit. The aggregate planning step comes just after the product and market planning and before the master production scheduling (MPS) procedure; thus the MRP process follows the MPS. The MPS can be defined as a short-term planning tool (around 3 months before production) that tries to optimize the production objectives in terms of product variations (vehicle models), per the production unit time, typically done on weekly basis. On the other hand, the MRP procedures specify the number and type of products needed for production per the MPS, and their delivery times. Hence, the long-term planning decides on the vehicle models to be built, their manufacturing locations and finally their long-term estimated volume. Intermediate planning is more concerned with the labor and subcontractor needs, the management of the production output into shipped vehicles or into controlled inventory levels. Finally, the short-term planning is dedicated to addressing the actual facility production capacity and its equipment loading and utilization, job assignments, and the production lot sizes and material orders.

The aggregate planning tries to optimize the resources' allocation to meet the targeted demand per the time period, typically on an annual basis. The types of resources can be classified into labor resources which can be described in terms of the number of workers needed and whether new hires on a full-time or part-time basis are required; other resources include the in-house production capacity, any subcontracting needed, and the current inventory levels.

The aggregate planning follows a very structured process that is composed of selecting an aggregate unit suitable to the product being made. The aggregate unit is selected to be generic in nature so that it increases confidence in the forecasting. For example, for automotive OEMs, if the market forecast demands are in terms of vehicle model A, B and C, then the aggregate unit will try to combine vehicles that require similar resources such as vehicle platforms, for example, the Toyota Camry, Highlander, RAV 4 and Avalon are all based on the same platform (chassis) and they share similar components and production processes, so they can all be combined in one aggregate unit, that is vehicle platform 1. Another example of the aggregate unit selection, if a company is processing different steel sheets and then forms them into different products, product C , D, E, etc., then a good aggregate unit will be the mass of steel processed in kg or Lbs, because the different products will go through similar processes and require similar resources, so it's more useful to describe the resources needed per 1 unit mass. However, one should be careful to note that the selected aggregate unit is related to the resources being planned.

The second step in aggregate planning is to decide on the aggregate demand per the time period, typically one or two years. This is done by aggregating the forecast numbers for the different products into the aggregate unit, so for the case of the aggregate unit defined in terms of unit mass, the products (steel) weights should be summed so that the demand is described in terms of unit mass of material that needs to be processed. This step starts with the demand forecast per month and aggregates it into aggregate unit per month. The third and most important step is actually deciding on the production strategy or plan that will be able to satisfy the demand; the production plan strategy is typically a combination of subcontracting, using existing inventory levels, and increasing the production capacity in terms of labor and work hours, if the extra production capacity is needed. On the other hand, if the smaller production capacity is enough to meet the demand, then the suggested plan might utilize building inventory levels, reducing the labor and labor hours, reducing subcontract percentages. After the production plan has been suggested, it should be evaluated based on the total cost; typically two or three plans are first introduced, then evaluated based on the plan overall cost. Also the company strategy might play a role in the evaluation process, so other considerations such as the importance of keeping a trained labor force in-house which reduces reliance on the reduction of labor through firing. The final step in the aggregate planning is disaggregating the plan from the aggregate unit into the original units so the MPS can be created. The disaggregation is simply translating the aggregate unit back into the original unit of the forecast.

To better illustrate the aggregate planning procedures, an example is discussed for an automotive supplier whose manufacturing components come in three different variations: models, A, B, and C. These models can be distinguished based on size and the amount of steel used. Model A is smallest with the lowest steel amount, while model B is the medium size, and C is the largest in size and has the largest steel amount. One of the customer OEMs provided their forecast based on their needs from the three different models for the coming year. This supplier's management team now needs to plan their production for the coming year based on this customer's demands from the three models. Model A is made up of around 2 kg of steel, while product B consumes 5 kg of steel, and finally C is the largest with 10 kg of steel. The forecast for each of these models per production month is shown in Table 8.1. Additionally, Table 8.1 shows the safety stock needs for each month. The safety stock is the inventory level needed to increase confidence in the forecast data.

So the first step in preparing an aggregate plan is to devise a suitable aggregate unit. Due to the fact that the three models are similar in their processing mechanism and they mainly differ in the amount of steel in each, one can use the unit mass of kg of steel as the aggregate unit, which is also suitable because it allows the supplier to plan the raw material needs, even though this is not the goal of the aggregate plan. So, if each model demand is replaced with its weight in kg, then the monthly demand can be translated into one column with the total kgs needed; also the same can be done for the safety stock levels; as in Table 8.2.

In addition to the forecast information, one needs the costs associated with the different resources and the limitations involved. For example, the cost to hire a new employee might be $40,000, however, the firing cost might be around $100,000, to include all the associated retirement and other costs. Additionally, one should know that, for example, the permissible number of employees the company can have is 100–200 people in that production location. Other limitations on the labor include the maximum productivity achieved per worker per production day, in this case, assumed to be 10 kg of steel. Other associated production costs include the raw material cost around $25 for each kg, which costs around $75/kg to be processed in the regular production hours and $80/kg if processed in overtime, which should not exceed 20% of the regular working hours. While subcontracting 1 kg costs the company $120, additionally the company in this example has a set limit on the subcontracting load per month of 150 kg. The inventory cost or holding cost reflects the

Table 8.1 The forecast for three product variations

Month	Working Days	Cumulative days	Forecast Demand		
			Small	Medium	Large
January	22	22	100	111	80
February	19	41	100	150	60
March	21	62	100	120	80
April	21	83	200	120	100
May	22	105	150	85	100
June	20	125	150	80	60
July	12	137	150	85	60
August	22	159	150	85	60
September	20	179	200	80	80
October	23	202	200	100	100
November	19	221	200	100	100
December	21	242	150	100	100

Table 8.2 Aggregate forecast of Table 8.1

Month	Working days	Cumulative days	Forecast demand			Aggregate forecast
			Small	Medium	Large	Demand
January	22	22	100	111	80	1555
February	19	41	100	150	60	1550
March	21	62	100	120	80	1600
April	21	83	200	120	100	2000
May	22	105	150	85	100	1725
June	20	125	150	80	60	1300
July	12	137	150	85	60	1325
August	22	159	150	85	60	1325
September	20	179	200	80	80	1600
October	23	202	200	100	100	1900
November	19	221	200	100	100	1900
December	21	242	150	100	100	1800

costs associated with keeping the products in inventory, and can be around $30/kg per year, with a minimum holding weight of 500 kg. Finally the backlog cost should be considered too, which are the losses endured by the company for any delays in meeting the customer demand, which can be assumed in this case to be $10/kg for each month of delay. After knowing all the associated costs and the production limitations, one can start compiling a production plan.

The first step in establishing a production plan is to compute the cumulative and net demands from the aggregated unit column. The net demand describes the monthly demand taking into consideration the inventory levels available from the previous month and the levels needed by the end of the current month. So, for example, the net demand for the first month is equal to the first month's (January) demand (1555 kg) + the inventory level needed for that month (900) – the inventory level available in stock from previous month (700), so the net demand is 1755 kg. On the other hand, the cumulative demand is the continuous accumulation of the monthly demands; so, for example, the net demand for January is 1700 and for February is 1500, then the cumulative demand by the end of February will be 3200 kg. Following this procedure, Table 8.2 can be converted into Table 8.3. Computing the net and cumulative demands helps estimate the production needs as the year progresses, thus the demand for certain months can be produced in advance. Additionally, this helps in giving the production engineers the big picture perspective of the expected demand, so they can anticipate any changes; the cumulative for this example is displayed in Figure 8.1.

The plot in Figure 8.1 helps when proposing a production plan to meet the cumulative demand; for example, a production plan can be suggested based on increasing

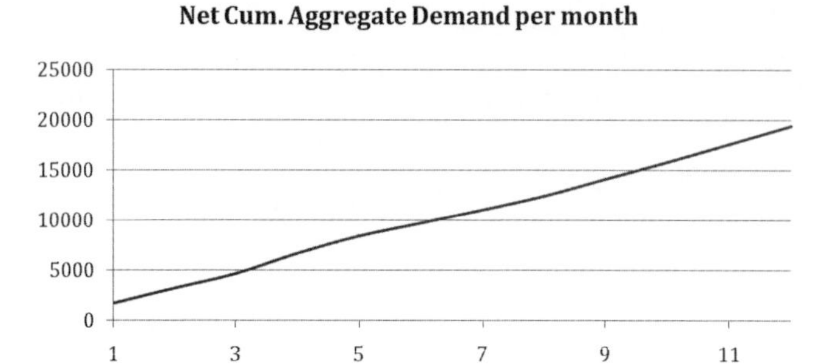

Figure 8.1 The net accumulative aggregate demand

Table 8.3 The aggregate demand information showing the net and net cumulative demands

Month	Working Days	Cum. days	Forecast demand			Aggregate Forecast	Safety stock	Net Demand	Net Cum.
			Small	Medium	Large	Demand	Inventory		
							700		
January	22	22	100	111	80	1555	900	1755	1700
February	19	41	100	150	60	1550	850	1500	3200
March	21	62	100	120	80	1600	700	1450	4650
April	21	83	200	120	100	2000	750	2050	6700
May	22	105	150	85	100	1725	750	1725	8425
June	20	125	150	80	60	1300	750	1300	9725
July	12	137	150	85	60	1325	750	1325	11050
August	22	159	150	85	60	1325	800	1375	12425
September	20	179	200	80	80	1600	900	1700	14125
October	23	202	200	100	100	1900	700	1700	15825
November	19	221	200	100	100	1900	600	1800	17625
December	21	242	150	100	100	1800	600	1800	19425

the number of employees from 150 to 165. The cumulative production based on this labor force allocation can be shown overlaid over the cumulative demand as in Figure 8.2.

Studying Figure 8.2 shows that the case of 165 workers, along with the production and cost computations from Tables 8.4 and 8.5, indicates that this plan can provide the expected demand along with the needed safety stock in accordance with the forecast information up to the month of June. After that, this plan tends to produce more than needed units, this results in an added holding cost. So planning engineers can modify this plan by reducing the number of workers after the month of June to reduce the over-production and hence the material and processing costs. However, if one considers firing 5 workers after June, this will reduce the holding, processing,

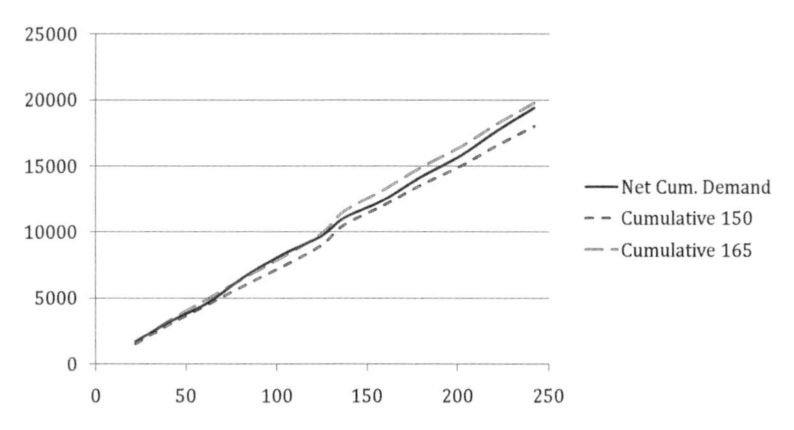

Figure 8.2 The net accumulative demand and production levels for different work levels

Table 8.4 Two production plans, 150 workers and 165 workers

Production Plan 1 (150 workers)			Production Plan 1 (165 workers)		
Net	Cumulative	Difference	Net	Cumulative	Difference
1500	1500	−200	1650	1650	−50
1500	3000	−200	1650	3300	100
1500	4500	−150	1650	4950	300
1500	6000	−700	1650	6600	−100
1500	7500	−925	1650	8250	−175
1500	9000	−725	1650	9900	175
1500	10500	−550	1650	11550	500
1500	12000	−425	1650	13200	775
1500	13500	−625	1650	14850	725
1500	15000	−825	1650	16500	675
1500	16500	−1125	1650	18150	525
1500	18000	−1425	1650	19800	375

and material costs, but it will add the firing cost, so the total cost for a plan that starts with 165 workers then drops to 160 workers will be around $8,954,062, which is more than the case of the 165 workers. Other modifications can be through hiring fewer workers as the case with 150 workers and then using subcontracting to satisfy the remaining demand. Other options include increasing the productivity from 10 kg/worker to 10.5 kg/worker through investing in new machinery or equipment; if one considers this case, then the total cost becomes $8,359,875, which means that planning engineers can invest around $600,000 in new equipment and improvements to raise the production capacity.

Table 8.5 The cost details for each of the production plans from Table 8.4

Cost plan 150		Cost plan 165		Cost plan 165/160		Cost plan 160 + improvements	
Hiring	6000000	Hiring	6600000	Hiring	6500000	Hiring	6400000
Proc.	1350000	Proc.	1485000	Proc.	1462500	Proc.	1464750
Mat.	450000	Mat.	495000	Mat.	487500	Mat.	488250
Hold	19687.5	Hold	812.5	Hold	812.5	Hold	1375
Delay	14250	Delay	3250	Delay	3250	Delay	5500
				Firing cost	500000		
Total.	7833937.5	Total.	8584062.5	Total.	8954062.5	Total.	8359875

Other approaches to compiling production plans involve the use of the transportation tables, as shown in the example in the matrix in Figure 8.3.

The main advantages of aggregate planning are its simplicity in terms of the resource allocations and final cost computations, because it does not rely on complicated programming codes or methodologies but can be done in simple programming environments such as Microsoft Excel. At the same time the aggregate planning does not require very accurate or specific forecast data. Also it tends to increase the confidence level in the forecast because of the generic nature of the aggregate unit. However, the aggregate planning process, as seen in previous examples, requires exact or near exact costs for each resource, this can be mitigated by using resources with a reasonable sensitivity range, but this might also change the suggested plan based on any changes in the OEM's operating environment, labor wages increases, etc. Additional disadvantages of the aggregate planning are due to the fact that it is based on specific demand and production periods.

8.3 Master Production Scheduling (MPS)

After completing the aggregate production plan, it should be further translated into shorter time periods and with shorter time horizons. The master production scheduling (MPS) uses the disaggregated data to create a weekly production plan with specifics on the model type, and volumes to be manufactured every week. The MPS is necessary so the weekly resource assignment and allocation are done, taking into consideration the specific production conditions, such as the change-over time needed to produce the product mix, in other words, the MPS decides on an approximate capacity plan for equipment and the production line in general. The MPS is hence a short-term planning tool used about 3–6 months before the start of production.

The first step in constructing an MPS is to start with the aggregate plan monthly production data, which is then mapped over the number of working days, an example of this is shown in Tables 8.6 and 8.7 for the month of January and using thc 160

	Period 1 (Mar)	Period 2 (Apr)	Period 3 (May)	Unused Capacity (Dummy)	T. Capacity Available (Supply)
Beginning Inventory	0 100	2	4	0	100
Regular	40 700	42	44	0	700
Overtime	50	52 50	54	0	50
Subcontract	70	72 150	74	0	150
Regular	X	40 700	42	0	700
Overtime	X	50 50	52	0	50
Subcontract	X	70 50	72	0 100	150
Regular	X	X	40 700	0	700
Overtime	X	X	50 50	0	50
Subcontract	X	x	70	0	130
Total Demand	800	1000	750	230	2780

Figure 8.3 The use of a transportation table for aggregate planning

Table 8.6 The demand information for the month of January

Mths	Work Days	Cum. days	Small	Med	L.	Demand	Inventory Level	Net Demand	Net	Cum 160
Jan.	22	22	100	111	80	1555	700 900	1755	1627.5	1627.5

workers plan with improved productivity, which helps in highlighting the non-production days.

The next step is to decide on the production strategy between the two main schemes; the former aims to meet the demand while keeping a level production volume, while the latter aims to meet the demand but with the minimum number of

Table 8.7 The production demand split over the working days within the month of January

	JANUARY				
	W1	W2	W3	W4	W5
M		7	14	21	28
T	1	8	15	22	29
W	2	9	16	23	30
TH	3	10	17	24	
F	4	11	18	25	
S	5	12	19	26	
S	6	13	20	27	

Table 8.8 The production demand based on mixed production

	January		
	Small	Medium	Large
Week 1	38	106	152
Week 2	48	132	190
Week 3	48	132	190
Week 4	48	132	190
Week 5	29	79	114

changes. The former is based on producing the demand mix on continuous terms, that is, every week the facility should be manufacturing the complete mix of products. On the other hand, the second strategy aims to reduce the number of changes, hence follows a batch production scheme, where one product type is produced for a complete week, then other product types will be made in the following weeks. As will be discussed in the Value Stream Mapping discussion, the first strategy enables the OEMs to keep production sequenced with the customer demand in volume and type. The implementation of these two strategies is shown in Tables 8.8 and 8.9.

Table 8.8 splits the production mix into the different weeks, keeping the production volume spread over the three types while keeping their percentages of production according to their demand mix. On the other hand, the second strategy spreads batches of one product type for each week, for example, the first two weeks are only dedicated to producing large products, while the small product type is only produced in the fourth week.

Table 8.9 The production demand based on based on batched production

	January		
	Small	Medium	Large
Week 1		296	
Week 2		285	85
Week 3			370
Week 4			370
Week 5	209		13

The second strategy exposes the company to unbalanced production mixes; in addition, it uncouples the production sequence from that of the customer, so if any customer changes take place in terms of product type or volume, then the OEM will not be able to respond by changing their product type because of the already produced stock of a single product type. Additionally, the second strategy increases the time to market because, the OEM cannot ship complete orders until they finish the complete product mix needed, which is spread over the whole month. For example, for any customer demand for the small product type, the OEM can only respond after one month. The first strategy enables the OEM to be more flexible and agile to any changes because any changes in terms of increasing or decreasing the demand of one of the product variations affects only the weekly batch which still contains all three model types. In addition, the first strategy allows for shorter production to market time, which leads to quick turnover of the OEM's invested cash in material and labor costs.

After the MPS is planned on a weekly basis, the production engineers should start preparing for the material and components acquisition plan, which is typically done through the use of one form of material requisition planning (MRP).

8.4 Material Requirement Planning (MRP)

The material requirement planning (MRP) activities are designed to manage the material and components orders in the plant; such activities should be driven by the production requirements in terms of material type, volume, and delivery timing. To achieve this, the MRP system relies on the MPS and the current inventory levels within the production facility. However, the MPS should be translated into materials and part counts, which is typically done using the Bill of Materials (BoM) for each product type; the BoM breaks down each product type into its basic components and parts. This translation into parts and components can be done through the very basic elements of the product; the OEM uses a BoM break-down level commensurate with

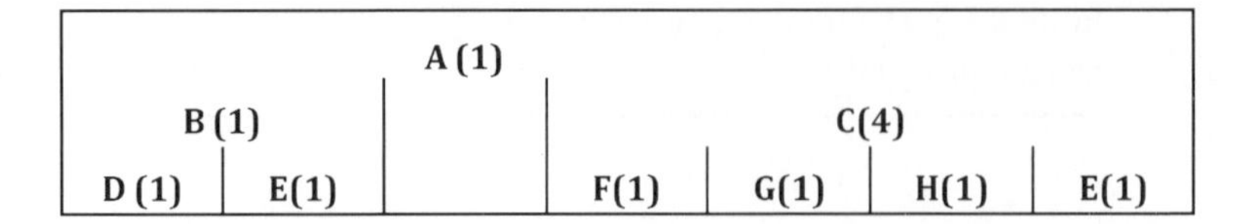

Figure 8.4 A simplified example of a BoM

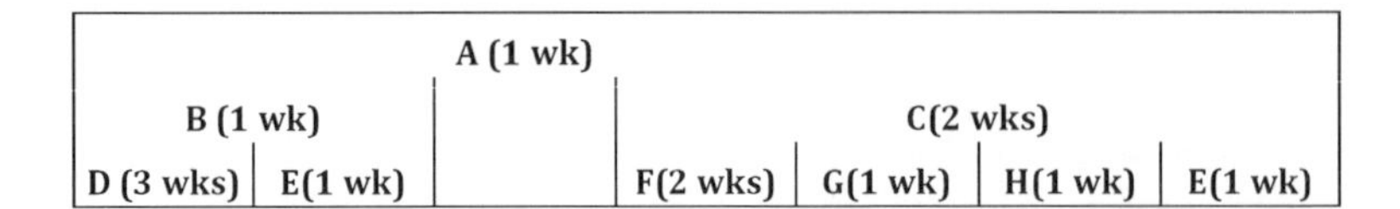

Figure 8.5 Lead time for each sub-component

Table 8.10 The product X required components

	A	B	D	E	C	F	H	E
UNITS	10	10	10	50	40	40	40	40

its supplier sub-assemblies. At the same time, the supplier utilizes a more specific BoM because the supplier needs to acquire the basic parts to build the sub-assemblies. Knowing the BoM for a certain product provides good information about the product, such as its basic working mechanism, through knowing the role of each sub-component. In addition, the BoM offers designers the opportunity to analyze the product for parts consolidation efforts, to reduce product total cost through simplifying its production sequence, or to improve the product performance through replacing its sub-components with ones that have better service behavior. Also, the BoM can be very useful for systems' hazard analysis studies, which replace each sub-component from the BoM with its function and then trace the different types of failures for each part and its effect on the overall product functionality.

In addition to describing the product inner configuration at the sub-component level, the BoM is typically complemented by information about the production lead time for each component, so that the OEM can estimate the order time for each part to ensure that all the parts and sub-assemblies are ready for delivery when needed. A simplified example of a BoM is shown in Figure 8.4. Thus, one can estimate the number of components needed for each one complete assembly, i.e. vehicle model H requires the components listed in Table 8.10.

So if one assumes that the lead time for each sub-component is shown circled in Figure 8.5, then the order time can be established in addition to the quantities ordered per one vehicle model. This process of computing the order time and the sub-components' quantities is the MRP process. The MRP process tabulates this informa-

Table 8.11 The basic MRP table

Product Information				
Period	1	2	3	4
Gross requirement				
Scheduled Receipts				
On Hand				
Planned Order Release				

Table 8.12 The MRP table for product X

Product Information: lead time 3 wks, lot size = 20						
Period	1	2	3	4	5	6
Gross requirement	15	20		20	10	
Scheduled Receipts		20				
On Hand = 27	12	12	12	12	2	2
Planned Order Release	20					

tion for each sub-assembly to compose the complete vehicle; to manage this process such tabulation can be illustrated in Table 8.11.

Table 8.11 defines the amount required in the row entitled Gross requirement, while the order quantity is described in the row entitled Planned order release, which is typically called the action row. The On hand row describes the inventory level quantities available or quantities expected to reach the production from previous issued orders. The time scale is typically displayed in weeks needed to manufacture the part, i.e. the production lead time, is the upper row entitled Period.

An example of a completed MRP table is shown in Table 8.12, which shows the MRP table for a product called X, its production lead time is 3 weeks and the typical lot size for this product is shipments of 20 units. From this table, in the first week, the needed quantity of this product is 15 units, while no scheduled material is expected to arrive in the first week. The starting inventory or available stock of product X is 27 units, while a 20-unit order is supposed to be released to order one lot size of 20 units. For the second week, the required quantity is 20 units, while the remaining stock available is 27 (starting level) – 15 (needed for first week production) = 12 units remaining for second week. No inventory is consumed for the second week, because an order of 20 units will be arriving according to the scheduled receipts row, for the inventory for the third week is 12 units, and all the arriving stock in the second week will be consumed in that week. The planned orders that were submitted in the first week will take three weeks (production lead time) to arrive.

The MRP table should be managed to adapt to any changes in the product's delivery time or quantities. In addition, it should change the order release timing and quantities based on the actual production requirements which might alter due to changes in customer demand.

The most commonly used technique to manage the MRP tables is based on the adjustment of the ordered lot size, because the lot size affects the inventory level and the cash flow. One possible adjustment for the lot size is based on the economy of scale, hence the larger the lot sizes, the less the cost of the sub-product components. However, this technique forces the OEM to follow a push production strategy that makes the product independent of the customer demand which affects the OEM's ability to respond to market changes. Other adjustments of the lot size might be based on the periodic order quantity (POQ) principle, which means that the lot size should be enough satisfy the production requirement for a certain, set number of production periods. In this case the lot size is specified based on the number of production periods, for example, the lot size is POQ 5 is the case displayed in Table 8.13 and Table 8.14, which show the application of the POQ principle for two products A and B with different lead times. Other method of controlling the lot sizes is to use a fixed

Table 8.13 The MRP table for A based on POQ of 5, with lead time of 2 wks

Product Information: lead time 2 wks, POQ = 5								
Period	1	2	3	4	5	6	7	8
Gross requirement	2	24	3	5	1	3	4	50
Scheduled Receipts								
On Hand = 28	26	2	13	8	7	4		
Planned Order Release	14					50		

Table 8.14 The MRP table for A based on POQ of 5, with lead time of 4 wks

Product Information: lead time 4 wks, POQ = 5								
Period	1	2	3	4	5	6	7	8
Gross requirement	14					50		
Scheduled Receipts	14							
On Hand = 28	2	2	2	2	2			
Planned Order Release		48						

value for the lot size. The MRP table for the production example from Figure 8.6 is shown in Table 8.15.

The current MRP systems have been undergoing continuous improvement so their management style can be optimized and made more flexible. One of the competing material acquisition management styles is the Just In Time (JIT) system adopted and developed as part of the Toyota Production System (TPS). The JIT is based on continuous (higher frequency orders) with small lot sizes to help the OEM reduce the inventory levels, synchronize the production demands with the raw material orders and deliveries to eliminate any accumulated Work In Process (WIP). This system also enables the OEM to receive the parts shipments and use them instantly in production, this means that any defective products will be exposed and the final product quality improved. However, the JIT system requires strong communication between the suppliers and the OEMs, in terms of shipment size, frequency, and the expected quality levels. The JIT system is typically best applied when a small number of suppliers are used to make the product. In addition, to achieve the JIT material delivery, the supplier production line should be synchronized to that of the OEM and the supplier processes are optimized so it is based on short lead times and is able to keep up the production rate of mixed products, even with high change-over frequency. The Toyota Motor Company have established a Toyota Supplier Support Center (TSSC) to work with the local suppliers to optimize their production lines and styles, so that they are able to meet the JIT requirements.

After the MPS, the MRP is completed and the final production plan should be evaluated to check the production capacity and whether the available resources will be able to meet the expected demand. This is typically done through a detailed capacity plan for man-hours, equipment capacity and material planning. If the detailed plan is approved, the execution of the MPS is commenced and the MRP system is executed.

Table 8.15 The complete MRP table for each of the different components

Type	Period	1	2	3	4	5	6	7	8
A	Requirement								10
	Order							10	
C	Requirement							40	
	Order					40			
B	Requirement							10	
	Order						10		
D	Requirement					40			10
	Order				40				
E	Requirement					40	10		
	Order				40	10			

8.5 Production Line Control and Management Style

The production line control is considered part of the operational management tasks, where the production engineers monitor the production levels and material deliveries to ensure the smooth execution of the MPS. This monitoring is typically done through a variety of tools, which include hardware tools such as the different sensory and communication systems established within the production line to ensure certain quality, production, and communication levels on the line. The communication tools vary from the basic wireless tools between the engineers, to tools that integrate the different machine cells with their programmable ladder controller (PLC) and to the master production PLC. Also, other systems can output quality and productivity signals from each PLC to a monitoring display to show the overall progress in terms of number produced to daily targets. In addition, this system outputs offline for line performance tracking and for reliability studies. The TPS have introduced an online display system to inform all the line personnel of the production status and of any stoppages; such a system is called the Andon system.

However, the most important control and monitoring strategy for a running production line is based on standardizing the work procedures in an exact way, so that all the workers are following best practice and delivering their products according to the production sequence. To achieve this level of operation, the production engineer should establish a set of documentation that describes the work standards, i.e. the best practices which not only include the specific steps in making the product and their timing, but also the exact tools to be used. In addition, the work standards include any safety concerns or specific issues to be aware of. An example of the work standard charts is shown in Figure 8.6.

Having the complete documentation is vital because it ensures that each process within the line is standardized and synchronized according to the production plan. Additionally, it ensures that the workers are following the best practices, i.e. the most optimized procedure and such a procedure can be captured and shared with new workers. This enables the OEM to accumulate the best practices and the skills set, independent of workers' years of experience. To highlight the role of such practices, the Toyota Motor Company estimates the time to train a new line worker to be around two days, which is extremely low when compared with other OEMs not practicing the standardized work procedures.

The standardized work sheets typically include the following types to better describe the work procedure:

- *capacity charts by parts*: to record the order of processes, process names, machine numbers, basic times, tool changing times, number of items and processing capacity;
- *standard task combination*: to determine the order in which individual workers' operations take place;

<table>
<tr><td colspan="6">WORK STANDARD SHEET</td><td>Prod type</td><td></td><td colspan="2">Task</td><td></td><td>DATE</td><td>USE</td></tr>
<tr><td colspan="6"></td><td></td><td></td><td colspan="2"></td><td></td><td></td><td></td></tr>
<tr><td colspan="6"></td><td>NO.</td><td>WO
RK
SE
Q</td><td>SYMBOL</td><td>FREQUENCY</td><td>IMPORTANT POINT</td><td colspan="2">NOTES</td></tr>
<tr><td colspan="6"></td><td></td><td></td><td></td><td></td><td></td><td colspan="2"></td></tr>
<tr><td>SAFETY</td><td>QUAL
CHECK</td><td>SAFETY
Other</td><td>STAND
STOCK</td><td>CONVE
CALL</td><td>ENVIR
O
IMPAC
T</td><td></td><td></td><td></td><td></td><td></td><td colspan="2"></td></tr>
</table>

Figure 8.6 A typical work standard sheet

- *task manuals*: to determine procedures for elements of the operations requiring special attention, e.g., tool changing, set-up changes;
- *task instrumentation manual*: to provide guidelines for correctly teaching standard operations;
- *standard operating sheets*: to describe the equipment layout diagrams, they also indicate cycle time, order of operations, work times, safety and quality checks.

8.5.1 Lean Manufacturing Management of Workers

This section will only discuss the lean manufacturing practices of some of its unique production control strategies; such strategies permit the labor force management to achieve a flexible operation, that is, to enable workers to move between stations and perform new tasks that relate to new products without drastic training efforts. So the discussion will focus on the Toyota production system methodology in analyzing the production activities that take place in an automotive manufacturing plant. First, the Toyota production system classifies the tasks into the following two stages: processes and operations [66]:

1. *Processes*. The TPS divides the processes into four main types: the actual transformational processing of materials and sub-assemblies, which is called processing. The second type is the inspection processes which are the steps conducted to ensure the conformance to the standards. The third type is the transportation of sub-assemblies, raw materials and finished products within the production facility. Finally, the delay is the fourth type involved in the manufacturing sequence; the delays can be related to two main causes: the processing delays and the lot delays, which is caused by the batch production strategy. From the above four types, one can see that the manufacturing processes can be improved through two main routes; the first relates to improving the product's manufacturability so that it is easier to manufacture. This can be achieved through applying the design for manufacturing practices to reduce the parts' count and hence consolidate the required processes. Additional design for manufacturing efforts should also focus on reducing the number of design features and instead try to utilize the functional or manufacturing features for styling, i.e. dual functionality. On the other hand, the second improvement route is based on improving the flow of the material within the plant so as to reduce the transportation effort, eliminate the lot delays, and furthermore improve the actual processing.

   In regard to the inspection efforts, it can be improved by applying an informative inspection scheme rather than a judgment inspection. The informative inspection helps increase the process understanding and the defects' formation mechanisms through providing a continuous feedback about when the defect is generated so that the root cause is better defined. The informative inspection can also be achieved without the need for specialized inspectors, but through a self or a successive mode. The self-inspection implies that the worker who is processing the part, inspects it too, even though this mode is highly instructive for the worker regarding the different deviations and their mitigation, it is considered a rather subjective method. On the other hand, the successive mode is based on initiating a chain of internal customers, that is considering the subsequent operator is a customer who first inspects the product and then processes it; this method is very useful because it is very effective in detecting the defects at the point of generation and also prevents any further value addition to a defective product. To enable these modes of inspection, the TPS introduced different tools such as the failproof (*poke yoke*) tooling and fixtures in addition to including the inspection task as part of the standardized work sheets. Additionally, the TPS introduced the Andon display system which helps expose the problem areas in production so defects are not passed along and/or hidden.

   To eliminate the transport delays, the TPS focused its improvement (*Kaizen*) activities on creating a smooth one-piece flow within the production line without any inventories to disrupt the flow. Also to remove the delays, the TPS analyzed the delays into three main types based on their root causes; the first is named the E-Storage based delays which is due to the imbalance in the production capacities between the stations and from product to product. The second type is the C-Storage which is typically created to improve the production robustness against small variations or line stoppages, and the third type is the S-storage which is caused by the management lack of confidence in the production ability to meet the market demand, so this type is created to shorten the time-to-market. The TPS approach to reducing these three types is based on balancing the machines and equipment capacity per the production demand not the actual machine productivity. In addition, the TPS increased the focus on the regular maintenance efforts to improve the production readiness and reliability, thus eliminating the need for the S-storage. Finally, the TPS introduced the JIT system to increase confidence in the material scheduling and reduce the delivered lot sizes.

The main issue with large lot sizes is that it forces longer production lead times, to illustrate this point with an example, if one considers the case where the product will pass through four stations with a lead time of 3 hours, then if a large lot size is processed, each piece will be processed but not transported to the next station until the complete lot is completed, thus the total time for one piece to reach to the customer is $4 \times 3 = 12$ hours. However, if a one piece flow is created where once the piece is processed it can move to the next station, the total time will be $= 3 + (4 - 1) \times$ processing time per station $= 3$ hours $+ 3$ for the processing time; typically the processing time (cycle time) is the in order of 1 minute. So the total time is 3 hours and 3 minutes compared with 12 hours.

2. *Operations*. The TPS defines the operations into set-up operations which take place before and after the actual processing, e.g. die exchange, and incidental operations, which covers the loading/feeding, unloading of parts and supplies into the machines. In addition, the operations cover the marginal allowances such as the lubrication, clearing, and parts stacking and sorting into pallets. Finally, the operations include the personal allowances which include the fatigue allowance, i.e. the breaks and rest periods. These classifications help distinguish between the value-added and the non-value-added operations, and whether these operations are necessary or unnecessary. Furthermore one looks at these operations according to their frequency (every time every product, once per shift, every time every change-over, etc.) which helps to understand their impact on production. The best and most famous example of how the TPS have addressed the operations is the single minute exchange of die (SMED) system which helped reduce the time expended on the set-up operations for the stamping dies. The SMED methodology is based on identifying the operations that should be done online (internal) which requires stopping the stamping press, and the offline (external) activities which can be done independently of the die operation. Then the online steps are improved to reduce their time requirements, or better convert some of the online steps into offline steps through the use of specialized fixtures that facilitate the mounting and calibration work. Finally, through the use of standardized shapes and fixture functions ,the overall time is reduced drastically. The SMED [66] have helped

reduce the time needed for die change-over from 6 hours to 9 minutes and hence the name single minute, i.e. single digit.

The above analyses of the different activities involved in automotive manufacturing have led the TPS to better define the best practices in managing and utilizing the labor force. For example, the separation of the delays into lot delays and process delays exposes the time wasted while workers are waiting for machines to finish their cycle. Also, the study of the different types of operations yielded further time and effort wasted in feeding and setting up the machines. These two facts motivated the TPS to separate the man work from the machine work and to further study the different modes of automation that can be useful in reducing the production delays and at the same time increase the product quality. These proposed automation routes included: the mechanization of the hand work through automating the feeding and unloading operations. Another type of automation is based on mechanizing the brain work through automating the defect detection [66, 67]. The combination of these two automation schemes is termed autonomation in the TPS terminology which can be translated as machine automation with a human touch. Another TPS management tool is the Nagara system which is based on the functional and sequential groupings of activities with the main focus on the product. This means that the workers should be flexible and multi-skilled, so that they can perform different sequential tasks. This concept has contributed not only to improving the line robustness against works absenteeism but also it increased the workers' skills set and their understanding of the product integration and the role of each process down the line. Further advantages included a noticeable reduction in work injuries related to repetitive tasks, hence the work ergonomics has improved.

To achieve the multi-skilled labor force, the TPS have adopted several strategies based on enriching the work content per worker which is achieved by adding more tasks to each worker through a job rotation system. Another way is based on expanding the work content by adding a few steps to the process just before and just after the worker station. Also empowering the workers to make decisions about the process performance and product quality has aided in increasing the worker skill set. The workers made decisions through the suggestion box system where a worker has the right and an effective platform to come up with new and better ways of doing the task. To illustrate with an example the final assembly area starts with the workers removing the doors for easy access, this procedure was suggested by the line workers at the Toyota Motor Company and then widely applied by almost all OEMs. Regarding worker control over the product quality, this can be explained through the self and successive inspection systems, where the worker has the right to reject a defective product and stop the line until a solution is developed.

However, the flow of people between the production line stations is more complicated than the flow of materials. The TPS recognizes this issue and has developed two models for workers' movement, one is called the swimming model and the second is termed the track and field model. The swimming model, as the name implies, means that a worker (swimmer) can't move/help in a new station until the previous worker has finished his/her work. However, in the track and field model, the fastest runner (the worker with lowest work content) can start running (helping) to compensate for the slowest runner, because the track and field is based on exchanging runners in a buffer area, while in swimming a swimmer from the same team cannot start until his team mate has finished his role. Additionally, the development of the cell-based layout for some sub-assembly areas enables better workers' movement and better information flow. The result of all these efforts enabled Toyota in 1960 to have 700 workers operate 3000 machines, which translates into 1 worker per 5 machines, while the industry standard at that time was 1 specialist worker for 1 machine.

Even though the static aspect is the one concerned with the plant layout and machines allocations into stations and cells, the operational aspect is the one that controls the flow of people, materials, and information across this layout. So the TPS labor management strategies are responsible for achieving their proposed people exchange between the stations in either the swimming or the track and field modes. One unique attribute of the way that TPS has controlled the people allocations to the different stations stems from using the worker productivity as a measure of performance not the production rate. In other words, the TPS has decided on the number of people within cells to achieve the best worker productivity in terms of processed parts/worker at the expense of the production rate so that the imbalance between the stations in terms of the time content is reduced. To explain this point with an example, if one consider the case where three stations have time contents of 40, 45, and 50 seconds. If the management allocated three workers to one per station, then the productivity of each worker can be computed as; the time required to produce one piece is the cycle time of the slowest station i.e. 50 seconds/product which translates into 3600 (seconds in one hour)/50 seconds/product = 72 pieces are produced each hour, so for three workers this translates into 72 units/3 = 24 units/worker for each hour. Now, let's consider the case that two workers are allocated to these three stations, and they are sharing the work content by moving in the three stations as seen in Figure 8.7, note the three worker case is depicted in Figure 8.8 (indicates the movement of people and the movement of material). So, for the second case, to produce one unit the two workers need 40 + 45 + 50 seconds = 135 seconds/unit, and because the work is conducted by two people, the time is actually 135/2 workers = 62.5 seconds/unit which means that each worker can produce 3600/62.5 = 57.6 units in each hour. Comparing this number with the three worker case of 24 units/worker for each hour, shows almost doubled the worker productivity, however, with smaller number of units produced per hour; first case 24 units × 3 = 72 units, while the second case 3600/135 = 26.67 units produced every hour. So shifting the focus on worker productivity can lead to better utilization of the workers' time. Also to solve the productivity rate issue, the improvement activities should target the actual

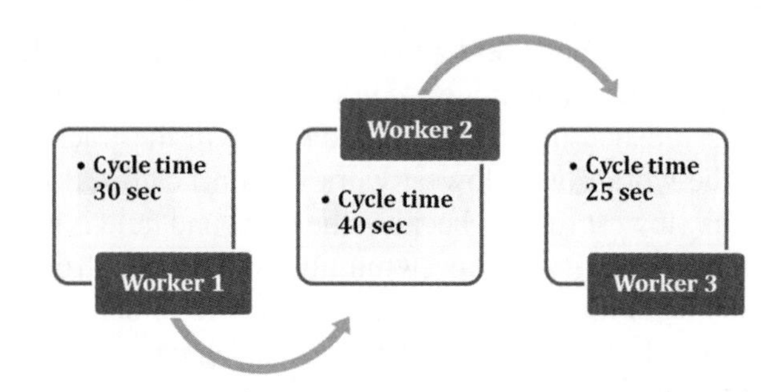

Figure 8.7 The workers' assignment/utilization before improvement

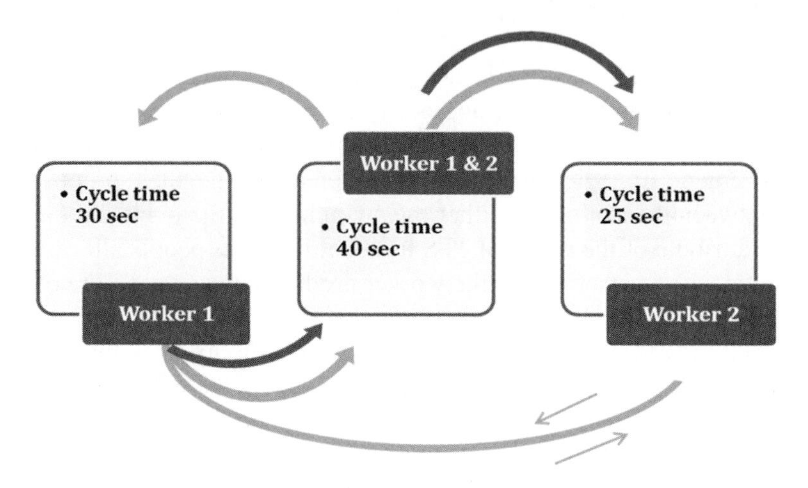

Figure 8.8 The workers' assignment/utilization after the improvements

processing time and aim to reduce it through the use of fixtures or through the improvement of the manufacturing technology. This approach is very unique to the TPS system because not only it focuses on better labor utilization but also it helps in balancing the work content between the stations by enriching and expanding the work content per worker.

8.6 Selection and Management of Suppliers

This section is dedicated to discussing the process of identifying and selecting the best supplier to support the automotive OEM manufacturing. First, the section introduces the different types and levels of automotive suppliers, then later in the section the systematic way of managing the suppliers through the quotation, the contract, and the monitoring processes are discussed.

The automotive industry relies on large numbers and levels of suppliers to provide component level, sub-assembly level and in some cases complete product level to the automotive OEMs. The traditional classification of those suppliers is based on the complexity of their parts, for example, the Tier 1 suppliers provide semi-complete assemblies to the automotive OEM plants directly to be integrated into the vehicle, while a tier 2 supplier provides parts at component level to the Tier 1, hence the supply chain is created all the way to the raw material extractor. However, such classification is not specific and does not reflect the actual role of the suppliers, so a better designation might be based on the following classifications: the raw material supplier is a supplier that feeds raw material either to the OEM or to another supplier. A supplier who sets the standards either on a global or a domestic basis for certain parts' performance or attributes can be classified as a standardizer. A supplier, whose role is to provide specific components or systems, can be called a component or system specialist supplier, while a supplier whose role is to integrate the components into the vehicle or assemble other semi-finished units can be called a system integrator. This classification describes each supplier based on their role and specialty, in addition, it provides information about the core competence for each supplier type; for example, the standardizer's core competence is based on the engineering research and design efforts in addition to testing and monitoring of parts and complete systems. On the other hand, the component specialist's main competence is the manufacturing of a certain component; at the same time, the component specialist might go through the standardizer to obtain a certificate of their component performance. So the role of the supplier is not only to provide physical materials and systems but also to service the OEMs with engineering and design information, or to help monitor or evaluate other type suppliers.

Examples of system integrators include the Dana Corporation which manufactures and integrates complete chassis systems, i.e. brakes, drive shafts, axles, etc. Other examples include Denso, and Magna Incorporated. Magna Inc. manufactures complete vehicles' BiW for some OEMs. These system integrators are sometimes called the Tier 0.5 suppliers. Because of the high level of competence required to integrate vehicular systems, the number of system integrators is the lowest compared to other types of suppliers, for example, their number is around 30 companies, while the component specialist can number up to 2000 suppliers.

However, before the OEM starts the supplier selection process, they should decide on a supply strategy which is typically formulated based on following issues;

- *In-house manufacturing versus out-sourcing, and for which components*: This issue is strongly related to the suggested production plan from the aggregate planning phase and to the OEM strategy in keeping certain core competences in-house. For example, when manufacturing the power-train components, most of the OEMs decide to keep it in-house because of its importance, and its cost percentage from the total vehicle cost.
- *The number of suppliers and their level of involvement*: This issue depends on the OEM production strategy and the manufacturing plant location. For example, for

an OEM using the JIT system, only one or two suppliers should be used because as the number of suppliers increase, their milk runs (lot sizes) is reduced to uneconomical levels. In addition, the OEM will be challenged to integrate and synchronize a large number of suppliers with different manufacturing technology, locations and cultures into their production sequence. At the same time, most of the OEMs would like to have two to three suppliers per product to avoid any major disruptions if one supplier's production line is stopped, also so no single supplier can control the cost of a certain component.

- Another important factor that controls the involvement and number of suppliers is the *location of the production plant.* If the plant is located in emerging markets where the OEM does not have adjacent production locations, the number of suppliers tends to increase and the share of the subcontracted work increases too. This also leads to an increase in the number of standardizers so that they can ensure the quality of incoming components and help train the workforce. To provide an example, the General Motors (GM) plant in Brazil is heavily dependent on suppliers even to assemble the final components into the vehicles. This strategy keeps GM immune against any variability in the Brazilian market or currency exchange rates due to their low investment cost, i.e. any changes in the market can be handled by manipulating the number of suppliers' personnel without carrying any long-term costs such as health insurance or retirement plans.

  However, having fewer suppliers tends to establish a better platform of long-term collaboration between the OEM and the supplier. In the OEM and major suppliers (Tier 1, system integrators) relationship, currently the automotive OEMs adopt different relationship strategies with their main suppliers, some are based on the vertical integration of the supplier into the OEM enterprise, which can be done through buying the supplier company and making it part of the OEM. This helps not only to reduce the cost but also in the exchange of proprietary engineering design and strategies. Examples of this include the GM relationship with Delphi, and the Ford Motor Company with Visteon. However, this scheme increases the OEMs' investment in the supplier core competence and increases the financial risks.
- Other relationship schemes include the *Keiretsu* strategy, which is typically found in the Japanese economy, where a group of companies of different levels (big and small) form a long-term relationship to exchange goods, services, and even cash loans. The Keiretsu typically include OEMs, banks, suppliers and heavy industry companies. Typically Keiretsu come in one of three types: the horizontally diversified business grouping; the vertical manufacturing networks; and the vertical distribution groups or networks. In Japan, several Keiretsu trees exist, for example, Mitsubishi Motor Company belongs to the Mitsubishi Keiretsu which also include the Bank of Tokyo, Nippon Oil, Nikon, and even Kirin Brewery; Mazda is a part of the Sumitomo Keiretsu which also include Sakura Bank, Japan Steel Works, and Toshiba. The Toyota Motor Company and Suzuki belong to the same Tokai Keiretsu which also include the Daido Steel, and Ricoh company.

8.6.1 Selection and Management Process

The process of the management of suppliers typically comes in five main stages; the first is the actual selection of the suppliers based on the OEM's defined strategy in terms of the number of suppliers sought and the metrics used in evaluating the suppliers. This process is shown in Figure 8.9. For example, the OEM might qualify suppliers based on the following criteria in regard to the product quality point of view;

- Is the supplier using a prevention-oriented approach to improve the quality of its parts? This criterion also means that the OEM is planning on involving the supplier early on in the process of designing a new vehicle. Also, this means that the OEM will require full documentation of the design review sessions that help in monitoring and evaluating any risks in terms of product quality. This documentation comes typically in the form of design review meeting minutes and the associated design plans.
- Is the supplier the Best in class in terms of product quality? This is typically applicable for well-established suppliers with long experience in making the specific parts.
- Is the supplier using a zero tolerance for defects regardless of the reject count? This typically means that the supplier should agree to have a quality team from the OEM to monitor the production and product quality levels
- Does the supplier have a worldwide presence in terms of its quality improvement and evaluation teams? This relates to large suppliers with well-established global companies.

Typically, the OEM will have to establish their priorities in regard to the above issues and then negotiate with the suppliers about the different agreements required. But the questions listed above help the OEMs to start the pre-selection process after inspecting the suppliers' profile which can be obtained from suppliers databases such as the SupplyOn database. SupplyOn is an online marketplace where all suppliers

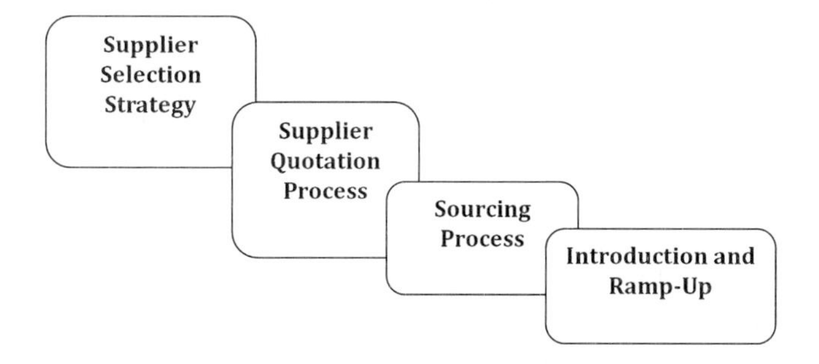

Figure 8.9 The supplier selection process

should be registered in order to be part of the quotation process of OEM and Tier 1. Also the supplier can provide their own self-assessment that can be verified by the OEM if needed.

The next step is composed of establishing the quotation process through negotiating the different agreements between the OEM and the selected candidate suppliers; such agreements might include a non-disclosure agreement, feasibility commitment agreement, and a general quality agreement. The general quality agreement (GQA) describes the quality level and strategy that both the OEM and supplier agree on and the different tools and metrics used in ensuring the GQA actual implementation. Typically the GQA should define the scope of the agreement, the audit system to be used, for example, in accordance with the German Association of the Automotive Industry VDA 6.3 regulations, the quality management system to be applied, and the different metrics/tests used to judge quality such as the run@rate and production parts approval process (PPAP). The quality management system requires the supplier to provide a certification in accordance with one of the quality standards; the following standards can be considered as examples:

- The ISO/TS 16949:2002 is considered a preferred requirement by most OEMs.
- The QS 9000; VDA 6.1 is considered an acceptable requirement by most OEMs.
- The ISO 9001:2000 is typically considered a minimum acceptable level.

Another item that is typically required in the GQA is the details of the Advanced Product Quality Planning (APQP), which is typically done in accordance with the AIAG/QS 9000 procedures and forms, or using the VDA 4.3. Other quality metrics used in the quality judgment process, including the run@rate, should define the tests' frequency and the reporting formats; for example, the run@rate is utilized to study the specific supplier's production process under the actual production conditions and to provide a validation of the supplier's process capability to produce quality parts and components within the production lead times provided.

The following step after compiling and negotiating all the documents and legal commitments in terms of production process and parts quality, the sourcing process, starts through reviewing suppliers' quotations. The review process helps the OEM narrow down the number of candidate suppliers who have passed previous stages. Additional reviews that take place at this stage include the supplier component review to verify all the OEM production conditions and requirements are clear and can be met by the supplier's current production practices. This is done based on a comprehensive review of the parts to be produced, their specifications and quality expectations. The supplier component review might also be done through a design review (DR) meeting between the OEM and the selected supplier, and once it is completed, the sourcing decisions and agreements are executed. A typical supplier component review format is shown in Figure 8.10.

SCR summary

GP 1 Supplier Component Review
- Mechanics -

Supplier (Name&No.)	Part (name)	Part No./Index(Rev):

Summary of the Supplier Component Review

Supplier's delegates / function	Co. delegates / department

Appropriated and handed over documentation:
- Drawing (No., Rev./Index):
- BOM:
- SV Specifications:
- SV Customer Specifications:

SOP date: 0 will be copied from logistics checklist!
VDA Bd.2 ISIR or PPAP: 0 will be copied from quality checklist!

	FY 02/03	FY 03/04	FY 04/05	FY 05/06
Quantity non-binding				

will be copied from logistics checklist!

The requirements of the discussed specifications in terms of engineering, quality, logistics, timing and integral components of the contract

☐ **are fulfilled without restrictions**

☐ **are fulfilled with the restrictions** as stated in Appendix ... / Checklist

☐ **are not fulfilled** as stated in Appendix ... / Checklist

by the supplier.

Remarks:

Production shall take place according to manufacturing procedures with assured conformance to the applicable national laws and requirements for safety.

Confirmed by compnay xx: Date(MMM.dd.yyyy e.g.:Sep.05.2002):

Engineering	Quality	Logistics	Purchasing			

Date / Supplier

Component Review

Engineering Checklist ☐

Quality Checklist ☐

Logistic Checklist ☐

Attachments of the Component Review:

	Required
Attach 1: Feasibility analysis	☐
Attach 2: Requirement of Production and Process Approval of VDA Bd.2 Req. 4.6	☐
Attach 3: Retention/Submission Requirement Table acc. QS 9000	☐
Attach 4: Status Report Supplier	☐
Attach 5: Retention/Submission Requirement Table acc. VDA 2	☐
Attach 6: PPAP-Run audit	☐
Attach 7: Tasklist	☐
Attach 8: Packaging data sheet	☐
Attach 9: List of toxic substances	☐

Date / delegate supplier   Date / SV delegate

Figure 8.10 A typical supplier component review

The final sourcing decisions are typically made based on the total cost of the product, which typically include the logistics, the engineering, and the quality costs associated with the development and manufacturing of the products. Based on the total cost of the part and other metrics such as the supplier/OEM past collaborations, each of the approved suppliers at the sourcing stage are evaluated. After a supplier has been selected, the OEM quality engineering team starts to verify and validate the actual production conditions for the release of the products through checking the PPAP, APQP, and other tests. The result from the above selection and approval process is to guarantee a smooth launch process for the new vehicle model and its parts at the suppliers' locations. One should also note that the supplier selection process involves the OEM, the supplier, and any Tier 1 suppliers involved.

If the supplier's production proves to be incapable of producing parts or components according to the agreed upon terms and specifications, different procedures are enforced. For example, if the supplier failed to provide a conforming product at first, then the OEM or Tier 1 supplier can extend the period to produce conforming parts so that the supplier improves their operations; however, if the supplier fails for the second time to provide conforming products, then it will typically be placed in a lower confidence level during the launch period, leading to more frequent audits and evaluations. Also as the supplier production delays the launch for the same or other reason, then confidence level in the supplier keeps declining. Typically, the OEM and Tier 1 suppliers apply a continuous evaluation and tracking procedure on the supplier parts and processes to ensure that the required manufacturing practices and quality expectations are met; these evaluations and tracking are then tabulated in the consequent supplier quality system. The severity of the supplier defects are typically judged based on the associated cost from the OEM point of view, for example, a level 1 is when the defect is generated but caught at the supplier location, leading to small loss from the OEM perspective, while a level 3 is when the defect results in warranty issues or customer recalls. On the other hand, some OEMs judge the supplier quality levels based on the probability that a supplier defect might lead to a production shut-down. Additionally the associated actions by the OEMs and Tier 1 suppliers are commensurate with the severity level; a level 1 issue requires a contamination plan with full mitigation actions, while a level 2 defect might lead the OEM to transfer the losses to the supplier, report the supplier to the quality certifying body, and also the OEM might ask for more frequent audits. The level 3 defects lead the OEM to conduct weekly audits, request weekly performance production data and share the costs with the supplier. The consequent supplier management system acquires all the data and audit results during the launch and serial production and uses it to evaluate the supplier for future quotation and sourcing agreements.

8.7 An Overview of the Automotive Quality Tools

This section addresses an overview of the different automotive quality methods and tools used within the industry to qualify or control the incoming products from suppliers or the produced parts within the OEM facility. The automotive quality methods help the OEM evaluate their production readiness for the release of new vehicle models composed of new parts and new BiW geometries; this time period is typically called the launch phase and is the most critical for automotive OEMs because it reduces the time to market, hence enabling the OEM to start getting returns in the form of selling vehicles, on their investment in developing the new model.

The quality tools discussed in this section include:

- the Production Part Approval Process (PPAP);
- the Advanced Product Quality Planning (APQP)
- the Failure Mode and Effect Analysis (FMEA).

8.7.1 The Production Part Approval Process (PPAP)

The PPAP process is a systematic methodology that helps the OEM to qualify the suppliers' components and production, in terms of its reliability. The PPAP process is part of the QS 9000 quality standard and is heavily used by General Motors GM, the Ford Motor Company and Chrysler to evaluate and qualify their suppliers' production practices. Typically the PPAP is applied to ensure customer (automotive OEM) release of: new products, existing products with new modifications, and for corrected parts or components.

The Automotive Industry Action Group (AIAG) PPAP manual specifies the production conditions when the PPAP can be conducted as at least 1–8 hours production, with a minimum of 300 produced parts in succession (or in accordance with the OEM request); with production conditions in accordance with the actual production cycle time, production rate, and tooling. The parts should be inspected separately from every single production step with additional representative parts inspections. The PPAP is based on 19 specific demands that should be fulfilled to qualify the suppliers' parts and processes; these demands are:

1. The design records (documents, CAD drawings) of each part should be available.
2. Any design change documents or engineering notes documenting any variations from original part plan should also be available.
3. Documents' release should be done under OEM written permission, because the PPAP might be conducted by a third party lab or consulting group.
4. The Design Failure Mode and Effect Analysis (DFEMA) for each supplier's engineered (designed) parts should be available; documenting the analyses conducted following the QS 9000 standards.
5. The process flow should be documented and explained through drain diagrams.
6. The Process Failure Mode and Effect Analysis (PFEMA) for each supplier's process should be conducted
7. Explain and document any measurement campaigns done at launch based on the OEM request or standards, with cross-reference to the design records and the engineering notes. Some referenced samples might be requested to verify the measurement process and sequence.
8. A material performance test should be conducted and documented. These tests are done to verify that the product complies with the standard specifications and that the sub-components have been acquired from a certified sub-supplier.
9. Short-term process capability showing the process ability to meet the actual production conditions using specific metrics such as the C_p and C_{pk} for each process; if the process is unstable, then a P_{pk} index should be used instead of the C_{pk}. The difference between the P_{pk} and the C_{pk} is mainly due to the method of calculating the standard deviation of the measured samples; the C_{pk} uses the average moving range to calculate the standard deviation assuming that this deviation is coming from a single source because the process is stable, while the P_{pk} sets the deviation based on the data at hand using a longer-term measurement

of the standard deviation because the process is unstable and so several deviation sources might be present, and should be taken into account. In other words, if a process is stable, then the standard deviation from a sub-group (short term) is exactly the same as the long-term deviation, but if the process is unstable then the standard deviation from a sub-group is related only to one source of variation and cannot be used to explain other variations so a long-term (from all collected data) standard deviation should be computed. To explain the difference in mathematical terms; the sequence of equations from 8.1 to 8.4 defines each of these terms starting from the C_p into the P_{pk}, respectively.

$$C_p = \frac{USL - LSL}{6\sigma_{sub}} \tag{8.1}$$

$$C_{pk} = \min\left(\frac{USL - \overline{\overline{X}}}{3\sigma_{sub}}, \frac{\overline{\overline{X}} - LSL}{3\sigma_{sub}}\right) \tag{8.2}$$

$$P_p = \frac{USL - LSL}{3\sigma_{long}} \tag{8.3}$$

$$P_{pk} = \min\left(\frac{USL - \overline{\overline{X}}}{3\sigma_{long}}, \frac{\overline{\overline{X}} - LSL}{3\sigma_{long}}\right) \tag{8.4}$$

10. Additionally, if the process is found to be unstable for the short-term capability study, then a specific plan of how to reach to stable conditions should be provided by the supplier, and a escalation plan should also be included if the stability targets are not reached.
11. Measurement system analysis procedures should also be described in detail, with focus on the data reliability and comparability, which describes the comparable precision in terms of the variations that result from repeated use of fixed work method, processing identical objects, and having the same operator perform the same tests using the same checking/measuring devices.
12. The availability of the documentation to testify to the laboratories' qualifications and accreditations, in terms of personnel, testing methods, and the usage of statistical methods.
13. The formulations of parts and processes control plan, which is the written description of how to control production processes. The control plan typically comes in three different levels:
    i. The Prototype level, which is a written description of the measurements of dimensions and the material performance tests during the prototype phase.
    ii. The Pilot level, which is the written description of measurement campagain after the prototype phase but just before series production.

iii. The Series or Full production level, which is an extensive documentation of product and process characteristics, the process control tests, and the measuring systems recorded during series production.
14. The Part Submission Warrant PSW, which should be done for each part.
15. The appearance release report, if the part has any aesthetic attributes such as painted surfaces or fabrics.
16. Check lists for special chemical products or OEM specific requirements
17. Production part samples, which should be identified and agreed on by the OEM
18. Reference part samples.
19. The Specific inspection devices, which should be listed for each part.
20. Finally, any customer (OEM) specific requirements should also be met in the PPAP.

In addition to the above 19 demands, the OEM has the right to be informed when certain production changes take place at the supplier location, such as when the supplier changes their subcontractor, or new materials are being used in the parts' manufacturing. The OEM can then request additional tests or samples to be included in the PPAP. Additionally, the OEM and supplier should agree and discuss the documents' storage and transmission policies, typically the PPAP documents should be kept as long as the part is still being produced plus one year, also the old PPAP documentations are integrated with any new documents for a new part from the same supplier.

8.7.2 The Advanced Product Quality Planning (APQP)

The APQP is a systematic planning methodology to ensure a high level of confidence in the final achieved quality level of planned products and components. The VDA 4.3 procedures specify a similar systematic way for the automotive project planning and management to secure meeting the OEM's expectation and requirements. The APQP or the VDA 4.3 can be thought of as designing the quality of parts and components into the automotive launch process. Also the QS 9000 quality standard is focused on the Quality Management practices within companies and industries and its evaluation process. In regard to the German Association for the Automotive Industry VDA standards, the VDA 6.1 is focused on the quality management demands and its evaluation, while the VDA 4.3 is focused on the product quality and the quality management planning.

The general APQP time chart can be displayed as in Figure 8.11, to show the different phases for the APQP, starting from the project initiation to the safe launch of the new product into series production. Figure 8.11 also describes the product's different levels and how it evolves from the prototype into the final series production level.

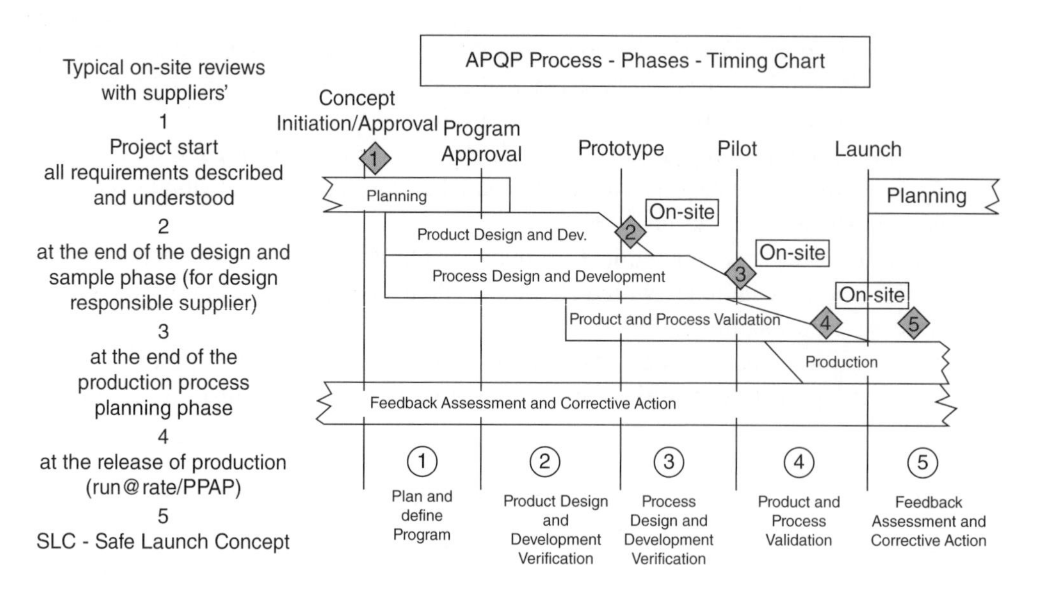

Figure 8.11 The different phases of the APQP process

The APQP aims to reduce the time to market for new products and ensures a safe and smooth launch process, where the OEMs and suppliers are following a specific set of steps and procedures to ensure product and process quality. The APQP, like any project management tool, is based on a predefined set of boundary conditions, assumptions, expectation of specific deliverables, time limit, and of course total cost. All these inputs should be discussed/negotiated and agreed upon by the OEM and the supplier, before the project is initiated. Once the project is initiated, the project can then be translated into smaller (in time and content) work steps, with time and cost estimates, to be handled by an inter-disciplinary team from the different departments within the OEM and the supplier. Additionally, the project structure should be planned from different perspectives; from the sought product functionality, from the different internal sub-components involved in making the product, and from the different development phases, i.e. the planning phase, the development phase, and the production phase. After these structures have been clearly defined, then for each work step, time, resource and cost allocations should be defined, for example, the required engineering resources and their capacities in terms of assigned personnel qualifications and number. Also, the time frame for each task, as well as estimates in regard to the earliest and the latest times the task can be completed. Finally, the critical path for the project's progress can be computed based on the time and cost estimates. Additionally to improve the quality of each work step, the Plan, Do, Check or Act (PDCA) procedure should be applied, where the Plan step corresponds to the re-design and/or required revisions of processes and/or components to improve its quality or reduce its cost or time. The Do action implements the suggested revisions

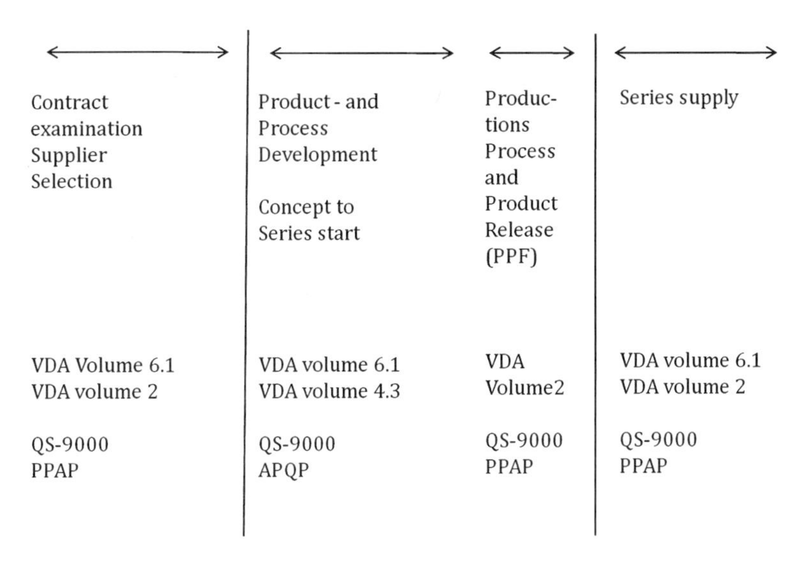

Figure 8.12 The supplier selection process and the associated standards

and evaluates its performance and cost/time savings, while the Check action assesses the measurements and reports the results to the group leaders or decision-makers. Finally, the Act action decides on changes needed to improve the process/product. The APQP procedures enforces certain forms and standards to be met at each work phase and within each work step. Figure 8.12 shows some of the standards and regulations that should be met as the project progresses.

A complete perspective of the APQP can be shown in the form in Figure 8.13, which describes all the items that should be checked through the project, starting with the General Quality Agreement (GQA) signed between the supplier and the OEM before the sourcing decision.

Also, an example check list for each phase can be further displayed in Figure 8.14, such a check list can be answered by Yes or No. Additionally, like any project management tool, the APQP includes risk analysis for each work step, which is composed of four stages: (1) identify the risk; (2) qualify the risk in terms of likelihood and impact; (3) then evaluate or prioritize the risks; and finally (4) mitigate the risk through one of four techniques: (a) Sharing the risk between the OEM and supplier; (b) Enduring the risk, for example, by increasing the production time; (c) Avoiding the risk, by outsourcing or consulting; and finally(d) Lessening the risk impact by negotiating a lower quality expectation.

Also, Figure 8.15 shows one more check list adopted by the Chrysler Corporation in their APQP product release or sign off process. Figure 8.16 shows an example of a control plan that can be used to address any of the issues if answered with No from the above check lists.

Advanced Product Quality Planning Status Report			Date Review No.:
Supplier		Program	
Location		Model Year	
Supplier Code		Lead Part No.	

Build Level	IPD Into Plant Date (MRD)	Status GYR		Quantity	Concurred		P.I.S.T.%	P.I.S.C.%
		Present	History		No. SC's	No. CC's		

APQP Elements	GYR Status		Risk[1] Evaluation per Element	Program Need Date	Supplier Timing Date	Closed Date	Resp. Engineer Initials	Remarks or Assistance Required
	Present	History						
1* Customer Input Requirements								
2 Agreed GQA								
3 Product Feasibility								
4 Sourcing Decision								
5 Change Management								
6 Environmental-Management/IMDS/ELV								
7 Prototype Build Control Plan								
8 Design Verification Plan								
9* Facilities, Tools, Gauges								
10* Design FMEA								
11* DFMA@Customer level (SV)								
12* Drawings and Specifications (Design Freeze)								
13 Information about Subcontractors								
14* Logistic Concept								
15* Packaging Specification (Serial & Spare Parts)								
16* Design Validation@Supplier Level								
17* Design Validation@Customer Level (SV)								
18* Measurement Syst. Analysis/ Measuring Agreem.								
....								
22 Safe Launch conceptincl. (SCL Or EPC)								
23* Manufacturing Process Flow Chart								
24 Process FMEA								
25 Working Instructions								
26 Requalification/Test Plan and Agreement Preliminary Process Capability								
27 Study								
28* Production Validation (PV) Testing								
30 Serial Control Plan								

Figure 8.13 A typical APQP standard sheet, form

8.7.3 The Failure Mode and Effect Analysis (FMEA)

The FMEA process is one of the most commonly used tools in the automotive industry to ensure the safe launch of new models and to qualify any product or process changes for current products or vehicle models. The FMEA can be done at the design level, hence DFMEA or at the process level, hence PFMEA. Additionally the FMEA is considered an integral part of the APQP and the PPAP procedures, because it analyzes the product issues from a functionality point of view then quantifies it using a Risk Priority Number (RPN). This section will only address the PFMEA because the DFMEA is more concerned with the design phase not the actual manufacturing process.

One should note two important attributes of the FMEA process: the first is that it is typically done for non-hazardous (not safety-related) failures, while the second is that FMEA is done to avoid building confidence based on inspection to prevent defect

Customer Input Requirements

Are the Product Requirements of Quotation (PRQ) from the customer available?
Are the technical requirements known? (mounting position, function and environmental influences)
Has the geographical location of use been defined? (e.g. Europe or world-wide; Note: quality of fuel)
Have the reliability targets been defined?
Have the timing milestones been defined? (design reviews, design freeze, prototype build phases, PSW, SOP)
If required: Does a Recycling Concept exist?
Have the warranty conditions been defined?

Agreed GQA
Did the supplier confirm the GQA?
Have the quality targets been confirmed?

Sourcing Decision
Has the supplier a QM system according to VDA 6.1 or QS9000 or ISO/TS 16949? If not: a strategy to reach it is required
Has the purchase department sent the APQP Template and the GQA to the supplier? Are the targets included in the purchasing documents
Has the Sourcing decision been communicated to the supplier?

Figure 8.14 Example check list for APQP

PQR Process (Supplier Source Selection)
Risk Assessment and Feasibility Issues
Supplier Source Approval Process
FMEA; D-FEMA → P-FMEA → Design for Manufacturing
Design Verification Planning and Reports
Prototype Tooling
Phase A Pre-production samples
Packaging R&D review
Determination of Engineering Standards and Review
Production Release of Drawings and Engineering Standards
Supplier System plan; control plan and P-FMEA
Die verification
Gages and Test stands check
Sub-supplier review
Gage Reliability and Repeatability Study
Phase B Pre-production
Production tooling
Preventive maintenance
Piece production run, machine capability study Cp and Cpk
Production trial run
Outer pilot review
.....
Launch

Figure 8.15 An example from the Chrysler APQP check list

Flow Chart	Critical Charac.	Class	Sample size and Freq.	Insp. or test proce	Control method	Gage	Report doc	Additional req.	Reaction Plan

Figure 8.16 A typical control plan document

escapes, or reactive design changes to avoid product failures, but is done to qualify (identify and prioritize) possible failures using the RPN number.

Also before conducting the FMEA, the engineer should discover the product or process history through previous FMEA (both P- FMEAs and D-FMEAs), Launch Logs, Bill of Materials, manufacturing routings, previous control plans, or supplier history. In addition, he should understand the product/process usage, failure modes, and its complete technical details.

The PFMEA process starts by identifying the analyses scope (Figure 8.17). The PFMEA process then defines the processing steps of the product, which is inserted in the FMEA form in the column entitled Process operations. For each processing step the product main or critical characteristics, which can include dimensions, alignment, or relative positions of the product sub-components, are included from the product characteristic matrix. These characteristics are inserted in the FMEA form in the column entitled Product characteristics with the target value for each of these items also included, along with the acceptable tolerance. The column entitled Dominant factor describes the main source of variation in processing for each of the product characteristics, it can include the operator, the raw material, or the equipment. A typical FMEA form is shown in Figure 8.18.

After that, the different failure modes are deducted from the product functionality point of view. However a failure mode relates to the failure in the manufacturing processes (assembling the product) not the product functionality, while the product functionality failures are called failure effects. For example, a failure mode is the incorrect placement of the adequate number of screws, while the failure effect is higher vibration level in the product. The failure modes can be classified according to the following: partial or incomplete processing, processing absence, incorrect processing. Note that the preliminary hazard analysis (PHA) process, which is typi-

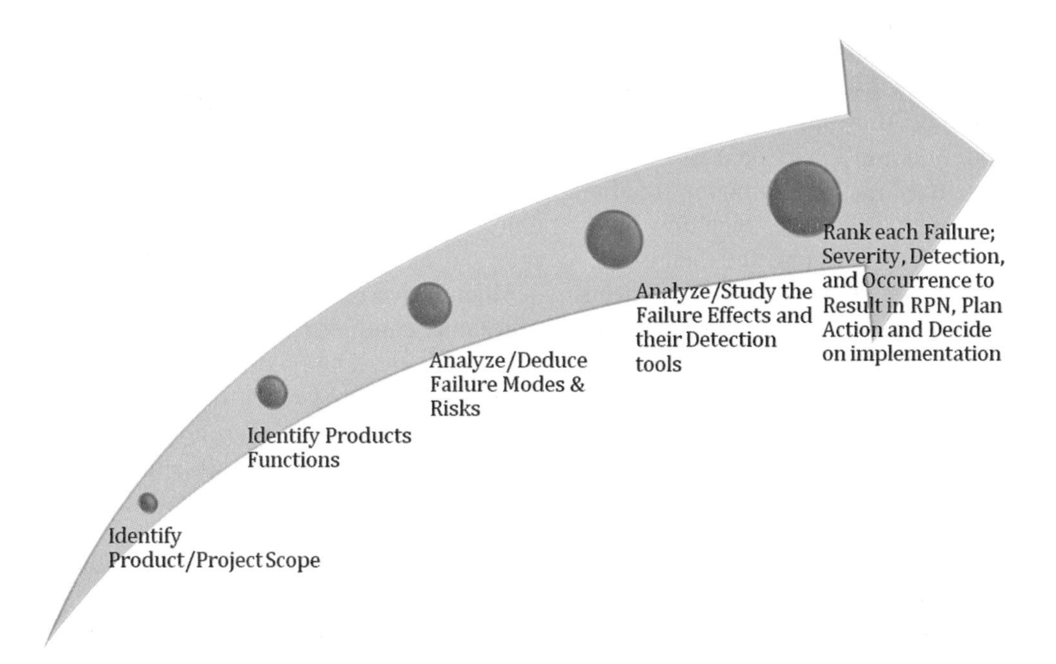

Figure 8.17 The PFMEA process flow diagram

<table>
<tr><td colspan="13">Failure Mode and Effect Analysis FMEA (Process FMEA)</td></tr>
<tr><td colspan="4">Process</td><td colspan="3">Process Responsibility</td><td colspan="6">FMEA number page __ of __</td></tr>
<tr><td colspan="4">Core team</td><td colspan="3">Date</td><td colspan="6">Prepared by</td></tr>
<tr><td>Func Requir</td><td>Poten Failure</td><td>Potential effect/impact</td><td>Severity</td><td>Potential Cause</td><td>Occurrence</td><td>Current process control</td><td>Detection</td><td>RPN</td><td>Recommended action</td><td>respon</td><td>Action results</td><td>NEW RPN</td></tr>
<tr><td></td><td></td><td></td><td></td><td></td><td></td><td></td><td></td><td></td><td></td><td></td><td></td><td></td></tr>
</table>

Figure 8.18 A typical FMEA form

cally done in a PPAP process, focuses on failures with respect to the product safety not product functionality. To provide an example, for FMEA failure effects for the assembly of automotive electric pumps, the failure effects can be: absence of function (pump doesn't start or doesn't stop), incomplete function (flow rate is lower than required), incorrect function (the pump starts without being started), and other failures such as leakage. Including all the failure effects and modes helps to predict the root causes of failure and their types, for example, internal causes and external causes with respect to the pump components. After identifying all the process failure modes and their effects on the product functionality, one can insert it in the FMEA form, under potential failure modes and effects respectively. The failure effects can then be used to classify each processing deviation in terms of severity, occurrence (frequency), and the possibility of its detection at the plant level. These three metrics; severity, occurrence and detection, help in computing the RPN for each failure mode. Additionally, the FMEA form contains another column for potential root causes of each failure mode, its prevention and current technologies for its detection. The prevention is more related to the failure modes while the detection is based on the failure effects. Each of these items, detection, prevention and occurrence, is based on a quantitative rank typically from 1–10, with 10 being, for example, no known technologies or means for detection. So for example, if the failure mode or effect cannot happen due to a design solution or set best practices, then the rank will be 1, while if the failure is detected post the design freeze stage (but still prior to launch) using a pass/fail test, then the rank might be 6–8. Finally, the RPN number is computed by multiplying the severity rank × occurrence rank × detection rank. The RPN number highlights the failure modes so it motivates an action plan, which can be added on to the FMEA form to specify the recommended action and the personnel or teams responsible. Also, the added action plan might also be complemented with the actual results (RPN) after the improvements.

Exercises

Problem 1

CEGC Incorporated just received the following forecast for the next year demand; knowing that CEGC main resource planning is focused on labor (# of workers) and they started the year with 100 workers, 100 units in storage and they want to end it with 100 workers, and 100 units in storage. Provide two aggregate plans: the first based on Leveled production scheme while the second is based on a Chase strategy; for each plan provide a table that describes the resource planning and a final cost estimate based on production data provided below.

Month	Sales forecast	Production	Cost $ and labor hrs
Jan.	750	Regular production cost	$2000 per unit
Feb.	760	Overtime production cost	$2062 per unit
Mar.	800	Average monthly holding cost	$40 per unit
Apr.	800	Average labor hours	20 hrs per unit
May	820		
June	840	Maximum regular production per mth	848 units
July	910	Allowable overtime per month	1/10 of regular production
Aug.	910		
Sept.	910	Hours worked per mth per employee	160 hrs
Oct.	880	Estimated cost of hiring	$1750
Nov.	860	Estimated cost of firing	$1500
Dec.	840		

Problem 2

Using the forecast information presented in the table below, suggest three aggregate plans to satisfy the expected production demand based on following strategies:

- PLAN 1: Plan based on leveled production of 50 units/day.
- PLAN 2: Plan based on leveled production based on min. period demand and subcontract when for other periods.
- PLAN 3: Plan based on hiring and firing to the exact monthly demand (Chase Strategy).

Then compute the cost associated with each plan using the cost information table provided.

Month	Expected Demand P_t	Production Days	Demand per Day	Cumulative Demand D_t
Jan.	900	22	41	900
Feb.	700	18	39	1600
Mar	800	21	38	2400
Apr	1200	21	57	3600
May	1500	22	68	5100
June	1100	20	55	6200

Cost Information		
Inventory carrying cost	\$5	/units / month
Subcontracting cost pre unit	\$10	per units
Average pay rate	\$5	per hr
Overtime pay rate	\$7	per hr
Labor hrs to produce a unit	1.6 hrs	per unit
Cost of increasing daily production rate	\$300	per unit
Cost of decreasing daily production rate	\$600	per unit

Problem 3

Solve the presented example above using the transportation method.

Problem 4

Compare and contrast the JIT system and the MRP system for material acquisition

Problem 5

Explain the SMED procedures when applied to automotive stamping presses.

Problem 6

What is the difference between *job enrichment* and *job enlargement* within the context of the Toyota Production System?

Problem 7

Explain the different stages of the supplier selection process.

Problem 8

What are the typical contents of the general quality agreement?

Problem 9

Explain the quality auditing system in qualifying suppliers' production.

Problem 10

What is the difference between the APQP and the PPAP processes?

Problem 11

Explain the differences and usages of the following statistical indices in automotive production: C_p, P_p, C_{pk}, P_{pk}.

Problem 12

What is the difference between the *failure mode* and the *failure effect* within the context of FMEA?

Problem 13

Conduct an FMEA for the process of assembling a mechanical pencil; knowing that its basic components are: the eraser, the barrel, the tube assembly, the clip, and the lead.

References

[1] Hitomi, Katsundo (1996) *Manufacturing Systems Manufacturing*, London: Taylor and Francis.

[2] Omar, M.A., Kurfess, T., Mears, L., and Kiggans, R. (2009) "Organizational learning in the automotive manufacturing; a strategic choice," *Journal of Intelligent Manufacturing*, ISSN 0956-5515, DOI 10.1007/s10845-009-0330-6.

[3] *Automotive News* (2004) *Market Data Book.*

[4] Xu, Yanwu (2006) *Modern Formability; Measurement, Analysis, and Applications*, Cincinnati, OH: Hanser Gardner.

[5] Keeler, S.P., Backofen, W.A. (1963) *ASM Trans. Q.*, 56, 25–48.

[6] Goodwin, G.M. (1968) "Application of strain analysis to sheet metal forming problems in the press shop", *La Metallurgica*, 60, 767–774.

[7] Yoshida, K. (1993) *Handbook of Ease Or Difficulty in Press Forming*, translated from Japanese, National Center for Manufacturing Science, Inc., Ann Arbor, MI.

[8] Greene, William H. Maximum likelihood estimation of stochastic frontier production models, ASTM E 646.

[9] Davies, G. (2003) *Materials for the Automobile Bodies*, Oxford: Elsevier Butterworth Heinemann.

[10] American Iron and Steel Institute (2003) "Formability characterization of a new generation of high strength steel", American Iron and Steel Institute Reports, March 2003, available at www.steel.org [accessed on August 5, 2010].

[11] American Iron and Steel Institute (2001) "Characterization of Fatigue and Crash Worthiness of New Generation of High Strength Steel FOR Automotive Applications (Phase I and Phase II)," American Iron and Steel Institute Reports, January 2001, available at www.steel.org [accessed on August 5, 2010].

[12] Aluminum Association (1998) *Aluminum Automotive Extrusion Manual*, Publication AT6, Washington, DC: Author.

[13] Cantor, Brian, Grant, Patrick, Johnston, Colin (2008) *Automotive Engineering, Lightweight, Functional and Novel Materials*, New York: Taylor & Francis.

[14] Kaufman, J., Gilbert, P. (2000) *Introduction to Aluminum Alloys and Tempers*, The Aluminum Association Inc, published by the ASM International Materials Park, OH.

[15] European Aluminum Association (2004) "Aluminum in commercial vehicles," Brussels, Belgium, available at www.eaa.net [accessed on August 5, 2010].

[16] Boljanovic, V. (2004) *Sheet Metal Forming Processes and Die Design*, New York: Industrial Press.

The Automotive Body Manufacturing Systems and Processes, First Edition. Mohammed A. Omar.

[17] Aoki, Itrau, Horita, Takashi, Herari, Toshio (1981) "Formability and applications of galvanized sheet steels", *Proc. Materials/Metal Working Technology Conference*, American Society of Metals, Michigan, April.
[18] Automotive Steel Partnership Report (2001) "Tailor Welded Blank Applications and Manufacturing," June, available at www.a-sp.org [accessed on August 5, 2010].
[19] Auto Steel Partnership Report (2001) "Blank Applications and Manufacturing," June available at www.a-sp.org [accessed on August 5, 2010].
[20] Koç, Muammer (2008) *Hydro-forming for Advanced Manufacturing*, Boca Raton, FL: CRC Press.
[21] Klaas, F. (1997) "Innovations in high-pressure hydro-forming", *Proc 2nd Int Conf on Innovations in Hydroforming Tech*, Columbus OH, 1–31.
[22] Automotive Steel Partnership Report (1999) "Automotive body measurement system capability," available at www.a-sp.org [accessed on August 5, 2010].
[23] Automotive Steel Partnership Report (2002) *Automotive Steel Design Manual*, available at www.a-sp.org [accessed on August 5, 2010].
[24] Carey, Howard. (1995) *Arc Welding Automation*, New York: Marcel Dekker.
[25] Omar, M.A., Kuwana, K., Saito, K., Couch, C. (2007) "Thermography for investigating industrial welding fires in automotive stations," *Journal of Fire Technology*, 43(4): 319–329.
[26] Kalpakjian, Serope, Schmid, Steven (2000) *Manufacturing Engineering and Technology*, New Jersey: Prentice Hall.
[27] Dickerson, D.W., Natale, T.V. (1981) "The effect of sheet surface on spot weldability, materials/ metal," *Proc. Working Technology Conference*, American Society of Metals. Michigan, April,
[28] Adams, A.D. (2005) *Adhesive Bonds: Science and Technology*, Cambridge: CRC Press.
[29] Cordes, E. (2003) "Adhesives in the automotive industry", in A. Pizzi, K. Mittal (Eds.), *Handbook of Adhesive Technology*, second edition, New York, Marcel Dekker, pp. 999–1015.
[30] Petrie, J. (2000) *Handbook of Adhesives and Sealants*, New York: McGraw-Hill.
[31] Zisman W.A. (1964) *Contact Angle, Wettability and Adhesion*, American Chemical Society.
[32] Nara, K.R., Omar, M.A. (2009) "Sensitivity analysis of non-load bearing 2K epoxy adhesive to hemming process variations, thermo-gravimetric, differential scanning calorimetry and FTIR analyses," *Journal of Progress in Organic Coatings*, 65, 104–108.
[33] Messler, R.W. Jr. (1993) *Principles of Welding: Processes, Physics, Chemistry, and Metallurgy*, New York: John Wiley and Sons, Inc.
[34] Messler, R.W. Jr. (1999) *Joining of Advanced Materials*, New York: Butterworth-Heinemann.
[35] Roobol, N. (2003) *Industrial Painting and Powder-coating; Principles and Practices*, Cincinnati, OH: Hanser Gardner.
[36] Lou, H.H., Huang, Y.L. (2000) "Integrated modeling and simulation for improved reactive drying of clearcoat," *Industrial Engineering Chemistry Res.* 39: 500.
[37] Xiao, Jie., Jia Li, Helen, Lou, H., Huang, Yinlun (2006) "Cure-window-based proactive quality control in topcoat curing", *Industrial Engineering Chemistry Res*, 45: 2351–2360.
[38] Zeng, F., Ayalew, B., Omar, M.A. (2008) "Robotic automotive paint curing using thermal signature feedback", *Journal of Industrial Robotics*, 36(4): 389–395.
[39] Hunter, Harold (1987) *The Measurement of Appearance*, New York: John Wiley Inter-science.
[40] Graedel, Thomas, Allenby, Braden (1995) *Industrial Ecology*, New Jersey: Prentice Hall.
[41] Clarke, G. (2005) *Automotive Production Systems and Standardization, from Ford to the Case of Mercedes-Benz*, Heidelberg: Physica-Verlag.
[42] Galitsky, C., Worrell, E. (2003) *Energy Efficiency Improvement and Cost Saving Opportunities for the Vehicle Assembly Industry*, Ernest Orlando Lawrence Berkeley National Laboratory, Environmental Energy Technologies Division.
[43] Patil, Y., Seryak, J., Kissock, K., Kissock, K. (2003) "Benchmarking approaches: an alternate method to determine best practice by examining plant-wide energy signature," ACEEE Summer Study on Energy-Efficiency in Industry, www.aceee.org [accessed January, 5, 2010].

[44] Ross, M., Thimmapuran, P. (1993) "Long term industrial energy forecasting LIEF model", Argonne National Laboratory. www.transportation.anl.gov [accessed July 20, 2009].

[45] Automotive Parts Manufacturer Association APMA, technical publications (2000), Obtained through the internet: www.apma.ca [accessed May 15, 2009].

[46] Graedel, Thomas, Allenby, Braden (1997) *Industrial Ecology and the Automobile*, New Jersey: Prentice Hall.

[47] Stodolsky, F., Vyas, A., Cuenca, R., Gaines, L. (1995) "Life cycle energy savings potential from Aluminum Intensive Vehicles," *Proc. Total Life Cycle Conference and Exposition*, October 16–19, 1995. Vienna, Austria.

[48] Greene (1993) *Journal of Econometrics Volume 18, Issue 2*, February 1982, Pages 285–289.

[49] Hicks, T., Dutrow, E. (2001) "Energy performance benchmarking for manufacturing plants." ACEEE 2001 Summer Study on Energy-Efficiency in Industry. www.aceee.org [accessed May 10, 2009].

[50] Boyd, G. (2005) *Development of a Performance-based Industrial Energy Efficiency Indicator for Automobile Assembly Plants*, Chicago: Argonne National Laboratory, Decision and Information Sciences Division.

[51] Kissock, K., Seyak J. (2004) "Lean energy analysis: identifying, discovering and tracking energy savings potential," *Proceedings of Society of Manufacturing Engineers: Advanced Energy and Fuel Cell Technologies Conference*, Livonia, MI, Oct. 11–13

[52] Kissock, K. and Eger, C. (2006) "Measuring Plant-wide Industrial Energy Savings", Society of Automotive Engineers World Congress and Exposition, Detroit, MI, April 3–6.

[53] IISI (1994) "Competition between steel and aluminum for the passenger car," International Iron and Steel Institute IISI report (1994) Brussels. www.worldstainless.org [accessed June, 10 2009].

[54] Klobucar, J. (2004) "Improving energy efficiency of automotive paint shops", Painting Technology Workshop, Lexington, KY.

[55] Newell, S. (1998) "Strategic evaluation of environmental metrics: making the use of life-cycle analysis", PhD Dissertation, MIT Department of Material Science and Engineering.

[56] Peters, T. (1997) "Environmental awareness in car design", SAE publication SP-1263, Feb.

[57] Hohmeyer, O., Ottinger, R.L. (eds.) (1992) *Social Costs of Energy*, New York: Springer Verlag.

[58] Zuckerman, B., Ackerman, F. (1995) "The 1994 update of the Tellus Institute Packaging study impact assessment method," *Proc. STEAC Impact Assessment Working Group Conference*, Washington DC.

[59] Rowe, R. (1995) *The New York Externality Study*, Empire State Electric Energy Research Corporation, New York: Oceana Publication.

[60] Steen, B., Ryding, S.-O. (1992) The EPS enviro-accounting method. IVL Report, Göteborg, Sweden.

[61] Ginley, D.M. (1994) "Material flows in the transportation industry, an example of industrial metabolism", *Resources Policy*, 20: 169–181.

[62] Roth, R., Diffenbach, J., Issacs, J. (1998) "economic analysis of the ultra-light steel auto. body," *Proceedings of the SAE IBEC*, Detroit. MI.

[63] Ashby, Michael, F. (2009) *Materials and the Environment*, Oxford: Elsevier Butterworth Heinemann.

[64] Muther, R. (1973) *Systematic Layout Planning*, 2nd edition, Boston: Cahners Books.

[65] Knox, Agnew (1998) *The Geography of the World Economy*, 3rd edition, London: Arnold.

[66] Shingo, Shigeo, Dillon, Andrew (1989) *A Study of the Toyota Production System: From an Industrial Engineering Viewpoint*, Portland, OR: Productivity Press.

[67] Shingo, Shigeo (1986) *Zero Quality Control, Source Inspection and the Poke-Yoke System*, Portland, OR: Productivity Press.

INDEX

The Automotive Body Manufacturing Systems and Processes, First Edition. Mohammed A. Omar.